Osmolytes and Plants Acclimation to Changing Environment: Emerging Omics Technologies

Noushina Iqbal • Rahat Nazar
Nafees A. Khan
Editors

Osmolytes and Plants Acclimation to Changing Environment: Emerging Omics Technologies

Springer

Editors
Noushina Iqbal
Department of Botany
Jamia Hamdard University
New Delhi, Delhi, India

Rahat Nazar
Department of Botany
Jamia Hamdard University
New Delhi, Delhi, India

Nafees A. Khan
Department of Botany
Aligarh Muslim University
Aligarh, UP, India

ISBN 978-81-322-2615-4   ISBN 978-81-322-2616-1 (eBook)
DOI 10.1007/978-81-322-2616-1

Library of Congress Control Number: 2015957584

Springer New Delhi Heidelberg New York Dordrecht London

Printed on acid-free paper

Springer (India) Pvt. Ltd. is part of Springer Science+Business Media (www.springer.com)

Acknowledgements

First and foremost we would like to thank the Almighty without whose blessings we are nothing. We would like to express our gratitude to the DS Kothari postdoctoral fellowship for providing the research grant. We also acknowledge the academic staff of Jamia Hamdard, New Delhi, and AMU, Aligarh, for their support.

We are also thankful to our parents for their love and trust and for being a continuous source of inspiration. We are thankful to our spouse for their continuous support and encouragement in spite of all the time it took us away from them. Thanks to our kids for bearing our absence.

And lastly, we take an opportunity to thanks Springer for providing us an opportunity to spread our work.

Contents

Contributors

Mohd Asgher Department of Botany, Aligarh Muslim University, Aligarh, UP, India

Monica Boscaiu Mediterranean Agroforestal Institute (IAM, UPV), Universitat Politècnica de València, Valencia, Spain

Mayukh Chakraborty Department of Biological Sciences, Presidency University, Kolkata, India

Mudasir Irfan Dar Environmental Botany Section, Department of Botany, Aligarh Muslim University, Aligarh, UP, India

Mehar Fatma Department of Botany, Aligarh Muslim University, Aligarh, UP, India

Masayuki Fujita Department of Applied Biological Science, Faculty of Agriculture, Laboratory of Plant Stress Responses, Kagawa University, Kagawa, Japan

Kamala Gupta Department of Biological Sciences, Presidency University, Kolkata, India

Department of Botany, Government General Degree College, Singur, Hooghly District, West Bengal, India

Bhaskar Gupta Department of Biological Sciences, Presidency University, Kolkata, India

Department of Zoology, Government General Degree College, Singur, Hooghly District, West Bengal, India

Mirza Hasanuzzaman Department of Agronomy, Faculty of Agriculture, Sher-e-Bangla Agricultural University, Dhaka, Bangladesh

Mohamad Al Hassan Institute for Plant Molecular and Cellular Biology (IBMCP, UPV-CSIC), Universitat Politècnica de València, Valencia, Spain

Sofi J. Hussain Department of Botany, Aligarh Muslim University, Aligarh, UP, India

Noushina Iqbal Department of Botany, Jamia Hamdard University, New Delhi, Delhi, India

Gurwinder Kaur School of Agricultural Biotechnology, Punjab Agricultural University, Ludhiana, India

M. Iqbal R. Khan Department of Botany, Aligarh Muslim University, Aligarh, UP, India

Nafees A. Khan Department of Botany, Aligarh Muslim University, Aligarh, UP, India

Fareed Ahmad Khan Environmental Botany Section, Department of Botany, Aligarh Muslim University, Aligarh, UP, India

Asim Masood Department of Botany, Aligarh Muslim University, Aligarh, UP, India

Kamrun Nahar Department of Agricultural Botany, Faculty of Agriculture, Sher-e-Bangla Agricultural University, Dhaka, Bangladesh

Mohd Irfan Naikoo Environmental Botany Section, Department of Botany, Aligarh Muslim University, Aligarh, UP, India

Fauzia Naushin Department of Botany, Women's College, Aligarh Muslim University, Aligarh, UP, India

Rahat Nazar Department of Botany, Jamia Hamdard University, New Delhi, Delhi, India

Bouzid Nedjimi Laboratory of Exploration and Valorization of Steppe Ecosystem, Faculty of Science of Nature and Life, University of Djelfa, Djelfa, Algeria

G.C. Nikalje Plant Stress Physiology and Biotechnology Section, Nuclear Agriculture and Biotechnology Division, Bhabha Atomic Research Centre, Trombay, Mumbai, India

Savitribai Phule Pune University, Pune, India

P. Suprasanna Plant Stress Physiology and Biotechnology Section, Nuclear Agriculture and Biotechnology Division, Bhabha Atomic Research Centre, Trombay, Mumbai, India

Tasir S. Per Department of Botany, Aligarh Muslim University, Aligarh, UP, India

A.N. Rai Plant Stress Physiology and Biotechnology Section, Nuclear Agriculture and Biotechnology Division, Bhabha Atomic Research Centre, Trombay, Mumbai, India

Faisal Rasheed Department of Botany, Aligarh Muslim University, Aligarh, UP, India

Farha Rehman Department of Botany, Faculty of Science, Mohammad Ali Jauhar University, Rampur, UP, India

Saroj Kumar Sah Department of Biochemistry, Molecular Biology, Entomology and Plant Pathology, Mississippi State University, Starkville, MS, USA

Jayita Saha Department of Biological Sciences, Presidency University, Kolkata, India

Atreyee Sengupta Department of Biological Sciences, Presidency University, Kolkata, India

Shahid Umar Department of Botany, Faculty of Science, Jamia Hamdard University, New Delhi, India

Oscar Vicente Institute of Plant Molecular and Cellular Biology (IBMCP, UPV-CSIC), Universitat Politècnica de València, Valencia, Spain

Shabir H. Wani Division of Genetics and Plant Breeding, SKUAST-K, Shalimar, Srinagar, JK, India

Osmolyte Accumulation and Implications in Plant Abiotic Stress Tolerance

1

P. Suprasanna, G.C. Nikalje, and A.N. Rai

Abstract

Plants being sessile are challenged by biotic and abiotic stresses. Osmotic adjustment has been shown to be an effective component of stress tolerance, and accumulation of osmoprotectants (proline, glycine betaine, gamma-aminobutyric acid and sugars) is a common response observed in different plant systems. Information on the metabolic pathways of these compatible solutes for their regulation, enzymes involved and compartmentalisation are well characterised in most important plant species. Such information has prompted researchers to devise strategies for improvement of plant tolerance through the accumulation of osmolytes in plants by expression of core biosynthetic enzymes or their derivatives and expression of related transporters. Pathway engineering for overproduction of compatible solutes is being attempted through stress-inducible and/or tissue-specific regulation in order to avoid primary metabolic costs and minimise the pleiotropic effects. Understanding the mechanisms by which plants perceive environmental signals and transmit them to cellular machinery to activate adaptive responses is crucial for the development of crop improvement strategies to impart stress tolerance.

1.1 Introduction

Plants are challenged by hostile environments of temperature, drought, salinity and heavy metals which disrupt cellular and developmental processes. One such mechanism is 'osmotic adjustment' by which plants accumulate small organic molecules called osmolytes to protect the cellular machinery against denaturing environmental stresses (Yancey 2005; Chen and Jiang 2010; Liang et al. 2013). Plants have developed com-

P. Suprasanna (✉) • A.N. Rai
Plant Stress Physiology and Biotechnology Section, Nuclear Agriculture and Biotechnology Division, Bhabha Atomic Research Centre, Trombay, Mumbai 400 085, India
e-mail: penna888@yahoo.com; prasanna@barc.gov.in

G.C. Nikalje
Plant Stress Physiology and Biotechnology Section, Nuclear Agriculture and Biotechnology Division, Bhabha Atomic Research Centre, Trombay, Mumbai 400 085, India

Savitribai Phule Pune University, Pune 411 007, India

N. Iqbal et al. (eds.), *Osmolytes and Plants Acclimation to Changing Environment: Emerging Omics Technologies*, DOI 10.1007/978-81-322-2616-1_1

plex and well-organised mechanisms to adopt and tolerate biotic and abiotic stress conditions (Munns and Tester 2008). Abiotic stresses such as drought, salinity, cold, freezing, high temperature, anoxia, high light intensity and nutrient imbalances affect plant growth, development and productivity. It is thus important to understand the mechanisms controlling different metabolic pathways and their interplay during abiotic stress and develop improved crop varieties for realising complete yield potential.

To combat stress, plant metabolism is altered in many different ways including compatible solute production (Fig. 1.1) to stabilise proteins and cellular structures and/or to maintain cell turgor by osmotic adjustment and redox metabolism to remove excess levels of ROS and re-establish the cellular redox balance (Chinnusamy and Zhu 2009; Janska et al. 2010; Krasensky and Jonak 2012). Osmotic adjustment in mediating stress tolerance and protecting subcellular structure has been considered as a central dogma in stress physiology (Hare et al. 1998); however, it is still debated whether increased osmolyte accumulation can benefit crop yield (Seeraj and Sinclair 2002).

Osmotic adjustment has been shown to be an effective component of stress tolerance, and accumulation of osmoprotectants such as proline, glycine betaine, gamma-aminobutyric acid (GABA) and sugars is a common response observed in different plant systems (Ashraf and Foolad 2007; Chen and Jiang 2010) (Fig. 1.2). Organic osmolytes found in plants include low molecular weight compounds (sugar and sugar alcohols), methylated tertiary N compounds (glycine betaine) and amino acids (proline and glutamate) and other low molecular weight metabolites (Chen and Jiang 2010). In addition, b-alanine betaine, proline betaine, dimethylsulphoniopropionate and pinitol have also been associated with adaptation in halophytic plant species (Slama et al. 2015). Osmolytes are involved in unique reactions that can protect plant cells in various ways other than osmotically (Yancey 2005). Table 1.1 presents some of the protective properties that have been demonstrated in several research reports. Genetic engineering of metabolic conduit for a number of compatible solutes such as proline, glycine betaine, sorbitol, mannitol and trehalose has led to successful demonstration that transgenic plants display increased resistance to drought stress, high salinity and cold stress (Bhatnagar-Mathur et al. 2008; Reguera et al. 2012). In this review, we provide an update on different osmolytes, their accumulation and transgenic approach to improve abiotic stress tolerance in plants.

1.2 Sugars and Sugar Alcohols

Sugarsnot only function as osmoprotectants during stress but as substrates for growth and regulators of gene expression (Koch 1996). Disaccharides (sucrose, trehalose), raffinose family oligosaccharides (RFOs) and fructans are essentially involved in plant stress tolerance (Keunen et al. 2013). It has been demonstrated that some stress conditions such as drought, salinity, low temperature and flooding, in general, result in increased soluble sugar levels, whereas low sugar levels are seen under other stress conditions like high light irradiance, heavy metals and nutrient deficiency (Strand et al. 1999; Gill et al. 2001). The initial response to drought stress appears to be associated with an increase in monosaccharides, while the more delayed response seems to be associated with an increase in fructan (Kerepesi and Galiba 2000). In many higher plants, carbohydrate metabolism is modified under dehydration stress to favour the conversion of other sugars to sucrose (Whittaker et al. 2001). In *Craterostigma plantagineum*, survival under almost complete tissue dehydration is associated with accumulation of sucrose in aerial tissues because of metabolism of 2-octulose (Norwood et al. 2000). During the development and maturation of seeds of some species, stachyose levels rise as water is lost from the seed (Saravitz et al. 1987). Trehalose possesses the unique feature of reversible water absorption capacity to protect biological molecules from desiccation-induced damage and appears to be superior to other sugars at conferring protection (Rontein et al. 2002; Delorge et al. 2014). In the plant kingdom, most species do not seem to accumulate detectable levels of trehalose, with the

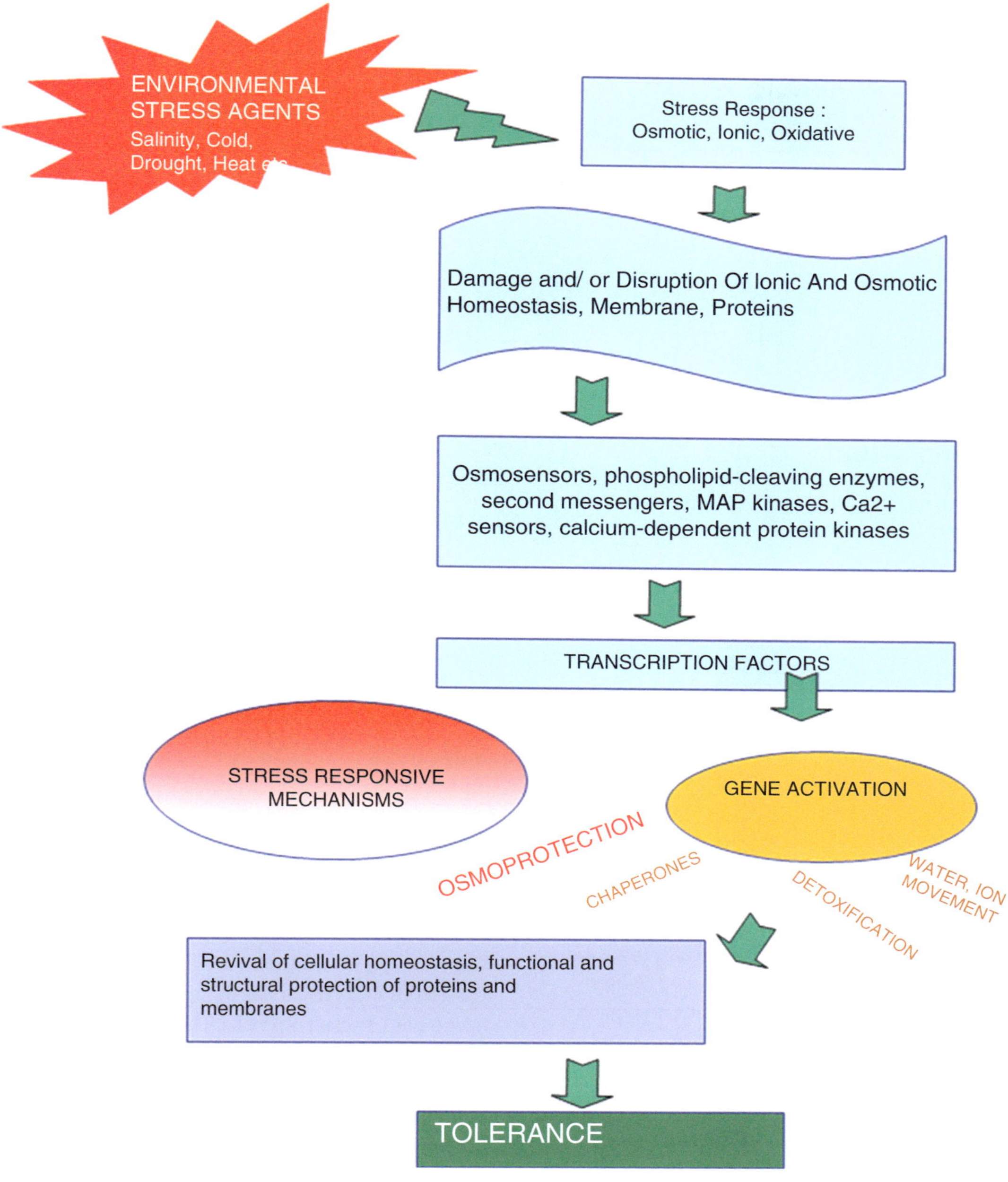

Fig. 1.1 Mechanisms of abiotic stress tolerance in plants. Different environmental stress conditions, such as drought, salinity and high/low temperatures induce signalling and regulatory networks leading to development of tolerance response

exception of the highly desiccation-tolerant 'resurrection' plants.

Raffinose family oligosaccharides (RFOs) have been implicated in mitigating the effects of environmental stresses such as cold, heat or dehydration (ElSayed et al. 2014). RFOs function in protecting cellular integrity during desiccation and/or imbibition, extending longevity in the dehydrated state and providing substrates for energy generation during germination. Taji et al. (2002) examined the functions of RFO in *Arabidopsisthaliana* plants under drought and

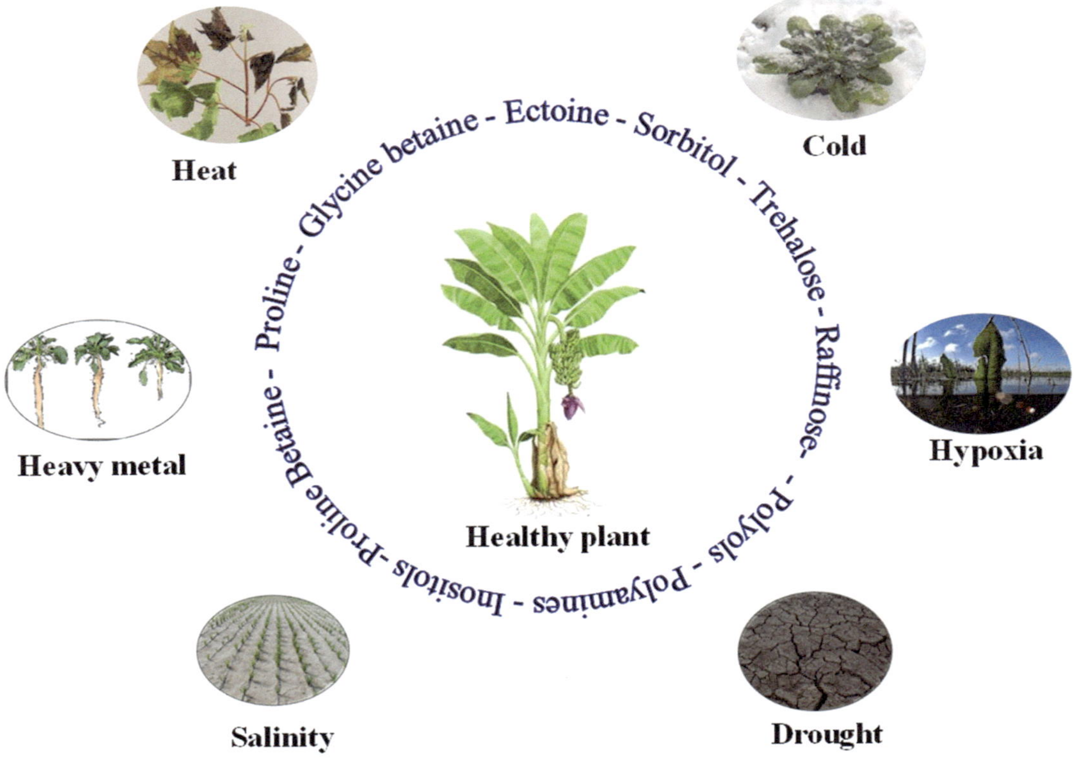

Fig. 1.2 Response of plants to different abiotic stresses and osmolyte production

Table 1.1 Compatible solutes and mechanism of protection

Compatible solutes	Mechanism of protection
Proline, glycine betaine	Protein integrity, ROS scavenging activity and singlet oxygen quenching ability
Trehalose	Reversible water absorption capacity, increase thermostability, membrane protection
Polyols	Stabilise enzymes or membrane structures, support redox control
GABA (γ-aminobutyric acid)	N metabolism, signal mediating plant-microbe interactions
Methylamines	Counteract inorganic ion inhibition

cold stress conditions based on the analyses of function and expression of genes involved in RFO biosynthesis. Sugar analysis showed that drought-, high salinity- and cold-treated Arabidopsis plants accumulate a large amount of raffinose and galactinol, but not stachyose, suggesting that raffinose and galactinol are involved in tolerance to drought, high salinity and cold stresses. Besides their role in abiotic stress, galactinol and RFOs are thought to act as signals to mediate stress responses (Valluru and Van den Ende 2011). Pathogen-induced systemic resistance has been shown to have signals mediated by these sugars possibly indicating a role in defence against biotic stresses (Kim et al. 2008; Valluru and Van den Ende 2011). Since biotic and abiotic stresses have integrating crosstalk, it is possible to assume such signalling mechanism operating in abiotic stress (Valluru and Van den Ende 2011).

Sugar alcohols also protect large molecules during dehydration and rehydration against rapid osmotic changes and oxidative stress (Tarczynski et al. 1993). Sugar alcohols could function as scavengers of activated oxygen species, thereby preventing peroxidation of lipids and resulting cell damage. Myo-inositol, D-pinitol and D-ononitol have been reported to be capable of scavenging hydroxyl radicals (Smirnoff 1998). Following stress in ice plant, enhanced inositol levels are utilised by the enzyme inositol O-methyl transferase to generate D-ononitol, which is then converted into D-pinitol. Myo-inositol accumulates under most abiotic stress conditions and is positively contributing to tolerance as a compatible solute (Tan et al. 2013). The sugar alcohol mannitol plays an important role in osmotic adaptation and confers enhanced tolerance to high salinity or water stress. In celery (*Apium graveolens*), mannitol is synthesised in equal amounts to that of sucrose and constitutes as much as 50 % of the translocated photoassimilate (Loester et al. 1992). Mannitol accumulation has also been shown to increase when plants are exposed to low water potential, and this accumulation is regulated by inhibition of competing pathways and decreased mannitol consumption and catabolism (Stoop et al. 1996).

Although accumulation of sugars is well associated with osmotic adaptation, sugars are also known to participate in specific signal transduction mechanisms during abiotic stress environment (Radomiljac et al. 2013). It is interesting to study the specificity of such sugar mediated signalling and derive a functional role for the individual sugars apart from their osmoprotectant roles (Radomiljac et al. 2013). In addition, it is also interesting to note that soluble sugars such as disaccharides, raffinose family oligosaccharides and fructans are strongly related to stress-induced ROS accumulation in plants.

1.3 Amino Acids

Accumulation of amino acids has been observed in plants exposed to abiotic stress (Rhodes et al. 1986; Lugan et al. 2010). During episodes of abiotic stress exposure, plants accumulate stress-related molecules with osmolyte role. Proline, glycine betaine and GABA) are involved and subsequently have become the candidates for genetic manipulation for improved stress tolerance (Huang et al. 2000; Giri 2011; Suprasanna et al. 2014).

1.4 Proline

Proline acts as a compatible solute to protect plant cells under osmotic stress (Csonka and Hanson 1991; Hare and Cress 1997; Kavi Kishor et al. 2005). As a molecular chaperone, proline has also been demonstrated to protect protein integrity and increase the enzyme activities. Proline is also suggested to bear an antioxidant role, having ROS scavenging activity and singlet oxygen quenching ability (Smirnoff and Cumbes 1989; Alia and Matysik 2001; Matysik et al. 2002). Proline is synthesised from (i) glutamate by Δ1-pyrroline-5-carboxylate (P5C) synthetase (P5CS) and P5C reductase (P5CR) and (ii) also from ornithine, which is converted into P5C/GSA via ornithine-δ-aminotransferase (Liang et al. 2013; Rai and Penna 2013). In plants, the pathway from glutamine is the primary route for the synthesis under osmotic stress conditions, and the pathway from ornithine is known to operate under nitrogen limitation (Delauney et al. 1993). In chloroplasts an enhanced rate of proline biosynthesis during stress can maintain the low NADPH: NADP+ ratio, contribute to sustaining the electron flow between photosynthetic excitation centres, stabilise the redox balance, and reduce photoinhibition and damage of the photosynthetic apparatus (Hare and Cress 1997).

During recovery from stress, proline is converted to Δ1-pyrroline-5-carboxylate (P5C) by proline dehydrogenase (PDH) and then to glutamate by P5C dehydrogenase (P5CDH). While proline accumulation under conditions of drought (low water potential), salinity and freezing (Verslues et al. 2006), heavy metal exposure (Sharma and Dietz 2006), and plant pathogen defence (Fabro et al. 2004) is reported, heat stress doesn't result in proline accumulation in tobacco and Arabidopsis plants, and induced proline accumulation renders plants more sensitive to heat (Rizhsky et al. 2004; Dobra et al. 2010; Lv et al. 2011).

A common adaptive response to various abiotic stresses in halophytes is seen through the accumulation of proline. The Aizoaceae family members accumulate large quantities of proline showing its role in osmoprotection (Deuschle et al. 2001; Lokhande et al. 2011). Such an osmotic adjustment mediated through proline accumulation is also evident in other halophyte species of different families (Slama et al. 2015, Table 1.2). High proline levels can improve the salt tolerance of the halophyte plant *Pancratium maritimum* by stabilising detoxifying enzymes and protein turnover machinery and stimulating the accumulation of stress-protective proteins (Khedr et al. 2003). Although proline hyperaccumulation correlates well with the halophytic behaviour in some plant species contributing to their salinity tolerance, it may not be an absolute requirement for them to adapt to extreme environmental conditions. For example, in some halophytes, such as *Camphorosma annua or Limonium* spp., instead of high proline, carbohydrate or betaine-derived osmolytes get accumulated under stress (Murakozy et al. 2003).

During multiple stress exposures (drought and heat stress), proline doesn't seem to accumulate while it gets accumulated under drought alone (Rizhsky et al. 2004). Heat stress has been shown to ameliorate the toxicity of proline to cells, and instead of proline, sucrose is accumulated in plants subjected to the stress combination, perhaps to protect hyperactive and susceptible mitochondria from the build-up of potentially toxic pyrroline-5-carboxylate (Rizhsky et al. 2004;

Table 1.2 Osmolytes, functions and distribution among angiosperm families

Type of osmolytes	Osmolytes	Role	Distribution in angiosperms
Amino acids	Proline, alanine, arginine, glycine, glutamine, asparagine, γ-aminobutyric acid, pipecolic acid, citrulline, ornithine	Precursors for most of the osmolytes, alleviation of cytoplasmic acidosis, prevents membrane damage and ion toxicity	Poaceae Cymodoceaceae Juncaginaceae Posidoniaceae Zosteraceae Asteraceae Brassicaceae Myrtaceae Plumbaginaceae Portulacea Aizoaceae Casuarinaceae Rhizophoraceae Fabaceae Plantaginaceae
Quaternary ammonium compounds	Glycine betaine, β-alanine betaine, proline betaine, choline-O-sulphate, hydroxyproline betaine, pipecolate betaine	Protection against damage of membrane, enzyme activity, regulation of ROS detoxification	Solanaceae, Fabaceae Acanthaceae Asteraceae Plumbaginaceae Amaranthaceae Aizoaceae Frankeniaceae
Tertiary sulphonium compounds	Dimethyl sulphoniopropionate (DMSP)	Osmoprotection, detoxifies excess sulphur	Poaceae
Sugars	Sucrose, trehalose, fructose, maltose, rhamnose	Osmotic adjustment and in stabilising membranes	Fabaceae, Plumbaginaceae Rhizophoraceae Cyperaceae Juncaceae, Poaceae Juncaginaceae Posidoniaceae Zosteraceae Asteraceae Brassicaceae
Sugar alcohols	Pinitol, mannitol, myo-inositol, ononitol, sorbitol	Facilitates osmotic adjustment and acts as signalling molecules	Cymodoceaceae Asteraceae Aizoaceae Combretaceae Plantaginaceae Fabaceae

Modified after Slama et al. (2015)

Mittler 2006). These findings suggest that during a combination of drought and heat stress, sucrose replaces proline in plants and functions as a major osmoprotectant. However, a recent study demonstrated the involvement of proline in the protection of plants against a combination of drought and heat stress (Cvikrova et al. 2013).

1.5 Glycine Betaine

Glycinebetaine (N,N,N-trimethylglycine) is widely accumulated in plants and other organisms (Chen and Murata 2002). In the case of osmoregulation, the compatible solute glycine betaine (GB), a small organic metabolite, can potentially play a crucial role in effective protection against salt, drought, and extreme temperature stress (Ashraf and Foolad 2007; Chen and Murata 2008, 2011; Giri 2011). GB is synthesised via two distinct pathways from two substrates: choline and glycine, respectively. The conversion of choline to GB involves two-enzyme pathway occurring naturally in various plants, animals and microorganisms. In this pathway, GB is formed as the result of the two-step oxidation of choline via the toxic intermediate betaine aldehyde. In higher plants, the reactions are catalysed by choline monooxygenase (CMO) and NAD+-dependent betaine aldehyde dehydrogenase (BADH), both of which are localised in the stroma of chloroplasts. The biosynthesis of GB is stress-inducible and the concentration of GB *in vivo* varies among plant species, ranging from 40 to 400 μ mol (g DW)$^{-1}$ (Rhodes and Hanson 1993).

In many crop plants, including sugar beet (Beta vulgaris), spinach (*Spinacia oleracea*), barley (*Hordeum vulgare*), wheat (*Triticum aestivum*) and sorghum (*Sorghum bicolor*), GB is known to accumulate in response to stress (McCue and Hanson 1990; Rhodes and Hanson 1993 and Yang et al. 2003). In these species, tolerant genotypes normally accumulate more GB than sensitive genotypes in response to stress. Chen and Murata (2008) listed plant species with stress-induced accumulation of GB. Natural accumulators of GB contain low levels of GB under non-stressed conditions, but accumulate larger amounts of GB under abiotic stress conditions (Rhodes and Hanson 1993).

Transgenic plants with overproduction of GB-synthesising genes exhibited increased production of GB and an enhancement in tolerance to salt, cold, drought or high temperature stress (Rhodes and Hanson 1993).

GB could be involved in inhibiting ROS accumulation, protection of photosynthetic machinery, activation of some stress-related genes and membrane protection (Chen and Murata 2008, 2011). The role of GB in ROS detoxification is also evident by reduced accumulation of ROS in transgenic plants under water-deficit stress as compared with control plants (Kathuria et al. 2009). Therefore, GB can provide tolerance to abiotic stresses even at low concentration by protecting photosynthesis under abiotic stress (Al-Taweel et al. 2007; Takahashi and Murata 2008). GB has also been implicated in protection of quaternary structure of proteins (thereby maintaining the enzyme activity) from damaging effects of environmental stresses. During salt or drought stress, synthesis of proteins involved in PSII repair is affected leading to photoinhibition (Al-Taweel et al. 2007; Takahashi and Murata 2008). GB antagonises the inhibition of protein biosynthesis and thus enhances the PSII repair, which leads to increased stress tolerance (Chen and Murata 2008).

Exogenous application of GB to low-accumulating or non-accumulating plants may help reduce adverse effects of environmental stresses (Yang and Lu 2005). GB accumulation could contribute to osmoregulation by stabilisation of the highly ordered structures of certain complex proteins, induction of the expression of specific genes that encode reactive oxygen species (ROS)-scavenging enzymes and subsequent depression of levels of ROS in plant cells, resulting in protection of the photosynthetic machinery from the combined effects of light stress and other kinds of stress, as well as of ion channel proteins and the integrity of cell membranes.

1.6 Gamma-Aminobutyric Acid (GABA)

GABA is a non-protein amino acid induced in response to biotic and abiotic stresses. The precursor for GABA is α-ketoglutarate which is converted by glutamate dehydrogenase to glutamate and by glutamate decarboxylase to GABA. Under adverse environmental conditions, GABA plays an important role in plant metabolism including osmotic and pH regulation and nitrogen metabolism, and it prevents accumulation of reactive oxygen species (Barbosa et al. 2010; Renault et al. 2010). Accumulation of GABA has been shown in response to salinity (Kinnersley and Turano 2000; Zhang et al. 2011; Renault et al. 2010). Flooding stress associated with marked decline in cytosolic pH and oxygen deprivation stress induces the higher accumulation of GABA (Kinnersley and Turano 2000). Other stresses, including cold, heat, salt and mild or transient environmental factors, such as touch, wind, rain, etc., rapidly increase cellular levels of Ca2+ which can stimulate calmodulin-dependent glutamate decarboxylase activity and GABA synthesis (Kinnersley and Turano 2000). Using *Caragana intermedia* (a legume shrub), exogenous GABA supply has been shown to alter the gene expression in roots under NaCl stress and activation of multiple mechanisms involved in signalling cascades, regulation of protein degradation, hormones biosynthesis, ROS production and PA metabolism (Shi et al. 2010).

1.7 Bioengineering for Development of Stress-Tolerant Crops

Progress in cellular and molecular biological tools has made it possible to clone important osmolyte biosynthetic genes and transfer them into crop plants (Wang et al. 2003). The elevated level of compatible osmolytes, radical scavengers and other transgene products has been demonstrated with the stress tolerance (Vinocur and Altman 2005). Several researchers have highlighted the genetic engineering efforts for enhanced osmolyte synthesis (for proline and glycine betaine) and stress tolerance (Wani et al. 2013). Overexpression of proline synthesis genes in tobacco plants have shown increased drought stress tolerance (Kavi Kishor et al. 1995). Overexpression of MfMIPS1 (*Medicago falcata myo-inositol phosphate synthase*) in tobacco resulted in an increased MIPS activity and myo-inositol, galactinol and raffinose, resulting in enhanced resistance to chilling, drought and salt stresses in transgenic tobacco plants (Tan et al. 2013).

Several of the glycine betaine non-accumulating and stress-susceptible crops have been used as targets for genetic engineering of osmotolerance by producing betaine in nonaccumulators (McCue and Hanson 1990), and transgenic plants with moderate tolerance to salt stress have been obtained (Jia et al. 2002). Transgenic sweet potato plants expressing *BADH* gene exhibited higher tolerance to various abiotic stresses, including salt, oxidative stress and low temperature (Fan et al. 2012). Exogenous choline addition increased the level of GB accumulation in *Arabidopsisthaliana, Brassica napus* and *Nicotiana tabacum*, suggesting that choline supplement is necessary for high GB levels in transgenic plants (Huang et al. 2000). Transgenic Brassica plants containing choline oxidase (codA) gene showed resistance to multiple abiotic stresses as a result of increased glycinebetaine accumulation (Wang et al. 2010). Overexpression of betaine by manipulation of *betaine aldehyde dehydrogenase* (BADH) via chloroplast genetic engineering may prove to be an important strategy in order to confer salt tolerance on desired crops (Kumar et al. 2004; Fitzgerald et al. 2009).

In order to genetically engineer plants for improved stress tolerance, trehalose biosynthesis pathway has served as the choice target system in several studies (Delorge et al. 2014). Initial studies used *otsA* (trehalose-6-phosphate synthase) and *otsB* (trehalose-6-phosphate phosphatase) genes from *E. coli* to raise transgenic tobacco plants more drought tolerance than controls, even though they exhibited morphological changes and higher levels of nonstructural carbohydrates

(Pilon-Smits et al. 1998). To overcome such phenotypic and growth penalties, overproduction of trehalose in rice transgenic plants was studied using stress-inducible or tissue-specific expression of a bifunctional trehalose-6-phosphate synthase/phosphatase (*TPSP*) fusion gene (comprising the *E.coli* trehalose biosynthetic genes *OtsAand OtsB* with TPS and TPP activity, respectively) (Garg et al. 2002). Cortina and Culianez-Macia (2005) explored a further possibility of genetic engineering for abiotic stress tolerance of tomato using the yeast *TPS1* gene. The *TPS1*-tomato plants had improved tolerance possibly through carbohydrate alterations produced via trehalose biosynthetic pathway. In some cases, although trehalose overproduction has resulted in plant growth abnormalities, such a condition can be circumvented by using stress-inducible promoters or in tissue-specific manner (Karim et al. 2007; Delorge et al. 2014). The Arabidopsis mutants deficient in GABA transaminase showed oversensitivity to salt stress (Renault et al. 2010), and the rice plants overexpressing the GAD gene accumulated extremely high amounts of GABA under salt stress treatment.

Ectoine, a common osmolyte mostly found in a halobacteria (Lentzen and Schwarz 2006) is shown to improve protein folding and protect biomolecules (Knapp et al. 1999). Three genes, ectB, ectA and ectC, which encode L-2, 4-diaminobutyric acid aminotransferase, L-2, 4-diaminobutyric acid acetyl transferase and L-ectoine synthase, respectively, are responsible for ectoine biosynthesis. Transformed cultured tobacco cell lines with ectA, ectB and ectC under the control of the constitutive CaMV 35S promoter resulted in increased resistance to mannitol-induced osmotic stress (Nakayama et al. 2000). Tobacco plants transformed with ectoine biosynthetic genes showed improvement in the maintenance of root function to enable water transport consistently to shoots under saline conditions and enhanced the nitrogen supply to leaves by increasing transpiration and by protecting RuBisCO proteins (Moghaieb et al. 2006).

1.8 Conclusions

In the face of climate change and impending recurrence of stress factors, agronomically important plants should be the targets for improving abiotic stress tolerance. Abiotic stress factors significantly affect plant growth and development and consequently productivity of crop plants. Since stress adaptive mechanisms include metabolic changes, gene expression and osmolyte synthesis, considerable research involving metabolites such as proline, mannitol, trehalose and glycine betaine has been done to validate their role in stress tolerance. Although genetic engineering efforts have mostly been conducted in model plants, this research has generated immense interest in the manipulation of crop plants for stress-related pathways of osmolyte production. This will undoubtedly require developing knowledge base of pathways and related networks in important crop plants.

In the past couple of decades, success has mostly been in improving tolerance to single stress factors (e.g. salt, drought, heat). Elucidation of multiple tolerance mechanisms through molecular detailing of the stress perception, signal transduction and genetic regulation and developing strategies for achieving tolerance can provide means to engineer tolerance to combined stresses. Understanding molecular crosstalk may lead to developing master controls to combat biotic and abiotic stresses towards sustainable agriculture.

References

Alia MP, Matysik J (2001) Effect of proline on the production of singlet oxygen. Amino Acids 21:195–200

Al-Taweel K, Iwaki T, Yabuta Y et al (2007) A bacterial transgene for catalase protects translation of d1 protein during exposure of salt-stressed tobacco leaves to strong light. Plant Physiol 145(1):258–265

Ashraf M, Foolad MR (2007) Roles of glycine betaine and proline in improving plant abiotic stress resistance. Environ Exp Bot 59:206–216

Barbosa JM, Singh NK, Cherry JH et al (2010) Nitrate uptake and utilization is modulated by exogenous -aminobutyric acid in Arabidopsis thaliana seedlings. Plant Physiol Biochem 48:443–450

Bhatnagar-Mathur P, Vadez V, Sharma KK (2008) Transgenic approaches for abiotic stress tolerance in plants: retrospect and prospects. Plant Cell Rep 27:411–442

Chen H, Jiang J-G (2010) Osmotic adjustment and plant adaptation to environmental changes related to drought and salinity. Environ Rev 18:309–319

Chen TH, Murata N (2002) Enhancement of tolerance of abiotic stress by metabolic engineering of betaines and other compatible solutes. Curr Opin Biotechnol 5:250–257

Chen TH, Murata N (2008) Glycinebetaine: an effective protectant against abiotic stress in plants. Trends Plant Sci 13:499–505

Chen TH, Murata N (2011) Glycinebetaine protects plants against abiotic stress: mechanisms and biotechnological applications. Plant Cell Environ 34(1):1–20

Chinnusamy V, Zhu JK (2009) Epigenetic regulation of stress responses in plants. Curr Opin Plant Biol 12:133–139

Cortina C, Culianez-Macia FA (2005) Tomato abiotic stress enhanced tolerance by trehalose biosynthesis. Plant Sci 169:75–82

Csonka LN, Hanson AD (1991) Prokaryotic osmoregulation: genetics and physiology. Annu Rev Microbiol 45:569–606

Cvikrova M, Gemperlova L, Martincova O et al (2013) Effect of drought and combined drought and heat stress on polyamine metabolism in proline-overproducing tobacco plants. Plant Physiol Biochem 73:7–15

Delauney AJ, Hu CA, Kishor PB et al (1993) Cloning of ornithine delta-aminotransferase cDNA from Vigna aconitifolia by trans-complementation in Escherichia coli and regulation of proline biosynthesis. J Biol Chem 268:18673–18678

Delorge I, Michal J, Sebastien C et al (2014) Fine tuning of trehalose biosynthesis and hydrolysis as novel tools for the generation of abiotic stress tolerant plants. Front Plant Sci 5:147

Deuschle K, Funck D, Hellmann H et al (2001) A nuclear gene encoding mitochondrial Δ1-pyrroline-5-carboxylate dehydrogenase and its potential role in protection from proline toxicity. Plant J 27:345–356

Dobra J, Motyka V, Dobrev P et al (2010) Comparison of hormonal responses to heat, drought and combined stress in tobacco plants with elevated proline content. J Plant Physiol 167:1360–1370

ElSayed AI, Rafudeen MS, Golldack D (2014) Physiological aspects of raffinose family oligosaccharides in plants: protection against abiotic stress. Plant Biol 16:1–8

Fabro G, Kovács I, Pavet V et al (2004) Proline accumulation and AtP5CS2 gene activation are induced by plant-pathogen incompatible interactions in Arabidopsis. Mol Plant Microbe Interact 17:343–350

Fan W, Zhang M, Zhang H, Zhang P (2012) Improved tolerance to various abiotic stresses in transgenic sweet potato (Ipomoea batatas) expressing spinach betaine aldehyde dehydrogenase. PLoS One 7(5):e37344

Fitzgerald TL, Waters DLE, Henry RJ (2009) Betaine aldehyde dehydrogenase in plants. Plant Biol 11(2):119–130

Garg AK, Kim J-K, Owens TG et al (2002) Trehalose accumulation in rice plants confers high tolerance levels to different abiotic stresses. Proc Natl Acad Sci U S A 99:15898–15903

Gill PK, Sharma AD, Singh P et al (2001) Effect of various abiotic stresses on the growth soluble sugars and water relations of sorghum seedlings grown in light and darkness. Bulg J Plant Physiol 27:72–84

Giri J (2011) Glycinebetaine and abiotic stress tolerance in plants. Plant Signal Behav 6:1746–1751

Hare PD, Cress WA (1997) Metabolic implications of stress-induced proline accumulation in plants. Plant Growth Regul 21:79–102

Hare PD, Cress WA, Staden J (1998) Dissecting the roles of osmolyte accumulation during stress. Plant Cell Environ 21:535–553

Huang J, Hirji R, Adam L et al (2000) Genetic engineering of glycine betaine production toward enhancing stress tolerance in plants: metabolic limitations. Plant Physiol 122:747–756

Janska A, Marsik P, Zelenkova S et al (2010) Cold stress and acclimation: what is important for metabolic adjustment? Plant Biol 12:395–405

Jia GX, Zhu ZQ, Chang FQ et al (2002) Transformation of tomato with the BADH gene from Atriplex improves salt tolerance. Plant Cell Rep 21:141–146

Karim S, Aronsson H, Ericson H et al (2007) Improved drought tolerance without undesired side effects in transgenic plants producing trehalose. Plant Mol Biol 64(4):371–386

Kathuria H, Giri J, Nataraja KN et al (2009) Glycinebetaine-induced water-stress tolerance in codA-expressing transgenic indica rice is associated with up-regulation of several stress responsive genes. Plant Biotechnol J 7:512–526

Kavi Kishor PB, Hong Z, Miao G et al (1995) Overexpression of Δ1-pyrroline-5-carboxylate synthetase increases proline overproduction and confers osmotolerance in transgenic plants. Plant Physiol 108:1387–1394

Kavi Kishor PB, Sangam S, Amrutha RN et al (2005) Regulation of proline biosynthesis, degradation, uptake and transport in higher plants: its implications in plant growth and abiotic stress tolerance. Sci Curl 88:3

Kerepesi I, Galiba G (2000) Osmotic and salt stress-induced alteration in soluble carbohydrate content in wheat seedlings. Crop Sci 40:482–487

Keunen E, Peshev DJ, Vangronsveld J et al (2013) Plant sugars are crucial players in the oxidative challenge during abiotic stress: extending the traditional concept. Plant Cell Environ 36:1242–1255

Khedr AHA, Abbas MA, Abdel Wahid AA et al (2003) Proline induces the expression of salt stress- responsive proteins and may improve the adaptation of *Pancratium maritimum* L. to salt-stress. J Exp Bot 54:2553–2562

Kim MS, Cho SM, Kang EY, Im YJ, Hwangbo H, Kim YC, Ryu CM, Yang KY, Chung GC, Cho BH (2008) Galactinol is a signaling component of the induced systemic resistance caused by Pseudomonas chlororaphis O6 root colonization. Mol Plant-Microbe Interact 12:1643–1653

Kinnersley AM, Turano FJ (2000) Gamma aminobutyric acid (GABA) and plant responses to stress. Crit Rev Plant Sci 19:479–509

Knapp S, Ladenstein R, Galinski EA (1999) Thermal stabilization of bovine ribonuclease A by naturally occurring osmolytesb-hydroxyectoine and betaine. Extremophiles 3:191–198

Koch KE (1996) Carbohydrate modulated gene expression in plants. Annu Rev Plant Physiol Plant Mol Biol 47:509–540

Krasensky J, Jonak C (2012) Drought, salt, and temperature stress-induced metabolic rearrangements and regulatory networks. J Exp Bot 63:1593–1608

Kumar S, Dhingra A, Daniell H (2004) Plastid-expressed betaine aldehyde dehydrogenase gene in carrot cultured cells, roots, and leaves confers enhanced salt tolerance. Plant Physiol 136(1):2843–2854. doi:10.1104/pp. 104.045187

Lentzen G, Schwarz T (2006) Extremolytes: natural compounds from extremophiles for versatile applications. Appl Microbiol Biotechnol 72:623–634

Liang X, Zhang L, Natarajan SK, Becker DF (2013) Proline mechanisms of stress survival. Antioxid Redox Signal 19:998–1011

Loester WH, Tyson RH, Everard JD et al (1992) Mannitol synthesis in higher plants: evidence for the role and characterization of a NADPH dependant mannose-6-phosphate reductase. Plant Physiol 98:1396–1402

Lokhande VH, Nikam TD, Patade VY, Ahire ML, Suprasanna P (2011) Effects of optimal and supra-optimal salinity stress on antioxidative defence, osmolytes and in vitro growth responses in Sesuvium portulacastrum L. Plant Cell Tiss Org Cult 104:41–49

Lugan R, Niogret MF, Leport L et al (2010) Metabolome and water homeostasis analysis of *Thellungiella salsuginea* suggests that dehydration tolerance is a key response to osmotic stress in this halophyte. Plant J 64:215–229

Lv WT, Lin B, Zhang M et al (2011) Proline accumulation is inhibitory to Arabidopsis seedlings during heat stress. Plant Physiol 156:1921–1933

Matysik J, Alia BB, Mohanty P (2002) Molecular mechanism of quenching of reactive oxygen species by proline under stress in plants. Curr Sci 82:525–532

McCue KF, Hanson AD (1990) Drought and salt tolerance: towards understanding and application. Trends Biotechnol 8:358–362

Mittler R (2006) Abiotic stress, the field environment and stress combination. Trends Plant Sci 11:15–19

Moghaieb RE, Tanaka N, Saneoka H, Murooka Y, Ono H, Morikawa H, Nakamura A, Nguyen NT, Suwa R, Fujita K (2006) Characterization of salt tolerance in ectoine-transformed tobacco plants (*Nicotiana tabacum*): photosynthesis, osmotic adjustment, and nitrogen partitioning. Plant Cell Environ 29(2):173–182

Munns R, Tester M (2008) Mechanism of salinity tolerance. Annu Rev Plant Biol 59:651–681

Murakozy E, Nagy Z, Duhaze C et al (2003) Seasonal changes in the levels of compatible osmolytes in three halophytic species of inland saline vegetation in Hungary. J Plant Physiol 160:395–401

Nakayama H, Yoshida K, Ono H et al (2000) Ectoine, the compatible solute of *Halomonas elongata*, confers hyperosmotic tolerance in cultured tobacco cells. Plant Physiol 122:1239–1247

Norwood M, Truesdale MR, Richter A et al (2000) Photosynthetic carbohydrate metabolism in the resurrection plant *Craterostigma plantagineum*. J Exp Bot 51(343):159–165

Pilon-Smits EA, Terry N, Sears T et al (1998) Trehalose-producing transgenic tobacco plants show improved growth performance under drought stress. J Plant Physiol 152:525–532

Radomiljac JD, Whelan J, van der Merwe M (2013) Coordinating metabolite changes with our perception of plant abiotic stress responses: emerging views revealed by integrative-omic analyses. Metabolites 3:761–786

Rai AN, Penna S (2013) Molecular evolution of plant P5CS gene involved in proline biosynthesis. Mol Biol Rep 40(11):6429–6435

Reguera M, Peleg Z, Blumwald E (2012) Targeting metabolic pathways for genetic engineering abiotic stress-tolerance in crops. Biochim Biophys Acta 1819:186–194

Renault H, Roussel V, Amrani E et al (2010) The Arabidopsis pop2-1 mutant reveals the involvement of GABA transaminase in salt stress tolerance. BMC Plant Biol 10:1–16

Rhodes D, Hanson AD (1993) Quaternary ammonium and tertiary sulfonium compounds in higher-plants. Annu Rev Plant Physiol Plant Mol Biol 44:357–384

Rhodes D, Handa S, Bressan RA (1986) Metabolic changes associated with adaptation of plant-cells to water-stress. Plant Physiol 82:890–903

Rizhsky L, Liang HJ, Shuman J et al (2004) When defense pathways collide. The response of Arabidopsis to a combination of drought and heat stress. Plant Physiol 134:1683–1696

Rontein D, Basset G, Hanson AD (2002) Metabolic engineering of osmoprotectant accumulation in plants. Metab Eng 4:49–56

Saravitz DM, Pharr DM, Carter TE (1987) Galactinol synthase activity and soluble sugars in developing seeds of four soybean genotypes. Plant Physiol 72:620–624

Seeraj R, Sinclair TR (2002) Osmolyte accumulation: can it really help increase crop yield under drought conditions? Plant Cell Environ 25:333–341

Sharma SS, Dietz KJ (2006) The significance of amino acids and amino acid-derived molecules in plant

responses and adaptation to heavy metal stress. J Exp Bot 57:711–726
Shi S-Q, Shi Z, Jiang ZP, Qi LW, Sun XM, Li CX, Liu JF, Xiao WF, Zhang SG (2010) Effects of exogenous GABA on gene expression of *Caragana intermedia* roots under NaCl stress: regulatory roles for H_2O_2 and ethylene production. Plant Cell Environ 33:149–162
Slama I, Abdelly C, Bouchereau A, Flowers T, Savoure' A (2015) Diversity, distribution and roles of osmoprotective compounds accumulated in halophytes under abiotic stress. Ann Bot 115(3):433–447
Smirnoff N (1998) Plant resistance to environmental stress. Curr Opin Biotechnol 9(2):214–219
Smirnoff N, Cumbes QJ (1989) Hydroxyl radical scavenging activity of compatible solutes. Phytochemistry 28:1057–1060
Stoop JMH, Williamson JD, Pharr MD (1996) Mannitol metabolism in plants: a method for coping with stress. Trends Plant Sci 1(5):139–144
Strand A, Hurry V, Henkes S et al (1999) Acclimation of Arabidopsis leaves developing at low temperatures. Increasing cytoplasmic volume accompanies increased activities of enzymes in the Calvin Cycle and in the sucrose-biosynthesis pathway. Plant Physiol 119:1387–1397
Suprasanna P, Rai AN, Hima Kumari P et al (2014) Modulation of proline: implications in plant stress tolerance and development. In: Anjum NA, Gill SS, Gill R (eds) Plant adaptation to environmental change: significance of amino acids and their derivatives. CABI, Oxfordshire, pp 68–96
Taji T, Ohsumi C, Luchi S et al (2002) Important roles of drought- and cold-inducible genes for galactinol synthase in stress tolerance in *Arabidopsis thaliana*. Plant J 29:417–426
Takahashi S, Murata N (2008) How do environmental stresses accelerate photoinhibition? Trends Plant Sci 13:178–182
Tan JL, Wang CY, Xiang B et al (2013) Hydrogen peroxide and nitric oxide mediated cold- and dehydration-induced myo-inositol phosphate synthase that confers multiple resistances to abiotic stresses. Plant Cell Environ 36:288–299
Tarczynski MC, Jensen RG, Bohnert HJ (1993) Stress protection of transgenic plants by production of the osmolyte mannitol. Science 259:508–510
Valluru R, Van den Ende W (2011) Myo-inositol and beyond – emerging networks under stress. Plant Sci 181:387–400
Verslues PE, Agarwal M, Katiyar-Agarwal S et al (2006) Methods and concepts in quantifying resistance to drought, salt and freezing, abiotic stresses that affect plant water status. Plant J 45:523–539
Vinocur B, Altman A (2005) Recent advances in engineering plant tolerance to abiotic stress: achievements and limitations. Curr Opin Biotechnol 16:123–132
Wang W, Vinocur B, Altman A (2003) Plant responses to drought, salinity and extreme temperatures: towards genetic engineering for stress tolerance. Planta 218:1–14
Wang GP, Li F, Zhang J et al (2010) Overaccumulation of glycine betaine enhances tolerance of the photosynthetic apparatus to drought and heat stress in wheat. Photosynthetica 48:30–41
Wani SH, Singh NB, Haribhushan A et al (2013) Compatible solute engineering in plants for abiotic stress tolerance- role of glycine betaine. Curr Genomics 14:157–165
Whittaker A, Bochicchio A, Vazzana C et al (2001) Changes in leaf hexokinase activity and metabolite levels in response to drying in the desiccation-tolerant species *Sporobolus stapfianus and Xerophyta viscosa*. J Exp Bot 52:961–969
Yancey PH (2005) Organic osmolytes as compatible, metabolic and counteracting cytoprotectants in high osmolarity and other stresses. J Exp Bio 208:2819–2830
Yang X, Lu C (2005) Photosynthesis is improved by exogenous glycine betaine in salt-stressed maize plants. Physiol Plant 124:343–352
Yang W-J, Rich PJ, Axtell JD et al (2003) Genotypic variation for glycine betaine in sorghum. Crop Sci 43:162–169
Zhang J, Zhang Y, Du Y et al (2011) Dynamic metabolomic responses of Tobacco (*Nicotiana tabacum*) plants to salt stress. J Proteome Res 10:1904–1914

Contribution of Osmolyte Accumulation to Abiotic Stress Tolerance in Wild Plants Adapted to Different Stressful Environments

2

Oscar Vicente, Mohamad Al Hassan, and Monica Boscaiu

Abstract

Abiotic stresses, mostly drought and salinity, are the major environmental factors which limit plant distribution in nature and reduce crop yields worldwide. The biotechnological improvement of crop stress tolerance would significantly contribute to the needed increase in food production, but requires a deep understanding of the mechanisms underlying plant responses to stress. Accumulation of osmolytes is one of those responses, which appears to be essential for tolerance in many species. Their main assumed role is to contribute to osmotic adjustment under conditions causing cellular dehydration, but they also have osmoprotectant functions as low-molecular-weight chaperons and reactive oxygen species (ROS) scavengers. Yet, important aspects of their mechanisms of action remain largely unknown, especially regarding the relevance and relative contribution of specific osmolytes to the stress tolerance of a given species. This gap in our knowledge is partly due to the experimental approaches commonly used to study those mechanisms, which have focused on non-tolerant model species and/or experiments performed under controlled – but artificial – laboratory or greenhouse setups.

In this review, we will summarise the (relatively scarce) data from field studies on the accumulation of different osmolytes in wild plants adapted to distinct stressful environments: saline, arid and gypsum habitats. We propose that more effort and resources should be invested on the study of the stress responses of wild plants in their natural habitats, as a comple-

O. Vicente • M. Al Hassan
Institute for Plant Molecular and Cellular Biology (IBMCP, UPV-CSIC), Universitat Politècnica de València, Camino de Vera s/n, 46022 Valencia, Spain

M. Boscaiu (✉)
Mediterranean Agroforestal Institute (IAM, UPV), Universitat Politècnica de València, Camino de Vera s/n, 46022 Valencia, Spain
e-mail: mobosnea@eaf.upv.es

N. Iqbal et al. (eds.), *Osmolytes and Plants Acclimation to Changing Environment: Emerging Omics Technologies*, DOI 10.1007/978-81-322-2616-1_2

ment to greenhouse experiments. We believe that this approach will significantly enhance our knowledge on this specific topic and could eventually be applied to the genetic improvement of crops.

2.1 Introduction

The great majority of wild plants and almost all cultivated species are sensitive to abiotic stress. The initial domestication and subsequent development of crop plants throughout history, by selection of traits such as growth rate, biomass accumulation, production of fruits and seeds, etc., did not enhance at all their stress tolerance. Quite on the contrary, cultivated plants are generally less tolerant than their wild predecessors, since inhibition of vegetative growth and reproductive development is actually the most general and immediate response to stress in plants, which use all their resources (energy, metabolic precursors, etc.) to survive the adverse environmental conditions (Serrano and Gaxiola 1994; Zhu 2001).

Yet, there are plants naturally adapted to different abiotic stresses, which grow in particular stressful habitats under harsh environmental conditions, such as halophytes, or salt-tolerant plants present on saline soils; gypsophytes or plants adapted to gypsum soils; and xerophytes, which are drought-resistant plants growing in arid and semiarid areas. In all these habitat types, an additional factor limiting plant development is generally the scarcity of nutrients in the soil. Only the specialised plants mentioned above are able to survive and complete their life cycle in such habitats, despite the adverse environmental conditions. Given the undoubted scientific interest of those environments, there are numerous botanical and ecological studies on such plants and their habitats. Their responses to stress have been also addressed in particular cases at the physiological and, to some extent, at the biochemical level. However, most of these studies have been performed in laboratory or greenhouse setups that, while allowing tighter experimental control, do not reflect the natural conditions of the plants. Moreover, very little is known about the molecular basis of the mechanisms of stress tolerance in these wild plants; paradoxically, most studies on stress responses and stress tolerance in plants have been conducted in model species that are not tolerant, especially in *Arabidopsisthaliana* and, to a much lesser extent, in some crops (Sanders 2000; Zhu 2000, 2001), which a priori do not seem to be the most appropriate experimental material for this type of research. Yet, the use of non-tolerant models could be at least partially justified, since there is abundant evidence that both sensitive and tolerant plants use the same basic mechanisms to respond to abiotic stress and that these mechanisms are overlapping or common to different types of environmental stress. The degree of tolerance to drought, salt, high temperatures, etc., which can reach *Arabidopsis* (or crop) plants is not at all comparable to those of wild plants adapted to each particular environment. This clearly indicates that the mechanisms of response to abiotic stress operating in the latter must be much more effective than those that function in non-tolerant models, although all may share the same molecular basis. Among the conserved mechanisms conferring stress tolerance in plants, the most important are based on: ***i)*** the control of ion homeostasis and cellular osmotic balance: regulation of water transport into the cell, compartmentalisation of toxic ions into the vacuole, synthesis and accumulation of compatible solutes, the so-called osmolytes (sugars, polyalcohols, amino acids or derivatives), in the cytoplasm; ***ii)*** the synthesis of protective low-molecular-weight compounds (the aforementioned osmolytes, in their 'osmoprotectant' role, see below) or proteins (heat shock proteins, LEA proteins, osmotin etc.); ***iii)*** the activation of antioxidant systems, both enzymatic (superoxide dismutase, catalase, ascorbate peroxidase, glutathione peroxidase, glutathione reductase) and non-enzymatic (reduced glutathione, vitamins C and E, carotenoids, flavonoids and other phenolic compounds, etc.) (Zhu 2001; Munns 2002; Vinocur and Altman 2005).

A possible explanation for the large differences observed between different plant species, with regard to their stress tolerance, is that certain proteins involved in stress responses – and playing a key role in the mechanisms of tolerance – have higher levels of expression and/or intrinsic activity in tolerant taxa as compared to the homologous proteins from non-tolerant species. These proteins, encoded by 'stress tolerance genes', may include, among others, plasma membrane and tonoplast ion transporters, enzymes involved in the synthesis of osmolytes, antioxidant enzyme systems or regulators of the expression or activity of any of the former, such as specific transcription factors, protein kinases, etc. Differences in gene expression levels could be due to the relative strength of the corresponding promoters or differences in their regulatory mechanisms; for example, the expression of a given 'stress tolerance' gene may be inducible under stress conditions in the tolerant species, but not that of its homologs in stress-sensitive plants. Therefore, stress-tolerant wild plants represent a convenient source – underutilised at present, in our opinion – of 'stress tolerance genes' and their regulatory elements. Once isolated and characterised, they can be used as biotechnological tools for the genetic improvement of stress tolerance in (transgenic) crop plants, tools that could hopefully be more effective than those presently available, isolated from non-tolerant models. The development of transgenic crops tolerant to drought, salinity and other abiotic stresses is one of the strategies that can significantly contribute to increasing agricultural productivity and thus food production, essential to feed a growing population in a global context of climate change, which will exacerbate the ongoing reduction of the area of cultivated land and increasing scarcity of water available for irrigation (DB Climate Change Advisors 2009).

Only in recent years several salt- and drought-tolerant species of the genus *Thellungiella*, somewhat related to *Arabidopsis*, as both genera belong to the same family (Brassicaceae), were proposed as extremophile models for abiotic stress tolerance studies (Inan et al. 2004; Kant et al. 2006). Comparative studies of these models are providing information on the molecular mechanisms of stress tolerance. A complementary approach, which could substantially contribute to the advance of knowledge in this field, would be based on the study of the stress response pathways in tolerant wild plants in their natural habitats; this will provide complementary information to that obtained using model systems subjected to controlled stress treatments under artificial laboratory/greenhouse conditions. Specifically, this strategy will allow establishing the relative contribution of different responses to stress tolerance mechanisms for a given species and under ecologically relevant conditions in the wild. For these studies, as mentioned before, we have a wide range of biochemical and enzymatic stress markers whose levels/activity can be determined in the plants; variations in these biomarkers could then be correlated with changes in the environmental conditions of the plants regarding, for example, soil properties and climatic conditions.

The synthesis of compatible organic solutes or osmolytes is one of the most common responses to stress in plants (Flowers and Colmer 2008; Hussain et al. 2008; Türkan and Demiral 2009). The accumulation of these compounds is induced in response to all environmental conditions leading to cellular dehydration, in all organisms, ranging from microorganisms to plants (Yancey et al. 1982; Burg et al. 1996), which represent a striking case of convergent evolution. Osmolytes are very soluble, low-molecular-weight organic compounds which do not interfere with normal metabolism, even at high concentrations. Chemically, they are diverse, including some amino acids, such as proline (Pro), and quaternary ammonium compounds (QACs), such as glycine betaine (GB), as well as polyols (glycerol, sorbitol, mannitol, several inositol isomers and derivatives) and soluble sugars (trehalose, sucrose, fructose or glucose) (Sairam and Tyagi 2004; Bartels and Sunkar 2005).

Besides their role in osmotic adjustment, osmolytes fulfil additional roles during the stress response, acting as 'osmoprotectants': they can directly stabilise proteins and membrane structures under dehydration conditions – acting as low-molecular-weight chaperons – and protect the cell against oxidative stress as scavengers of 'reactive oxygen species' (ROS), as it has been largely discussed in many reviews (Shen et al. 1997; Hong et al. 2000; Akashi et al. 2001; Zhu 2001; Chen and Murata 2002; Ashraf and Foolad 2007; Szabados and Savouré 2010).

In this chapter, we will focus on the responses of wild plants to stress in the field, based on the analysis of changes in the levels of different osmolytes in plant material collected in natural habitats and under different environmental conditions. Although conditions such as soil salinity, humidity, temperature or rainfall cannot be controlled in nature, they can be measured and correlated with the plant responses. The number of publications dealing with this specific topic is limited, which is not surprising considering the complexity of such studies, which require a multidisciplinary approach including plants' taxonomic identification, soil and climate analyses, determination of physiological responses and biochemical assays. In addition, the analysis of field data is much more complicated than the interpretation of laboratory results, since in their natural ecosystems, plants must respond simultaneously to different types of stress, which may activate similar or different mechanisms, either independent, showing additive effects or interacting in more complex ways (Krasensky and Jonak 2012; Ben Hamed et al. 2013).

2.2 Plant Stress Responses in Saline Environments

Salt stress, with its two well-known components – osmotic stress and ion toxicity – has a clear inhibitory effect on plant growth. The 'osmotic stress' component is generated not only by excessive soil salinity but also by drought, cold or high temperatures; all these conditions cause dehydration and loss of turgor at the cellular level (Greenway and Munns 1980; Serrano and Gaxiola 1994). Intracellular accumulation of Na^+ and/or Cl^- at high concentrations, on the other hand, is specific for salt stress; these ions have a toxic effect, since they inhibit the activity of many enzymatic systems and some basic cellular processes, such as protein synthesis or mRNA processing (Serrano 1996; Yeo 1998; Zhu 2001; Forment et al. 2002). NaCl also affects plant mineral nutrition, by interfering with K^+ and Ca^{2+} uptake, and promotes oxidative stress through generation of reactive oxygen species (Serrano and Gaxiola 1994; Yeo 1998; Zhu 2001). All these effects, in addition to inadequate photosynthesis due to stomatal closure and the subsequent reduction of CO_2 uptake, produce the inhibition of growth in the presence of salt and eventually plant death if the limits of tolerance are surpassed.

Halophytes are plants adapted to grow and complete their life cycle in habitats with soil salinity equivalent to, at least, 200 mM NaCl (Flowers and Colmer 2008). They represent only a small fraction (ca. 0.25 %) of all angiosperms, including about 350 species belonging to a small number of genera and families (Flowers et al. 2010). They grow in different saline ecosystems, including wet littoral and inland salt marshes, but also dry habitats such as saline deserts or dunes and cliffs in coastal areas. Their salt tolerance depends mainly on their ability to accumulate Na^+ and Cl^- in the vacuole, so that their levels in the cytoplasm are substantially lower, thus avoiding the inhibition of metabolic processes (Flowers et al. 1986; Serrano and Gaxiola 1994). The concomitant synthesis and accumulation of compatible solutes (osmolytes) in the cytoplasm is a fundamental mechanism for the maintenance of cellular osmotic balance.

Although in recent years the analysis of stress responses in halophytes is getting increasing attention, the overwhelming majority of these studies have been conducted in artificial laboratory or greenhouse conditions, which do not reflect at all the environment of plants in their natural habitats. Nevertheless, there are also publications reporting measurements of biochemical stress markers, and specifically of different osmolytes, using field-collected plant material from halophytes, which have provided information regarding the different mechanisms of response to salt stress of these salt-tolerant taxa. Some of these studies are summarised in the following paragraphs.

In earlier studies on the chemical composition of different halophytes, collected in the area of Neusiedler Lake in Austria, Albert and Kinzel (1973), Albert (1975) and Albert and Popp (1977, 1978) introduced the term 'physiotypes'; it refers to a peculiar chemical composition of halophytes, based mostly on ion contents and soluble carbohydrates accumulation in their leaves. A few years later, this concept was further developed by Gorham et al. (1980), who extended the measurements to other types of compatible solutes. A specific study on halophytes' osmolytes was published by Briens and Larher (1982), who quantified the contents of soluble carbohydrates, polyols, betaines and free proline in different organs of plants collected from coastal salt marshes. They identified three main groups of halophytes: those producing high levels of soluble carbohydrates only; those which accumulated mostly nitrogenous compounds, such as proline and glycine betaine; and species which synthesised both classes of osmolytes under high salinity conditions. An exhaustive study on betaines, in 63 species collected in coastal areas in Europe, Asia and South America, was reported by Adrian-Romero et al. (1998), who found high levels of these compounds in plants of the Chenopodiaceae family. The biochemical diversity of 51 halophytes from a salt marsh in Turkey was analysed by Tipirdamaz et al. (2006), who found that these species can be discriminated according to their capacity to accumulate in the field, either proline, glycine betaine or both compounds.

All these studies were extremely valuable for the understanding of the strategies adopted by different halophytes, present in the same habitat, to respond to specific environmental conditions. However, they did not provide much information regarding the contribution of osmolyte accumulation to the salt tolerance (or, more generally, stress tolerance) of those species, as the measurements were based on single samplings of plant material; that is, those experiments represent 'snapshots' under a particular set of environmental conditions, but did not correlate variations in osmolyte levels with changes in the degree of stress affecting the plants. This kind of analysis can be performed by determining osmolyte contents in two or more populations of the same halophytic species, growing in different geographic locations with distinct edaphic and climatic characteristics, but there are very few published studies using this experimental approach (Youssef 2009; Bankaji and Sleimi 2012). Nevertheless, even if differences in the levels of specific osmolytes are detected, they may not be due to the different external conditions of the plants, but rather the result of genetic variability between the populations.

Therefore, the correlation of osmolyte accumulation and the level of environmental stress in a given species, to assess its possible role in stress tolerance mechanisms, is more appropriately addressed by determining seasonal fluctuations in successive samplings of material from plants growing in the same location. Several publications report studies based on this strategy. For example, Doddema et al. (1986) analysed seasonal variations in water content, toxic ions (Na^+ and Cl^-) and mineral nutrients in soil and plants, as well as those of soluble carbohydrates, proline and proteins in roots and shoots of *Arthrocnemum fruticosum*, in a saline area by the Dead Sea (Jordan). Sodium and total soluble carbohydrate contents in plant roots showed a sudden sixfold increase in June, as compared to the values determined in previous months, in parallel to a drop in soil water content; proline was detected, at low concentrations, only in plants sampled in March. Seasonal variations of soluble carbohydrates levels were analysed by Murakeözy et al. (2002) in *Limonium gmelini* subsp. *hungarica* growing in a saline steppe in Hungary; these authors detected a fivefold increase in summer in the leaf contents of pinitol, the major osmolyte in this taxon, compared with the lowest values recorded in mid-April, but additional peaks were obtained in winter/early spring. These higher pinitol values corresponded to the most intense levels of water stress and lower temperatures, respectively. *Lepidium crassifolium* and *Camphorosma annua* from this habitat were also studied by the same group (Murakeözy et al. 2003), who detected the highest levels of reducing sugars, sucrose and pinitol in early spring, the period of maximum soil salinity and lowest atmospheric temperature. Variation in ion and osmolyte levels in three endemic *Limonium* species from Central Anatolia were analysed, together with soil characteristics, by Furtana et al. (2013), using samples collected in summer and autumn (2007–2008) from three locations around Tuz Lake. In this case, however, no clear positive correlation could be found between soil salinity and the levels of ions and the major osmolytes (proline, carbohydrates or quaternary ammonium compounds) in the plants. A significant increase in summer and autumn, as compared with winter and spring, was found by Mouri et al. (2012) in total soluble sugars and proline contents in *Ammophila arenaria* from sand dunes in Algeria, probably due to the intense drought and high temperatures affecting plants in both seasons.

Seasonal variation of compatible solutes in five different halophytes was systematically analysed by our group in a littoral salt marsh located near the city of Valencia, SE Spain, in a study that extended over a period of 2 years. EC and ion contents of soil samples, collected simultaneously to plant material in each sampling, together with meteorological data, were used to assess the level of environmental stress affecting the plants in the studied area. The most stressful season in Mediterranean habitats is generally summer, characterised by high temperatures and lack of rain. In salt marshes, and in addition to drought, salinity increases because of water evaporation and concentration of salts. Statistical analyses of the results revealed a significant correlation of environmental conditions associated to water stress and soil salinity with sorbitol levels in *Plantago crassifolia* and with those of sucrose, glucose and fructose in *Juncus acutus* and to a lesser extent in *J. maritimus* (Gil et al. 2011). On the contrary, in *Sarcocornia fruticosa* and *Inula crithmoides*, which were the most salt-tolerant taxa under study, there were no significant seasonal variations of toxic ions and glycine betaine, their major osmolyte, suggesting that their tolerance mechanisms are constitutive and relatively independent of external conditions. Proline levels were too low in all selected species to possibly have any effect on osmotic adjustment, but except for *P. crassifolia*, they showed a very good correlation with environmental stress, indicating that proline probably plays a role in stress tolerance based on its 'osmoprotectant' functions (Gil et al. 2014).

As mentioned above, most studies on halophytes' responses to salt stress, including osmolyte accumulation, have been carried out under controlled, but artificial, laboratory or greenhouse conditions. Therefore, the biological relevance of these responses should be confirmed in field experiments; there are, however, very few reports on this kind of comparative analyses. We have addressed this question by measuring osmolyte levels in plants grown in their natural habitat and in plants obtained by germination of seeds from the same location but grown in experimental controlled conditions; two salt-tolerant monocots of the genus *Juncus* (*J. acutus* and *J. maritimus*) and two halophytic dicots (*Plantago craasifolia* and *Inula crithmoides*) were selected for these studies (Boscaiu et al. 2013a; Pardo-Domènech et al. 2015). A great discrepancy was generally found between osmolyte contents in plants sampled in the field and those grown in artificial conditions, although EC measurements of the soil surface and the pots substrate may indicate a similar degree of salinity. The results of these studies reinforced the idea that direct quantitative comparisons between such data sets are not possible for several reasons. First, the developmental stages of the plants are different: seedlings or young plants are used in the growth chamber or greenhouse, while field material is sampled from fully grown adult individuals, and it is well known that responses to stress depend to a large extent on the developmental stage of the plants (Vicente et al. 2004). In addition, the roots of the potted plants are found in a limited and reduced environment with homogeneous salinity, whereas roots in the field can explore a more heterogeneous and considerably larger volume of soil. EC usually is higher in the upper soil layers that are sampled for soil analysis, but the root system can spread through deeper and less saline soil. Therefore, the degree of salt stress could be considerably lower for plants growing in the field than for plants submitted to controlled salt treatments, even at moderate concentrations of NaCl. In short, responses of a particular halophytic species to salt stress under controlled artificial conditions may differ considerably from its behaviour in nature, if only because of the impossibility to mimic accurately environmental conditions in laboratory setups. Fieldwork is extremely important to assess the biological relevance and the relative contribution of osmolyte accumulation to salt tolerance mechanisms in halophytes.

2.3 Plant Stress Responses in Arid Environments

Drought is a climatic condition specific to deserts, arid and semiarid habitats, which cover a huge surface throughout the world. Water availability is not the only restrictive factor for plant growth in these zones, where nutrients are also scarce, and there is an excess of solar radiation, all contributing to additional stress for plants living in those ecosystems (Valladares 2003). Arid and semiarid areas of the world are among the most sensitive to global climate change. Even the least pessimistic models predict for such zones an increase of average temperatures; decrease in rainfall; higher frequency; intensity and duration of periods of drought, 'heat waves' and other extreme meteorological phenomena; and changes in the seasonal weather patterns (Epstein and Mills 2005) in a relatively short period of time. Therefore, the knowledge of how xerophytes, plants naturally adapted to arid environments, can survive is of great importance. Among the many complex and intricate responses of plants to water stress, a relevant aspect is osmotic adjustment, providing the means to avoid cellular dehydration, which is essential for maintaining cellular activity (Bartels and Ramanjalu 2005). Although initially it was thought that osmotic adjustment was specific for plants subjected to high salinities, later studies proved that this response is also common in plants grown in conditions of water stress (Cushman 2001; Steudle 2000). Inorganic cations, organic acids, carbohydrates and free amino acids are the solutes known to accumulate as a response to drought (Munns et al. 1983).

There is difficulty in gathering information on the responses of xerophytes to stress in their natural environments, since often drought is associated with increased salinity, and many plants are at the same time water- and salt-stress tolerant.

In several recent studies, osmolyte levels were compared in plants from sites with different degrees of aridity. Sayed et al. (2013) analysed soil characteristics and plant contents of soluble sugars, total free amino acids and soluble proteins, as well as of several ions (Ca^{2+}, Mg^{2+}, Na^{+}, K^{+}, Cl^{-} and SO_4^{2-}) in *Zilla spinosa*, *Citrullus colocynthis* and *Morettia philaeana* from Wadi Natash (Eastern Desert, Egypt), but could not establish a clear pattern of correlation between soil properties and osmolytes accumulation in the plant species.

Water status, chlorophyll fluorescence, carotenoids and proline were analysed in combination with soil properties in two extreme xerophytes from three sites in China, with different ecological conditions (Zhou et al. 2010). *Reaumuria soongorica* had tenfold more proline than *Salsola passerina*, but while in the latter species proline levels correlated positively with the degree of aridity in the different sites, in the first taxon proline contents did not change significantly among sites.

Proline is one of the commonest and best studied compatible solutes in plants from arid regions, accumulating in response to water stress (Mohammed and Sen 1987). Proline concentrations were reported to increase with increasing water stress in three desert species from Pakistan (*Calotropis procera, Senna holosericea, Aerva japonica*) (Khan and Beena 2002). In the same three taxa and additionally in *Abutilon indicum*, Aziz and Khan (2003) analysed their water status and proline levels before and after the rainy season and found that proline decreased after rainfalls.

A further study on seasonal changes in proline in *Artemisia scoparia*, *Juniperus exceltata*, *Onobrychis cornuta* and *Sophora alopecuroides* growing in an arid region in Ziarat valley in

Pakistan (Aziz 2007) confirmed the good correlation of proline levels and water stress. After the snow melting and spring showers in April, proline concentrations decreased in all taxa. The same group determined seasonal variations of water status and proline contents in *Suaeda fruticosa*, *Heliotropium curassavicum*, *Haloxylon stocksii* and *Atriplex stocksii* from a natural area near Karachi, Pakistan, detecting again a peak in proline at increased osmotic potential, especially in *H. curassavicum* followed by *A. stocksii* (Aziz et al. 2011).

Proline accumulation has also been reported in two Mediterranean shrubs, *Halimium halimifolium* and *Pistacia lentiscus*, in both natural and experimental water deficit conditions (Ain-Lhout et al. 2001). In *H. halimifolium*, proline accumulation was induced only upon severe dehydration and was hardly reversible with plant mortality of 50 % under natural conditions. In *P. lentiscus*, on the contrary, proline accumulated gradually in response to drought and also to cold. These results indicated that in *P. lentiscus*, proline rather than an osmotic agent seemed to have protective role in cases of severe abiotic stress.

The levels of proline and soluble carbohydrates were determined in plants of several xerophytic Mediterranean species, collected in the wild in winter and spring, in a semiarid area near Valencia, Spain (Boscaiu et al. 2009). The results obtained showed a general reduction in both sugar and proline contents from the winter to the spring sampling, in agreement with a strong reduction of the environmental stress due to the heavy rains registered in early spring in the year of sampling.

A 5-month study, with samplings every month of plant material from four native species (*Sophora davidiana*, *Bauhinia faberi* var. *microphylla*, *Convolvulus tragacanthoides* and *Artemisia gmelinii*), was conducted by Yanqiong et al. (2007) in an arid valley in the area of Minjiang River in China. The content of chlorophyll, free proline and soluble sugars in the samples increased with increasing drought stress, whereas evaporation ratio showed a parallel reduction.

An additional difficulty when analysing data from field studies is that the endogenous concentration of osmolytes may change quite quickly in response to rapid changes in the environmental conditions of the plants, which could mask changes taking place within much longer timeframes, such as seasonal variations. For instance, as reported by Batanouny and Ebeid (1981) for two desert species, *Zygophyllum quatarense* and *Francoeuria crispa*, proline contents, which were relatively low before sunrise for both species – ca. 10 and 6 μmoles/g fresh weight, respectively – increased during the day, as the temperature and the rate of evapotranspiration rose to about double values in the afternoon than in the early morning.

2.4 Plant Stress Responses in Gypsum Habitats

Although geological formations bearing gypsum are widely spread in the world, gypsiferous soils are common only in arid or semiarid climates, since abundant rainfall removes the gypsum by leaching (Parsons 1976) and gypsum crystals occur on the surface only in those regions where there is a dry season. Soils with gypsum accumulation extend over 100 million ha in the world and are widespread in Africa and Asia (Verheye and Boyadgiev 1997). Within Europe, Spain is the country with some of the largest deposits of gypsum, which cover more than 30,000 km^2, mostly in the dry south-eastern part of the Peninsula (Mota et al. 2004). The vascular flora colonising these gypsum deposits is extremely diverse and includes many endemic and rare taxa. Iberian gypsum habitats are considered as priority habitats (Pueyo et al. 2007) and shelter about 30 endemic gypsophytes (Merlo et al. 1998). The high rate of endemicity is explained by the fact that gypsum soils represent an especially stressful environment, providing the selective force for the evolution of endemics (Meyer 1986).

Several physical constraints act in this habitat type: limited water retention in the soil and therefore an accentuated aridity, often associated with the formation of a hard soil surface crust which can limit seedling establishment, and a mechanical instability and lack of plasticity and cohesion, structural deterioration and low porosity, which interfere with root growth (Verheye and Boyadgiev 1997; Palacio et al. 2007; Martínez-Duro et al. 2010). Gypsum soils have also some chemical characteristics which negatively affect the development of plant communities. They are generally poor in organic matter – which ranges from between 0.4 and 1 % for surface layers to less than 0.2 % in the subsurface layers – and have very low levels of nitrogen and phosphorous (FAO 1990). The relatively high concentration of soluble calcium unbalances the availability of other macronutrients such as phosphorus, potassium and magnesium. Similarly, the interaction between calcium and several mineral micronutrients (zinc, iron and manganese) may cause a drop in their availability to plants. Another relevant aspect is the toxic effect of sulphates at high concentrations (Duvigneaud 1968; Ruíz et al. 2003).

An additional factor which affects the plant communities developing on gypsum is the marked irregularity in the distribution of their water content. In soils with a high amount of gypsum, moisture is uniform only on the surface layer, whereas below 10–15 cm, water is restricted to channels of 10–20 cm in size and the other parts of soils are dry. Therefore, water stress represents an important limiting factor for plants in gypsum areas, especially when combined with the aforementioned additional restrictive characteristic of this type of habitat.

Although the gypsicolous flora is of great scientific and conservationist interest, only a few studies have been published focusing on the responses to stress of plants from such habitats. Several authors have analysed the pattern of germination of plants present in gypsum zones (Caballero et al. 2003; Escudero et al. 1997; Ferriol et al. 2006; Moruno et al. 2011), and special attention has received the effect of the hard crusts which form on the surface of gypsum soils and restrain seedling establishment, thus limiting the presence of many species in such areas (Escudero et al. 1999; Meyer 1986; Meyer and García-Moya 1989; Romao and Escudero 2005). The effects of terrain topography, such as the presence of a slope and its orientation, have also been discussed by several authors (Meyer et al. 1992; Pueyo et al. 2007).

The mineral composition of plants living on gypsum was studied in earlier studies (Duvigneaud 1968; Duvigneaud and Denaeyer-De Smet 1966). More recently, Palacio et al. (2007, 2014) reported differences in the chemical composition of ash and in the concentrations of several ions, between gypsophytes (specific for gypsum environments) and gypsovags (plants that grow on gypsum but are frequent also on other soil types). Alvarado et al. (2000) analysed nitrogen metabolism – nitrate and nitrite reductase activities and nitrate, total free amino acids, soluble proteins, proline and organic N contents – in a study carried out on five gypsophytes from the Iberian Peninsula and found that *Lepidium subulatum* L., a shrubby gypsophyte of the Brassicaceae family endemic to the Iberian Peninsula and North of Africa, showed the highest levels of proline and the highest nitrogen metabolism. Although these authors sampled plant material in different seasons, they presented only average values, without addressing possible correlations of changes in the levels of these compounds with ecological factors.

All these studies, however, did not contribute relevant information regarding the mechanisms of response to environmental stress in plants adapted to gypsum habitats. The analysis of possible changes in the levels of biochemical stress markers, and specifically of different osmolytes, in plants living in gypsum soils, according to spatial or temporal (seasonal) variations in environmental conditions, has been undertaken only recently, by Boscaiu et al. (2013b) and Llinares et al. (2015), on several species – including gypsophytes and gypsovags – from a gypsum area in the province of Valencia (SE Spain).

There are several factors that could affect osmolyte contents in plants from gypsum habitats, especially those related to salt and water stress. If gypsum itself were the most important stressful factor for the plants, one should expect to detect higher osmolyte levels in plants present in those areas with higher gypsum content in the soil. However, the conclusion of these two studies is that drought (and not the osmotic or toxic effects of gypsum) represents the major ecological factor inducing the synthesis and accumulation of osmolytes in plants from such habitats. Proline was shown to be a common osmolyte in plants from gypsum habitats and a good indicator of environmental water deficit; its concentrations strongly varied among samples collected in different seasons, with generally higher levels in summer, the most stressful period of the year due to higher temperatures and lack of rain (Boscaiu et al. 2013b). When considering the topographic gradient, higher levels of proline were detected in plants collected in the driest plot, which was that with lower gypsum content in the soil. Salinity levels in the gypsum soils are moderate and steady along the year, since the soil solution is gypsum saturated independently of the humidity. It could be concluded that proline biosynthesis is mostly triggered by water deficit in the soil, not by the presence of high gypsum concentration (Boscaiu et al. 2013b).

Glycine betaine was also analysed in the same experimental plot in the gypsum zone, in two gypsophytes and two gypsovags, but only low levels were found in all but the gypsophyte *Ononis tridentata*, which had double amounts in comparison to the other selected taxa (Llinares et al. 2015). However, those levels were also much lower than the values detected in halophytes that are clear GB accumulators (Khan et al. 2000; Tipirdamaz et al. 2006). There was little spatial and seasonal variation of this compound in the four analysed species, and it did not correlate with soil water availability or other environmental factors that could affect the synthesis of osmolytes. It was also not possible to correlate the levels of total sugars with the degree of environmental stress, thus supporting the notion that proline is probably the major osmolyte in most of the analysed gypsum-adapted species.

2.5 Conclusions

Drought and salinity are the most important environmental factors responsible for the reduction of crop yields worldwide, and these adverse conditions will worsen in the near future in many regions due to the foreseeable effects of global climate change, thereby exacerbating the problem, especially in the most productive arable lands of the world, those cultivated under irrigation in arid and semiarid zones. There is an urgent need of increasing food production to feed a growing human population, and the genetic improvement of the stress tolerance of our major crops, both by traditional breeding methods and by modern biotechnological approaches, would significantly contribute to this goal. Yet this requires a deep understanding of the molecular mechanisms underlying the responses of plants to environmental stress, which – apart from the undeniable academic interest of this topic – has stimulated the study of these mechanisms over the last few decades.

One of the most general responses of plants to abiotic stresses, which appear to be essential for stress tolerance in many species, is based on the synthesis and accumulation of osmolytes. Osmolytes are compatible solutes that contribute to the maintenance of cellular osmotic balance under all those environmental conditions causing cellular dehydration (such as drought, soil salinity, cold or high temperatures), but also play 'osmoprotectant' roles by acting as low-molecular-weight chaperons and reactive oxygen species (ROS) scavengers. Despite intensive research, important aspects of their mechanisms of action remain largely unknown, especially regarding the biological relevance and the relative contribution of specific osmolytes to the stress tolerance mechanisms of a given species under natural conditions. This gap in our knowledge is partly due to the common approach used in these studies, which have focused on non-tolerant model species and/or experiments performed under controlled – but artificial – laboratory or greenhouse setups, which cannot mimic the conditions plants encounter in their natural habitats.

In this short review, we have commented some of the relatively few published data reporting field studies on the accumulation of osmolytes in wild plant species adapted to specific stressful environments: saline habitats (halophytes), arid or semiarid environments (xerophytes) and gypsum soils (gypsophytes). The available information support the notion that the responses of plants to controlled stress treatments in the laboratory do not reflect their behaviour in their natural habitats, where they must react simultaneously to different, uncontrollable and continuously changing environmental stress factors, which can interact in complex ways rather than show simple additive effects. Aware of these limitations, and as a complementary approach to laboratory and greenhouse experiments, we propose that more effort and resources should be invested on the study of the abiotic stress responses of wild plants growing in their natural, stressful habitats. We believe that this fieldwork will significantly contribute to enhance our knowledge on the general mechanisms of environmental stress tolerance in plants, which could eventually be applied to the genetic improvement of stress tolerance in crop species.

References

Adrian-Romero M, Wilson SJ, Blunden G, Yang M, Carabot-Cuervo A, Bashir AK (1998) Betaines in coastal plants. Biochem Syst Ecol 26:535–543

Ain-Lhout F, Zunzunegui M, Diaz Barradas MC, Tirado R, Clavijo A, Garcia Novo F (2001) Comparison of proline accumulation in two Mediterranean shrubs subjected to natural and experimental water deficit. Plant Soil 230:175–183

Akashi K, Miyake C, Yokota A (2001) Citrulline, a novel compatible solute in drought-tolerant wild watermelon leaves, is an efficient hydroxyl radical scavenger. FEBS Lett 508:438–442

Albert R (1975) Salt regulation in halophytes. Oecologia 21:57–71

Albert R, Kinzel H (1973) Unterscheidung von Physiotypen bei Halophyten des Neusiedlerseegebietes (Österreich). (Distinction of physiotypes in halophytes from the Neusiedler Lake region, Austria). Z Pflanzenphysiol 70:138–158

Albert R, Popp M (1977) Chemical composition of halophytes from the Neusiedler Lake region in Austria. Oecologia 27:157–170

Albert R, Popp M (1978) Zur Rolle der löslichen Kohlenhydrate in Halophyten des Neusiedlersee-Gebietes (Österreich). (On the role of soluble carbohydrates in halophytes from the Neusiedler Lake region, Austria). Oecol Plant 13:27–42

Alvarado JJ, Ruiz JM, López-Cantarero I, Molero J, Romero L (2000) Nitrogen metabolism in five plant species characteristic of gypsiferous soils. Plant Physiol 156:612–616

Ashraf MY, Foolad MR (2007) Roles of glycine betaine and proline in improving plant abiotic stress resistance. Environ Exp Bot 59:206–216

Aziz I (2007) Seasonal flux in water potential, chlorophyll and proline content in plants at Ziarat Valley Balochistan, Pakistan. Pak J Bot 39:1995–2002

Aziz I, Khan MA (2003) Proline and water status of some desert shrubs before and after water rains. Pak J Bot 35:902–906

Aziz I, Gul B, Gulzar S, Khan MA (2011) Seasonal variation in plant water status of four desert halophytes from semi-arid region of Karachi. Pak J Bot 43:587–594

Bankaji I, Sleimi N (2012) Polymorphisme biochimique chez quelques halophytes autochtones du nord Tunisien (Chemical polymorphism of some North Tunisian autochthonous halophytes). Rev Écol 67:29–39

Bartels D, Ramanjulu S (2005) Drought and salt tolerance in plants. Plant Sci 24:23–58

Bartels D, Sunkar R (2005) Drought and salt tolerance in plants. Crit Rev Plant Sci 24:23–58

Batanouny KH, Ebeid MM (1981) Diurnal changes in proline content of desert plants. Oecologia 51:250–252

Ben Hamed K, Ellouzi H, Zribi Talbi O, Hessini K, Slama I, Ghnaya T, Munné Bosch S, Savouré A, Abdelly C (2013) Physiological response of halophytes to multiple stresses. Funct Plant Biol 40:883–896

Boscaiu M, Mora E, Fola O, Scrion S, Llinares J, Vicente O (2009) Osmolyte accumulation in xerophytes as a response to environmental stress. Bull UASVM Hort 66:96–102

Boscaiu M, Bautista I, Lidón A, Llinares J, Lull C, Donat P, Mayoral O, Vicente O (2013a) Environmental-dependent proline accumulation in plants living on gypsum soils. Acta Physiol Plant 35:2193–2204

Boscaiu M, Lull C, Llinares J, Vicente O, Boira H (2013b) Proline as a biochemical marker in relation to the ecology of two halophytic *Juncus* species. J Plant Ecol 6:177–186

Briens M, Larher F (1982) Osmoregulation in halophytic higher plants: a comparative study of soluble carbohydrates, polyols, betaines and free proline. Plant Cell Environ 5:287–292

Burg MB, Kwon ED, Kültz D (1996) Osmotic regulation of gene expression. FASEB J 10:1598–1606

Caballero I, Olano JM, Loidi J, Escudero A (2003) Seed bank structure along a semi-arid gypsum gradient in Central Spain. J Arid Environ 55:287–299

Chen THH, Murata N (2002) Enhancement of tolerance of abiotic stress by metabolic engineering of betaines and other compatible solutes. Curr Opin Plant Biol 5:250–257

Cushman JC (2001) Osmoregulation in plants: implications for agriculture. Am Zool 41:758–769

DB Climate Change Advisors (2009) Investing in agriculture: far-reaching challenge, significant opportunity. An asset management perspective. (Deutsche Bank report). Whitepaper available online at: http://www.dbcca.com/research

Doddema H, Eddin RS, Mahasneh A (1986) Effects of seasonal changes of soil salinity and soil nitrogen on the N-metabolism of the halophyte *Arthrocnemum fruticosum* (L.) Moq. Plant Soil 92:279–293

Duvigneaud P (1968) Essai de classification chimique (éléments minéraux) des plantes gypsicoles du bassin de l'Ebre (Test of chemical classification (mineral elements) of gypsophytes from the Ebro basin). B Soc Roy Bot 101:279–291

Duvigneaud P, Denaeyer-De Smet S (1966) Accumulation du soufre dans quelques espèces gypsophiles d'Espagne (Sulphur accumulation in several species of gypsophytes from Spain). B Soc Roy Bot 99:263–269

Epstein PR, Mills E (eds) (2005) Climate change futures. Health, ecological and economic dimensions. The Center for Health and the Global Environment, Harvard Medical School, Harvard

Escudero A, Carnes LF, Pérez García F (1997) Seed germination of gypsophytes and gypsovags in semi-arid central Spain. J Arid Environ 36:487–497

Escudero A, Somolinos RC, Olano JM, Rubio A (1999) Factors controlling the establishment of *Helianthemum squamatum*, an endemic gypsophila of semi-arid Spain. J Ecol 87:290–302

FAO (1990) Management of gypsiferous soils, FAO Soils Bull 62. FAO, Rome: FAO Land and Water Development Division

Ferriol M, Pérez I, Merle H, Boira H (2006) Ecological germination requirements of the aggregate species *Teucrium pumilum* (Labiatae) endemic to Spain. Plant Soil 284:205–216

Flowers TJ, Colmer TD (2008) Salinity tolerance in halophytes. New Phytol 179:945–963

Flowers TJ, Hajibagheri MA, Clipson NJW (1986) Halophytes. Q Rev Biol 61:313–335

Flowers TJ, Galal HK, Bromham L (2010) Evolution of halophytes: multiple origins of salt tolerance in land plants. Funct Plant Biol 37:604–612

Forment J, Naranjo MA, Roldán M, Serrano R, Vicente O (2002) Expression of *Arabidopsis* SR-like splicing proteins confers salt tolerance to yeast and transgenic plants. Plant J 30:511–519

Furtana GB, Duman H, Tipirdamaz R (2013) Seasonal changes of inorganic and organic osmolyte content in three endemic *Limonium* species of Lake Tuz (Turkey). Turk J Bot 37:455–463

Gil R, Lull C, Boscaiu M, Bautista I, Lidón A, Vicente O (2011) Soluble carbohydrates as osmolytes in several halophytes from a Mediterranean salt marsh. Not Bot Horti Agrobo 39:9–17

Gil R, Bautista I, Boscaiu M, Lidón A, Wankhade S, Sánchez H, Llinares J, Vicente O (2014) Responses of five Mediterranean halophytes to seasonal changes in environmental conditions. AoB Plants. doi:10.1093/aobpla/plu049

Gorham J, Hughes L, Wyn Jones RG (1980) Chemical composition of salt-marsh plants from Ynys Môn (Anglesey): the concept of physiotypes. Plant Cell Environ 3:309–318

Greenway H, Munns R (1980) Mechanisms of salt tolerance in non-halophytes. Annu Rev Plant Physiol 31:149–190

Hong Z, Lakkineni K, Zhang Z, Verma DPS (2000) Removal of feedback proline accumulation of Δl-pyrroline-5-carboxylate synthetase results in increased proline accumulation and protection of plants from osmotic stress. Plant Physiol 122:1129–1136

Hussain TM, Chandrasekhar T, Hazara M, Sultan Z, Saleh BK, Gopal GR (2008) Recent advances in salt stress biology – a review. Biotechnol Mol Biol Rev 3:8–13

Inan G, Zhang Q, Li P, Wang Z, Cao Z, Zhang H, Zhang C, Quist TM, Goodwin SM, Zhu J, Shi H, Damsz B, Charbaji T, Gong Q, Ma S, Fredricksen M, Galbraith DW, Jenks MA, Rodees D, Hasegawa PM, Bohnert HJ, Joly RJ, Bressan RA, Zhu JK (2004) Salt stress. A halophyte and cryophyte *Arabidopsis* relative model system and its applicability to molecular genetic analyses of growth and development of extremophiles. Plant Physiol 135:1718–1737

Kant S, Kant P, Raveh E, Barak S (2006) Evidence that differential gene expression between the halophyte, *Thellungiella halophila*, and *Arabidopsis thaliana* is responsible for higher levels of the compatible osmolyte proline and tight control of Na^+ uptake in *T. halophila*. Plant Cell Environ 29:1220–1234

Khan MA, Beena N (2002) Seasonal variation in water relations of desert shrubs from Karachi, Pakistan. Pak J Bot 34:329–340

Khan MA, Ungar IA, Showalter AM (2000) The effect of the salinity on the growth, water status, and ion content of a leaf succulent perennial halophyte, *Suaeda fruticosa* (L) Forssk. J Arid Environ 45:73–84

Krasensky J, Jonak C (2012) Drought, salt, and temperature stress-induced metabolic rearrangements and regulatory networks. J Exp Bot 63:1593–1608

Llinares JV, Bautista I, Donat MP, Lidon A, Lull C, Mayoral O, Shantanu W, Boscaiu M, Vicente O (2015) Responses to environmental stress in plants adapted to Mediterranean gypsum habitats. Not Sci Biol 7:37–44

Martínez-Duro E, Ferrandis P, Escudero A, Luzuriaga AL, Herranz JM (2010) Secondary old-field succession in an ecosystem with restrictive soils: does time from abandonment matter? Appl Veg Sci 13:234–248

Merlo ME, Mota JF, Cabello J, Alemán MM (1998) La gipsofilia en plantas: un apasionante edafismo. Investigación y Gestión 3:103–112

Meyer SE (1986) The ecology of gypsophyle endemism in the eastern Mojave desert. Ecology 67:1303–1313

Meyer SE, García-Moya E (1989) Plant community patterns and soil moisture regime in gypsum grasslands of north central Mexico. J Arid Environ 16:147–155

Meyer SE, García-Moya E, Lagunes-Espinoza LC (1992) Topographic and soil surface effects on gypsophila plant community patterns in central Mexico. J Veg Sci 3:429–438

Mohammed S, Sen DN (1987) Proline accumulation in arid zones plants. J Arid Environ 13:231–236

Moruno F, Soriano P, Vicente O, Boscaiu M, Estrelles E (2011) Opportunistic germination behaviour of *Gypsophila* (Caryophyllaceae) in two priority habitats from semi-arid Mediterranean steppes. Not Bot Horti Agrobo 39:18–23

Mota JF, Sola AJ, Jiménez-Sánchez ML, Pérez-García F, Merlo ME (2004) Gypsicolous flora, conservation and restoration of quarries in the southeast of the Iberian Peninsula. Biodivers Conserv 13:1797–1808

Mouri C, Benhassaini H, Bendimered FZ, Belkhodja M (2012) Variation saisonnière de la teneur en proline et en sucres solubles chez l'oyat (*Ammophila arenaria* (L.) Link) provenant du milieu naturel de la côte ouest de l'Algérie. Acta Bot Gallica 159:127–135

Munns R (2002) Comparative physiology of salt and water stress. Plant Cell Environ 25:239–250

Munns R, Greenway H, Kirst GO (1983) Halotolerant eukaryotes. In: Lange OL, Nobel PS, Osmond CB, Ziegler H (eds) Encyclopedia of plant physiology III, vol 12c. Physiological plant ecology. Springer, Berlin, pp 59–135

Murakeözy ÉP, Smirnoff N, Nagy Z, Tuba Z (2002) Seasonal accumulation pattern of pinitol and other carbohydrates in *Limonium gmelinii* subsp. *hungarica*. J Plant Physiol 159:485–490

Murakeözy ÉP, Nagy Z, Duhazé C, Bouchereau A, Tuba Z (2003) Seasonal changes in the levels of compatible osmolytes in three halophytic species of inland saline vegetation in Hungary. J Plant Physiol 160:395–401

Palacio S, Escudero A, Montserrat-Martí G, Maestro M, Milla R, Albert M (2007) Plants living on gypsum: beyond the specialist model. Ann Bot 99:333–343

Palacio S, Aitkenhead M, Escudero A, Montserrat-Martí G, Maestro M, Robertson AH (2014) Gypsophila chemistry unveiled: Fourier transform infrared (FTIR) spectroscopy provides new insight into plant adaptations to gypsum soils. PLoS One. doi:10.1371/journal.pone.0107285

Pardo-Domènech LL, Tifrea A, Grigore MN, Boscaiu M, Vicente O (2015) Proline and glycine betaine accumulation in two succulent halophytes under natural and experimental conditions. Plant Biosyst. doi:10.1080/11263504.2014.990943

Parsons RF (1976) Gypsophily in plants. A review. Am Midl Nat 96:1–20

Pueyo Y, Alados CL, Maestro M, Komac B (2007) Gypsophila vegetation patterns under a range of soil properties induced by topographical position. Plant Ecol 189:301–311

Romao RL, Escudero A (2005) Gypsum physical soil crusts and the existence of gypsophytes in semi-arid central Spain. Plant Ecol 181:127–137

Ruíz JM, López-Cantarero I, Rivero RM, Romero L (2003) Sulphur phytoaccumulation in plant species characteristic of gypsiferous soils. Int J Phytorem 5:203–210

Sairam RK, Tyagi A (2004) Physiology and molecular biology of salinity stress tolerance in plants. Curr Sci 86:407–421

Sanders D (2000) The salty tale of *Arabidopsis*. Curr Biol 10:486–488

Sayed SA, Gadallah MAA, Salama FM (2013) Ecophysiological studies on three desert plants growing in Wadi Natash, Eastern Desert, Egypt. J Biol Earth Sci 3:B135–B143

Serrano R (1996) Salt tolerance in plants and microorganisms: toxicity targets and defence responses. Int Rev Cytol 165:1–52

Serrano R, Gaxiola R (1994) Microbial models and salt stress tolerance in plants. Crit Rev Plant Sci 13:121–138

Shen B, Jensen RG, Bohnert HJ (1997) Mannitol protects against oxidation by hydroxyl radicals. Plant Physiol 115:527–532

Steudle E (2000) Water uptake by roots: effects of water deficit. J Exp Bot 51:1531–1542

Szabados L, Savouré A (2010) Proline: a multifunctional amino acid. Trends Plant Sci 15:89–97

Tipirdamaz R, Gagneul D, Duhazé C, Aïnouche A, Monnier C, Özkum D, Larher F (2006) Clustering of halophytes from an inland salt marsh in Turkey according to their ability to accumulate sodium and nitrogenous osmolytes. Environ Exp Bot 57:139–153

Türkan I, Demiral T (2009) Recent developments in understanding salinity tolerance. Environ Exp Bot 67:2–9

Valladares F (2003) Light heterogeneity and plants: from ecophysiology to species coexistence and biodiversity. Progress Bot 64:439–471

Verheye WH, Boyadgiev TG (1997) Evaluating the land use potential of gypsiferous soils from field pedogenic characteristics. Soil Use Manag 13:97–103

Vicente O, Boscaiu M, Naranjo MA, Estrelles E, Bellés JM, Soriano P (2004) Responses to salt stress in the halophyte *Plantago crassifolia* (Plantaginaceae). J Arid Environ 58:463–481

Vinocur B, Altman A (2005) Recent advances in engineering plant tolerance to abiotic stress: achievements and limitations. Curr Opin Biotechnol 16:123–132

Yancey PH, Clark ME, Hand SC, Bowlus RD, Somero GN (1982) Living with water stress: evolution of osmolyte systems. Science 217:1214–1222

Yanqiong L, Xingliang L, Shaowei Z, Hong C, Yongjie Y, Changlong M, Jun L (2007) Drought-resistant physiological characteristics of four shrub species in arid valley of Minjiang River, China. Acta Ecol Sin 27:870–878

Yeo A (1998) Molecular biology of salt tolerance in the context of whole-plant physiology. J Exp Bot 49:915–929

Youssef AM (2009) Salt tolerance mechanisms in some halophytes from Saudi Arabia and Egypt. Res J Agric Biol Sci 5:191–206

Zhou H, Tan H, Zhang Z-S, Jia X, Fan H, Yuan J (2010) Physiological responses and adjustment mechanisms of the dominate species of natural vegetation of Eastern Tengger Desert. Sci Cold Arid Reg 2:455–463

Zhu J-K (2000) Genetic analysis of plant salt tolerance using *Arabidopsis*. Plant Physiol 124:941–948

Zhu J-K (2001) Plant salt tolerance. Trends Plant Sci 6:66–71

Salinity Tolerance: Growth, Mineral Nutrients, and Roles of Organic Osmolytes, Case of *Lygeum spartum* L., A Review

3

Bouzid Nedjimi

Abstract

Soil salinity is a major environmental problem that limits plant growth, productivity, and survival. Proper drainage and the application of high quality water can although solve the problem; however, these measures are very costly and cannot be applied in extensive agriculture. An alternative strategy for sustainable agriculture in saline marginal lands is to select plants that can tolerate salinity. The plant's ability to tolerate salinity depends on multiple biochemical pathways that enable retention and/or acquisition of water, protect photosynthetic functions, and maintain homeostasis of ions. *Lygeum spartum* L. is a pioneer grass species, used for sand dune fixation, desalination, and rehabilitation of degraded arid lands. The plant tolerates abiotic constraints such as salinity and drought and is also used to provide standing feed browse for livestock. The present paper reviews *L. spartum* plant responses to salinity stress with emphasis on the biochemical and physiological mechanisms of salt tolerance.

3.1 Introduction

Soilsalinityis a principal environmental restriction to crop production, affecting an estimated 45 million hectares of irrigated land, and is expected to increase due to global climate changes and as a consequence of many irrigation practices (Rengasamy 2010). The main reasons for increasing soil salinity is decreased precipitation, high evaporation from surface, nature of the native rocks, use of brackish water for irrigation, and poor cultural practices. The damaging effects of salinity on the growth of plants are associated with (1) low osmotic potential (water stress), (2) nutritional imbalance, (3) specific ion effect (salt stress), or (4) a combination of these factors (Ashraf 1994).

Salinity has diverse effects on the plant growth and development, and the biological effects can be detected at physiological, biochemical, and molecular levels (Munns 2002). Generally, the negative effects of salinity are assigned to the increased content of Na^+ and Cl^- ions in different

B. Nedjimi (✉)
Laboratory of Exploration and Valorization of Steppe Ecosystem, Faculty of Science of Nature and Life, University of Djelfa, Cité Aîn Chih, P.O. Box 3117, Djelfa 17000, Algeria
e-mail: bnedjimi@yahoo.fr

N. Iqbal et al. (eds.), *Osmolytes and Plants Acclimation to Changing Environment: Emerging Omics Technologies*, DOI 10.1007/978-81-322-2616-1_3

parts of plants. These ions intercept with different plant mechanisms and produce the critical conditions for the survival of plant.

Salt-tolerant plants exhibit multiple adaptation features in both morphology and/or structure and also in metabolic and physiological strategies that enable them to survive under harsh saline conditions. The osmotic adjustment (OA) mechanism is an important strategy that can help plants to avoid toxicity of ions and maintain water uptake and retention by accumulating large amount of osmolytes (Ashraf and Foolad 2007). In salt-stressed plants, OA can occur by accumulating either low molecular weight organic solutes or high concentrations of either inorganic ions. Although the role of both organic and inorganic osmolytes is crucial in higher plants grown under saline conditions, however, their relative contribution varies between species and cultivars and even between different compartments in the same plant (Shannon 1998).

The soil and water resources in many arid and semiarid regions in the world are too saline for majority of the common economic crops. Increased use of brackish groundwater in many areas for landscape irrigation has resulted in a need for breeding salt-tolerant plant species or selection of suitable halophytes (Nedjimi 2014).

In the soils with high salt concentration, the halophytes and salt-tolerant plants emerge as highly evolved and specialized organisms with well-adapted morphological and physiological characteristics that allow them to tolerate these extreme conditions. Exploitation of these plants for the better utilization of saline water and for the rehabilitation of highly saline soils can be an important strategy for sustainable agriculture (Khan and Duke 2001). However, it is important to know the mechanisms these plants adapt to complete their life cycle under high salinity. Among the operational mechanisms that these plants use to adapt themselves to extreme saline conditions, the foremost important is one that helps to maintain the content of Na and Cl ions at a much lower level in the cytoplasm than in the soil solution, which protects the cytoplasmic organelles from toxic effect of ions (Glenn et al. 1999). The second mechanism should reduce their water potential, thus increasing the water supply to plants under high soil salinity (Flowers and Colmer 2008). Munns (2002) realized that maintenance of low Na and Cl concentrations in young plant parts with active metabolism, such as the meristem and generative organs, is a strategy of ionic homeostasis (Munns 2002).

Among the different halophytic grasses, *Lygeum spartum* L., commonly called "Albardine," is an important grass of the *Poaceae* family. This plant has a distinct biochemical physiological flexibility that enables it to adapt and survive under conditions of abiotic stress (Pugnaire and Haase 1996; Conesa et al. 2007; Nedjimi et al. 2010). *L. spartum* constitutes an essential element for soil protection of semiarid and arid zones, its aboveground biomass, and its extensive root system makes them an efficient tool against desertification and in the rehabilitation of degraded lands. Its remarkable ability to accumulate green fodder over several seasons has contributed to its traditional use as a source of browsing fodder for livestock (Nedjimi 2009). This chapter provides a comprehensive review of the major responses of *L. spartum* plants to saline environments and the mechanism adaptation of this species against salt stress.

3.2 Salt Stress Tolerance

Salt-tolerant plants can reduce the harmful effects of salts (e.g., toxicity of ions, osmotic stress, and nutritional disorder) by modifying morphological, anatomical, and physiological mechanisms (Munns and Tester 2008). The plant salt tolerance may involve many strategies like controlled uptake of salts, ions sequestration in the vacuoles and specialized organs, and production of compatible solutes.

3.2.1 Germination

Seed germination is one of the most fundamental and vital phases in the growth cycle of plants that determine plant establishment and the yield of the crops. In salt conditions, the germination rates and percentage of germinated seeds at a particular time vary considerably among species and cultivars (Baskin and Baskin 1998).

Under salt stress conditions, the critical factors to crop production are the rapid germination of seed and stand establishment. In several species, germination of seed and early seedling growth are the most crucial stages to salinity. Salinity may reduce plant growth and yield by delaying germination, reducing the rate of germination, and increasing the dispersion of germination events (Khan and Gul 2006).

Seed germination is also affected by change in temperature. They may affect the permeability of membrane and the activity of membrane-bound and cytosolic enzymes (Probert 1992). Soil moisture is another important factor determining germination and the growth of seedling and plays an important role in determining the pattern of distribution of species (Bewley 1997). Ungar (1995) reported that under salinity, stress temperature affects germination in saline dry areas. Although high salinity reduces germination, the deleterious effect of salinity is generally less harmful at the suitable thermal optimum (Tlig et al. 2008; Nedjimi 2013). Seed germination under saline conditions usually occurs after high precipitation, when soil salinity is reduced due to leaching (Ungar 1995).

One of the first physiological disorders taking place during seed germination under salt stress is a decreased seed imbibition due to low water potential of the germination medium. In addition slow rate of imbibition may lead to a chain of metabolic changes, including regulation of enzyme activities (Guerrier 1988), perturbation in the mobility of inorganic nutrients (Ashraf and Wahid 2000), imbalances in the levels of plant growth regulators (Khan et al. 2006), and reduction in hydrolysis and utilization of food reserves (Promila and Kumar 2000).

In a recent study, Nedjimi (2013) reported that *L. spartum* seeds showed the highest percentage of germination under non-saline conditions. As salinity increased, germination was inhibited; however, about 20 % of the seeds germinated even at 150 mM NaCl (Fig. 3.1). This suggests that *L. spartum* is a moderately salt-tolerant grass at germination stage compared to the other grasses found at the Algerian steppes.

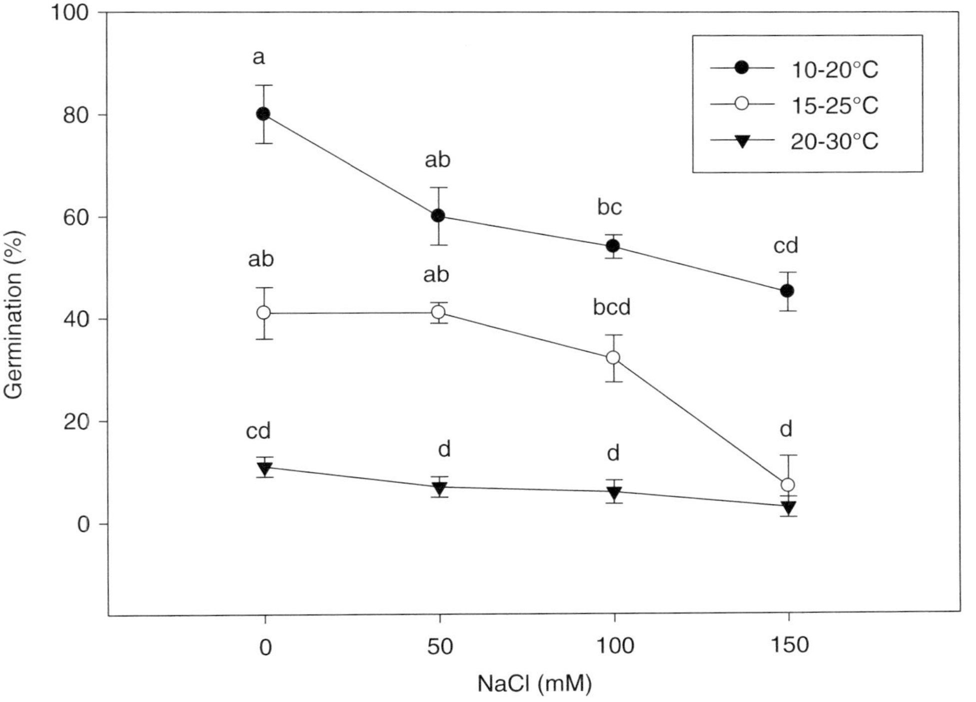

Fig. 3.1 Effect of salinity (NaCl) and temperature on final germination percentages of *L. spartum*. *Bars* represent mean ± SE ($n = 4$) (Nedjimi 2013)

3.2.2 Growth

Salt tolerance involves a variety of mechanisms and is a complex phenomenon. Flowers and Colmer (2008) defined salt tolerance as the ability of a plant to complete their growth cycle under saline conditions with appreciable or acceptable growth and yield. Leaf surface expansion is reduced as the first response of salt stress (Munns 2002). Salt stress also decreases the fresh and dry weights of boot roots and shoots (Abdul Qados 2011). Under salt stresses, *L. spartum* exhibits various adaptations through growth and water-use efficiency. The growth observed in terms of fresh and dry weight of root and shoots is found to significantly decrease under higher salt exposure (Nedjimi 2011). However, improved growth of *L. spartum* under lower concentrations of salt has been correlated with delicate balance among the accumulation of ions, osmotic adjustment, maintenance of pressure potential, and osmolytes synthesis. Thus, *L. spartum* may be conserving growth and nutrition acquisition during moderate salt stress as a consequence of decreased water potential and increased water-use efficiency (Nedjimi et al. 2013).

L. spartum, being a moderate salt-tolerant grass, grows well under salinity up to 50 mM NaCl. Data on tolerance of the plant exposed to salinity reveal that *L. spartum* maintains its growth by sequestration of Na and Cl ions into the vacuoles to maintain the osmotic balance between vacuole and cytoplasm (Parida and Das 2005).

3.2.3 Photosynthesis

Salt stress reduces photosynthesis by decreasing the water potential. There are some other factors that reduce photosynthetic rates under salt stress: enhanced senescence, changes in enzyme activity, and alterations of chloroplasts (Sudhir and Murthy 2004). Photosynthesis is also reduced due to reduction in stomatal conductance which restricts the availability of CO_2 for carboxylation reactions (Suárez 2011). Under saline conditions, the decreased stomatal conductance helps to avoid water losses and to minimize Na ions uptake. Under salt stress, leaf pigment contents decrease in general (Nedjimi 2014). Nedjimi (2009) showed that stomatal conductance (g_s), Net CO_2 (A_{CO2}) assimilation rate, and photosynthetic pigments (Chl *a* and *b*) did not differ significantly at 50 mM NaCl (Fig. 3.2); however, these tree parameters decreased when salinity reached the value of 100 mM NaCl. At high salinity, the rate of salt loading might be reduced through decreased transpiration rates and reduced stomatal conductance. Reduction in transpiration rate would increase the longevity of leaf by maintaining salts at sub-toxic levels for longer duration compared to conditions when transpiration rates were not decreased (Ramani et al. 2006).

3.2.4 Mineral Nutrients

Mineral nutrients play an important role in osmotic adjustment. Among the organic and inorganic nutrients, in some plants, inorganic ions play more important roles than that of compatible solutes. Among the mineral ions, Na ion appears to be the main ion and a low-cost source of osmoticum to maintain the osmoregulation under salt stress (Blumwald 2000). Under high salinity, K/Na is altered by the unrestrained influx of Na through the K pathways. Accumulation of Na ion and its sequestration into the vacuole consequently restrict the K uptake by roots, therefore decreasing K level in the shoots (Maathuis and Amtmann 1999).

Excess accumulation of minerals and salts contributes to increased ash content in *L. spartum* which was found to be higher in the shoots than in the roots and increased with increase in salinity (Nedjimi et al. 2010). The accumulation of Na ions in shoots was reported to be two times higher than in the roots (Nedjimi 2009). The root system of this grass is adjusted to enable absorption of large amounts of ions and the transfer of bulk amounts of Na and Cl to the aboveground organs. These results suggest that *L. spartum* is able to accumulate Na in their shoots and to maintain turgor and osmotic adjustment. It appeared that salinity tolerance was maintained in this species

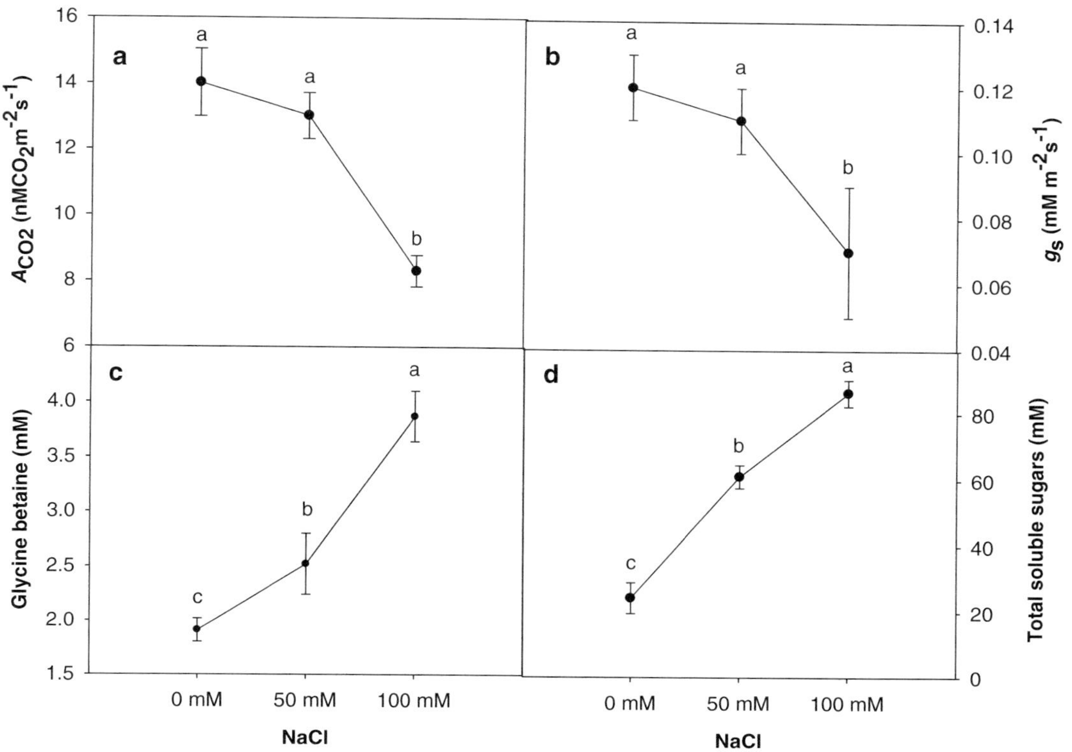

Fig. 3.2 Effects of salinity (NaCl) on net CO_2 assimilation rate (A_{CO2}) (**a**), stomatal conductance (g_s) (**b**), glycine betaine, (**c**) and total soluble sugars (**d**) of *L. spartum* grown in hydroponic conditions. Values represent means ± SE ($n = 5$) (Nedjimi 2009)

essentially by using Na as an osmoticum and compartmentalizing it within leaf cells away from the cytosol, perhaps by Na/H antiport at the vacuolar membranes (Hasegawa et al. 2000).

In contrast, increasing the external NaCl concentration was accompanied by a concomitant decrease in K content. Increased concentration of Na can reduce the uptake of K and result in K deficiency and growth reduction. Therefore, instead of low Na or high K, it is the relative K and Na concentration that is a key factor in determining plant salt tolerance (Maathuis and Amtmann 1999). Potassium is important for the activation of enzyme, synthesis of protein, and photosynthesis, and it helps in osmoregulation during cell expansion, stomatal movements, and tropisms. Besides, it is required for solute transport through phloem and for cation/anion balance in the cytosol as well as in the vacuole. The ability of plants to prevent salinity stress depends strongly on the status of their K nutrition. Reduction in K content in the salt-stressed *L. spartum* plants may be due to a downregulation of the genes involved in the transport of K (Zhu 2003) and/or return of K to the root media due to increased salt-induced plasma membrane damages (Shabala et al. 2005). Further, salinity-induced reduction in the K content in tissues of *L. spartum* might also be merely due to competition of K with Na at the root plasma membrane (Tester and Davenport 2003). Many researchers have reported that the reduction in K accumulation of salt-stressed plants is usually attributed to the competition between Na and K during uptake (reducing the influx of K) and returning back of K to the bathing solution (increasing the efflux of K), mostly due to loss in plasma membrane integrity (Patel et al. 2009; Nedjimi 2014). All of these factors either alone or in combination strongly reduce the net uptake of the K nutrient.

3.2.5 Supplementary Calcium Enhances Salinity Tolerance

Calcium is an essential plant nutrient that has a role in metabolic activities, including membranes stabilization, transduction of signals through second messenger, and control of the activity of enzymes (White and Broadley 2003). It is known to ameliorate the deleterious effects of salinity in plants, probably by facilitating higher K/N selectivity (Hasegawa et al. 2000). Calcium helps in maintaining the integrity of membrane and hydraulic conductivity of roots involving aquaporins (Martinez-Ballesta et al. 2006). Elevated concentration of Ca in nutrient solution alleviates the adverse effects of NaCl by inhibiting the uptake of Na (Epstein 1998) and reducing the leakage of membrane (Kaya et al. 2002). Application of Ca protects cell membranes from the adverse effect of Na and reduces the leakage of cytosolic K. In this manner it may help in the restoration of the K concentrations to an adequate level in the root that is reduced by salinity (Tuna et al. 2007). Thus, maintaining an adequate supply of Ca in saline soil solutions is an important factor in controlling the severity of specific ion toxicities, particularly in crops which are susceptible to Na and Cl injury (Epstein 1998). For example, Nedjimi et al. (2010) found that root applied 40-mM Ca as $CaCl_2$ counteracted salt-induced growth inhibition in *L. spartum*. This ameliorative effect of Ca on growth was suggested to be due to increased leaf K and Ca coupled with decrease in leaf Na. Likewise, Nedjimi and Daoud (2009a) found that the growth promoting effect of supplemental Ca on salt-grown two saltbush species was due to the ameliorative effect of Ca on root hydraulic conductivity as well as water uptake. Similarly, Nedjimi and Daoud (2009b) found that addition of $CaCl_2$ to a saline nutrient medium helped in maintaining membrane permeability and thus increasing concentrations of Ca and K and reducing concentration of Na, which could offer an economical and simple solution to *Atriplex halimus* under saline conditions. The presence of adequate Ca in the substrate influences the selectivity of K/Na by shifting the uptake ratio in favor of K at the expense of Na. Improvement in Ca-mediated membrane integrity invariably leads to reduction of K leakage from root cells and a more favorable root-K status (Nedjimi and Daoud 2009b).

3.2.6 Osmolytes Accumulation

Plants exposed to osmotic stress accumulate active compounds called osmolytes that lower their osmotic potential. Osmolytes are also referred to as compatible metabolites since they do not apparently interfere with the normal cellular metabolism of the cell. The main role of compatible solutes is to maintain the cell turgor and to provide the driving gradient for water uptake (Parida and Das 2005).

The accumulation of inorganic and organic solutes helps in osmotic adjustment, which is necessary for growth in a saline environment. Inorganic ions are believed to be sequestered in the vacuoles, while organic solutes are assumed to be compartmentalized in the cytoplasm to balance the low osmotic potential in the vacuole (Munns and Tester 2008). Plants synthesize a variety of compatible solutes such as proline, glycine betaine, and soluble sugars. These compounds protect plants from oxidative damage caused by free radicals and maintain the enzyme activities under salt stress (Ashraf and Harris 2004).

Glycine betaine (GB) is the most widely and extensively studied compatible compound due to its versatile functions. Several studies have shown that the level of GB is increased in the naturally accumulating plants under stressful conditions. GB has not only protected the plants against various stresses but also protected various enzymes involved in stress protection, including the quaternary structure of proteins and other macromolecules under stressful conditions (Chen and Murata 2002).

Glycine betaine in plants is synthesized in the chloroplasts from serine via ethanolamine, choline, and betaine aldehyde. It is mainly localized in chloroplasts and plays a vital role in chloroplast adjustment and protection of thylakoid membranes, thereby maintaining photosynthetic efficiency (Ashraf and Foolad 2007). *L. spartum* accumulates large quantities of GB, a quaternary ammonium compound that often contributes to osmoprotection, when exposed to osmotic stress (Fig. 3.2). Accumulation of this organic solute under various abiotic stresses in *L. spartum* indicates that GB plays a key role in osmoregulation to retain its growth potentialities in saline conditions (Nedjimi 2009). Higher salinity also causes increased accumulation of valine (Nedjimi 2011). The accumulation of free amino acid in plants under salt stress has often been attributed to the change in biosynthesis and degradation processes of amino acids and proteins (Hare et al. 1998). Valine has also been reported to accumulate in plants in response to salt stress such as strawberry (Keutgen and Pawelzik 2008) and *Phragmites australis* (Hartzendorf and Rolletschek 2001). Despite involvement of GB and valine in OA, the contribution of total soluble sugar accumulation to osmotic adjustment was significant. The total soluble sugar content was observed to increase with an increase in salinity (Fig. 3.2). It is widely believed that synthesis and accumulation of soluble sugars contribute to OA, carbon storage, radical scavenging, and stabilization of the structure of proteins such as RuBisCO (Dubey and Singh 1999). In several plants, sugars are main osmolytes for OA. For example, Cha-um et al. (2009) and López et al. (2008) found that high salinity increased the total soluble sugars in rice cultivars and *Medicago* species, respectively.

3.2.7 Root Hydraulic Conductivity (L_0)

Under abiotic stress, such as water deficit or salinity, stomata closure is considered to be the first mechanism that the plant employs to preserve water (Munns 2002). However, water absorption by the roots is among the main mechanisms by which plants can maintain their water content under stress conditions (Javot and Maurel 2002). The discovery of aquaporins in plants has resulted in a paradigm shift in the understanding of plant water relations (Maurel and Chrispeels 2001). These proteins provide an organism with the possibility to accelerate water movement across membranes, but diffusion will still occur in parallel. Furthermore, the ability to increase or decrease the water permeability of a cell seems to justify the enormous effort in expressing large amounts of these proteins (Schäffner 1998).

Under salinity stress, growth reduction could be caused by inhibition of the uptake of water due to reduction in hydraulic conductivity of roots cells. This in turn may result in a reduced overall hydraulic conductivity of roots (Martinez-Ballesta et al. 2006). L_0 in *L. spartum* seedlings declined significantly with the increase of salinity (Nedjimi et al. 2013). Effects of increased salinity on L_0 have been reported also in other plants (Martinez-Ballesta et al. 2003; Nedjimi and Daoud 2009a), and it has been suggested that they are due to the high concentrations of Na and Cl in the cytoplasm that reduce water transport through the plasma membrane aquaporins (Carvajal et al. 1999). We assume that the large reduction in root hydraulic conductivity of salinized plants was probably related to the decrease in activity or concentration of aquaporins in the root plasma membrane, this effect being due mainly to the specific toxicity of Na and Cl (Martinez-Ballesta et al. 2006).

3.3 Conclusion

Lygeum spartum L. (Poaceae) is a pioneer plant species, used for sand dune fixation, desalination, and phytoremediation along steppe regions. The plant tolerates abiotic constraints such as salinity and drought. This perennial grass is also used as a fodder for domestic cattle. *L. spartum* grows luxuriantly at 50 mM NaCl concentration. The present review analyses research undertaken during the last decade in physiology and biochemistry to abiotic stresses understands plant tolerance mechanism of this species. *L. spartum* having the ability to survive in the salt environment accumulates organic solutes and/or inorganic ions as osmolytes to alleviate or eliminate the salinity environmental stresses. Osmolytes for OA are accumulated in this plant by synthesizing in cells and/or absorbing the exogenous solutes from the environment. Large-scale multiplication of *L. spartum* in the arid and semiarid regions should reduce the load of salt soils and provide a fodder for livestock.

Acknowledgments All researches were funded by MESRS (National Program of Research, Project No. 1/U7/7606 and CNEPRU Project code F–02820140009).

References

Abdul Qados AMS (2011) Effect of salt stress on plant growth and metabolism of bean plant *Vicia faba* (L.). J Saudi Soc Agri Sci 10:7–15

Ashraf M (1994) Breeding for salinity tolerance in plants. Crit Rev Plant Sci 13:17–42

Ashraf M, Foolad MR (2007) Roles of glycine betaine and proline in improving plant abiotic stress resistance. Environ Exp Bot 59:206–216

Ashraf M, Harris PJC (2004) Potential biochemical indicators of salinity tolerance in plants. Plant Sci 166:3–16

Ashraf M, Wahid S (2000) Time-course changes in organic metabolites and mineral nutrients in germinating maize seeds under salt (NaCl) stress. Seed Sci Technol 28:641–656

Baskin CC, Baskin JM (1998) Seeds: ecology, biogeography, and evolution of dormancy and germination. Academic, San Diego

Bewley JD (1997) Seed germination and dormancy. Plant Cell 9:1055–1066

Blumwald E (2000) Sodium transport and salt tolerance in plants. Curr Opin Cell Biol 12:431–434

Carvajal M, Martinez V, Alcaraz CF (1999) Physiological functions of water channels as affected by salinity in roots of paprika pepper. Physiol Plant 105:95–101

Cha-um S, Charoenpanich A, Roytrakul S, Kirdmanee C (2009) Sugar accumulation, photosynthesis and growth of two indica rice varieties in response to salt stress. Acta Physiol Plant 31:477–486

Chen THH, Murata N (2002) Enhancement of tolerance of abiotic stress by metabolic engineering of betaines and other compatible solutes. Curr Opin Plant Biol 5:250–257

Conesa HM, Robinson BH, Schulin R, Nowack B (2007) Growth of *Lygeum spartum* in acid mine tailings: response of plants developed from seedlings, rhizomes and at field conditions. Environ Pollut 145:700–707

Dubey RS, Singh AK (1999) Salinity induces accumulation of soluble sugars and alters the activity of sugars metabolising enzymes in rice plants. Biol Plant 42:233–239

Epstein E (1998) How calcium enhances plant salt tolerance. Science 40:1906–1907

Flowers TJ, Colmer TD (2008) Flooding tolerance in halophytes. New Phytol 179:964–974

Glenn EP, Brown JJ, Blumwald E (1999) Salt tolerance and crop potential of halophytes. Crit Rev Plant Sci 18:227–255

Guerrier G (1988) Comparative phosphatase activity in four species during germination in NaCl media. J Plant Nutr 11:535–546

Hare PD, Cress WA, Van Staden J (1998) Dissecting the roles of osmolyte accumulation during stress. Plant Cell Environ 21:535–553

Hartzendorf T, Rolletschek H (2001) Effects of NaCl-salinity on amino acid and carbohydrate contents of *Phragmites australis*. Aquat Bot 69:195–208

Hasegawa PM, Bressan RA, Zhu J-K, Bohnert HJ (2000) Plant cellular and molecular responses to high salinity. Annu Rev Plant Physiol Plant Mol Biol 51:463–499

Javot H, Maurel C (2002) The role of aquaporins in root water uptake. Ann Bot 90:301–313

Kaya C, Kirnak H, Higgs D, Saltali K (2002) Supplementary calcium enhances plant growth and fruit yield in strawberry cultivars grown at high (NaCl) salinity. Sci Hortic 93:65–74

Keutgen AJ, Pawelzik E (2008) Contribution of amino acids to strawberry fruit quality and their relevance as stress indicators under NaCl salinity. Food Chem 111:642–647

Khan MA, Duke NC (2001) Halophytes – a resource for the future. Wetl Ecol Manag 6:455–456

Khan MA, Gul B (2006) Halophyte seed germination. In: Khan MA, Weber DJ (eds) Ecophysiology of high salinity tolerant plants. Springer, Dordrecht, pp 11–30

Khan MA, Ahmed MZ, Hameed A (2006) Effect of sea salt and L-ascorbic acid on the seed germination of halophytes. J Arid Environ 67:535–540

López M, Tejera NA, Iribarne C, Lluch C, Herrera-Cervera JA (2008) Trehalose and trehalase in root nodules of *Medicago truncatula* and *Phaseolus vulgaris* in response to salt stress. Physiol Plant 134:575–582

Maathuis FJM, Amtmann A (1999) K^+ nutrition and Na^+ toxicity: the basis of cellular K^+/Na^+ ratios. Ann Bot 84:123–133

Martinez-Ballesta MC, Martinez V, Carvajal M (2003) Aquaporin functionality in relation to H^+–ATPase activity in root cells of *Capsicum annuum* grown under salinity. Physiol Plant 117:413–420

Martinez-Ballesta MC, Silva C, Lopez-Berenguer C, Cabãnero FJ, Carvajal M (2006) Plant aquaporins: new perspectives on water and nutrient uptake in saline environment. Plant Biol 8:535–546

Maurel C, Chrispeels MJ (2001) Aquaporins: a molecular entry into plant water relations. Plant Physiol 125:135–138

Munns R (2002) Comparative physiology of salt and water stress. Plant Cell Environ 25:239–250

Munns R, Tester M (2008) Mechanisms of salinity tolerance. Annu Rev Plant Biol 59:651–681

Nedjimi B (2009) Salt tolerance strategies of *Lygeum spartum* L.: a new fodder crop for Algerian saline steppes. Flora 204:747–754

Nedjimi B (2011) Is salinity tolerance related to osmolytes accumulation in *Lygeum spartum* L. seedlings? J Saudi Soc Agri Sci 10:81–87

Nedjimi B (2013) Effect of salinity and temperature on germination of *Lygeum spartum* L. Agric Res 2:340–345

Nedjimi B (2014) Effects of salinity on growth, membrane permeability and root hydraulic conductivity in three saltbush species. Biochem Syst Ecol 52:4–13

Nedjimi B, Daoud Y (2009a) Effects of calcium chloride on growth, membrane permeability and root hydraulic conductivity in two *Atriplex* species grown at high (sodium chloride) salinity. J Plant Nutr 32:1818–1830

Nedjimi B, Daoud Y (2009b) Ameliorative effect of $CaCl_2$ on growth, membrane permeability and nutrient uptake in *Atriplex halimus* subsp. *schweinfurthii* grown at high (NaCl) salinity. Desalination 249:163–166

Nedjimi B, Daoud Y, Carvajal M, Martinez-Ballesta MC (2010) Improvement of the adaptation of *Lygeum spartum* L. to salinity under the presence of calcium. Commun Soil Sci Plant Anal 41:2301–2317

Nedjimi B, Guit B, Toumi M, Daoud Y (2013) Water status of *Lygeum spartum* L. seedlings subjected to salt stress. Bio Ressources 3:1–5

Parida AK, Das AB (2005) Salt tolerance and salinity effects on plants: a review. Ecotoxicol Environ Saf 60:324–349

Patel AD, Bhensdadia H, Pandey AN (2009) Effect of salinisation of soil on growth, water status and general nutrient accumulation in seedlings of *Delonix regia* (Fabaceae). Acta Ecol Sin 29:109–115

Probert RJ (1992) The role of temperature in germination ecophysiology. In: Fenner M (ed) Seeds: the ecology of regeneration in plant communities. CAB International, Wallingford, pp 327–348

Promila K, Kumar S (2000) *Vigna radiata* seed germination under salinity. Biol Plant 43:423–426

Pugnaire FI, Haase P (1996) Comparative physiology and growth of two perennial tussock grass species in a semi-arid environment. Ann Bot 77:81–86

Ramani B, Reeck T, Debez A, Stelzer R, Huchzermeyer B, Schmidt A, Papenbrock J (2006) *Aster tripolium* L. and *Sesuvium portulacastrum* L.: two halophytes, two strategies to survive in saline habitats. Plant Physiol Biochem 44:395–408

Rengasamy P (2010) Soil processes affecting crop production in salt-affected soils. Funct Plant Biol 37:613–620

Schäffner AR (1998) Aquaporin function structure and expression: are there more surprises to surface in water relations? Planta 204:131–139

Shabala L, Cuin TA, Newman IA, Shabala S (2005) Salinity-induced ion flux patterns from the excised roots of Arabidopsis SOS mutants. Planta 222:1041–1050

Shannon MC (1998) Adaptation of plants to salinity. Adv Agron 60:75–119

Suárez N (2011) Effects of short-and long-term salinity on leaf water relations, gas exchange, and growth in *Ipomoea pes-caprae*. Flora 206:267–275

Sudhir P, Murth SDS (2004) Effects of salt stress on basic processes of photosynthesis. Photosynthetica 42:481–486

Tester M, Davenport R (2003) Na^+ tolerance and Na^+ transport in higher plants. Ann Bot 91:503–527

Tlig T, Gorai M, Neffati M (2008) Germination responses of *Diplotaxis harra* to temperature and salinity. Flora 203:421–428

Tuna AL, Kaya C, Ashraf M, Altunlu H, Yokas I, Yagmur B (2007) The effects of calcium sulphate on growth, membrane stability and nutrient uptake of tomato plants grown under salt stress. Environ Exp Bot 59:173–178

Ungar IA (1995) Seed germination and seed-bank ecology of halophytes. In: Kigel J, Galili G (eds) Seed development and germination. Marcel Dekker, New York, pp 599–629

White PJ, Broadley MR (2003) Calcium in plants. Ann Bot 92:487–511

Zhu JK (2003) Regulation of ion homeostasis under salt stress. Curr Opin Plant Biol 6:441–445

Roles of Osmolytes in Plant Adaptation to Drought and Salinity

4

Kamrun Nahar, Mirza Hasanuzzaman, and Masayuki Fujita

Abstract

Drought and salinity are an immense threat to crop production worldwide, and crop growth models predict that due to global climate change the frequency and severity of drought and salinity stresses will be more severe in the future to cause substantial damages. Drought, salinity and other abiotic stresses are the reasons for crop losses worldwide as much as 50 %. So, it is a burning concern to protest and prevent the destructive global effects of drought and salinity. To cope with the stresses, plants, however, adopt a variety of physiological and biochemical mechanisms at the cellular and whole organism levels. Under these stress conditions, many plants and organisms synthesize and accumulate compatible solute compounds termed as osmolytes or osmoprotectants. Osmoprotectants are small, electrically neutral non-toxic molecules at molar concentrations and highly soluble organic compounds that efficiently maintain osmotic balance and stabilize proteins and membranes under salt, drought or other stress conditions. Researches in metabolomic, polyomic, transcriptomic and transgenic levels explored that plants overexpressing osmolyte biosynthesis or metabolic genes showed enhanced stress tolerance. This chapter presents information highlighting the vital roles of osmoprotectants against the devastating effects of drought and salt stresses.

K. Nahar
Department of Agricultural Botany, Faculty of Agriculture, Sher-e-Bangla Agricultural University, Sher-e-Bangla Nagar, Dhaka 1207, Bangladesh
e-mail: knahar84@yahoo.com

M. Hasanuzzaman
Department of Agronomy, Faculty of Agriculture, Sher-e-Bangla Agricultural University, Sher-e-Bangla Nagar, Dhaka 1207, Bangladesh
e-mail: mhzsauag@yahoo.com

M. Fujita (✉)
Department of Applied Biological Science, Faculty of Agriculture, Laboratory of Plant Stress Responses, Kagawa University, Miki-cho, Kita-gun, Kagawa 761-0795, Japan
e-mail: fujita@ag.kagawa-u.ac.jp

N. Iqbal et al. (eds.), *Osmolytes and Plants Acclimation to Changing Environment: Emerging Omics Technologies*, DOI 10.1007/978-81-322-2616-1_4

Keywords
Abiotic stress • Drought • Osmotic adjustment • Oxidative stress • Proline • Salinity

4.1 Introduction

Drought and salinity are immense threat for crop production worldwide, and crop growth models predict that due to global climate change the frequency and severity of drought and salinity stresses will be more severe in the future to cause substantial damages. Drought stress is supposed to cause global crop production losses of up to 30 % by 2025, compared to present yield [according to "Water Initiative" report of "The World Economic Forum (2009)" at Davos; Zhang 2011]. On the other hand, worldwide, more than 45 million hectares (M ha) of irrigated land have been aggravated by salt which account for more than 20 % of total land. Every year, more than 1.5 M ha area is taken out of production due to salinity problem (Pitman and Läuchli 2002; Munns and Tester 2008), and increasing salinity problem is expected to result in up to 50 % loss of cultivable lands by the middle of the twenty-first century (Mahajan and Tuteja 2005; Hasanuzzaman et al. 2013a, b). Drought, salinity, and other abiotic stresses are the reasons for crop losses worldwide as much as 50 % (Rodríguez et al. 2005; Acquaah 2007). So, it is a burning concern to protest and prevent the destructive global effects of drought and salinity. Drought and salinity prejudice normal growth of plant. These stresses impair water relations, water use efficiency, osmotic adjustment, and ionic balance in plants and disrupt membrane integrity and consistency and activity of proteins/enzymes. As a result, physiological processes including stomatal conductance, transpiration, photosynthesis, and assimilate translocation are distorted to severely affect plant biomass accumulation and yield (Peñuelas et al. 2013; Hasanuzzaman et al. 2014a, b). To cope with the stresses, plants, however, adopt a variety of physiological and biochemical mechanisms at cellular and whole-organism levels. Plants often alter gene expression and metabolism toward regulation of concentrations of several antistress compounds which help to change their physiology, phenology, growth, and reproduction to cope up with stress conditions (Peñuelas et al. 2013). Under stress conditions, many plants and organisms synthesize and accumulate compatible solute compounds termed as osmolytes or osmoprotectants. Osmoprotectants are small, electrically neutral nontoxic molecules at molar concentrations and highly soluble organic compounds which efficiently maintain osmotic balance and stabilize proteins and membranes under salt, drought, or other stress conditions (Yancey 1994). Considering the chemical basis, osmoprotectants are of three types (Fig. 4.1): betaines (Bet) (*N*-methylated amino acid derivatives) and related compounds such as dimethylsulfoniopropionate (DMSP) and choline-O-sulfate, amino acids such as proline (Pro) and ectoine, and polyols and nonreducing sugars such as trehalose (Tre), mannitol, and sorbitol (Rhodes and Hanson 1993; Bohnert et al. 1995). Numerous research findings reported the stress-induced changes of osmoprotectants within plant cell. Reports exist describing the beneficial roles of exogenously applied osmoprotectants under drought, salt, and other stress conditions. Researches in metabolomic, polyomic, transcriptomic, and transgenic levels explored that plants overexpressing osmolyte biosynthesis or metabolic genes showed enhanced stress tolerance (Chen and Murata 2002, 2008; Rontein et al. 2002; Ashraf and Foolad 2007; Peñuelas et al. 2013). This chapter is our little effort to gather and present information highlighting the vital roles of osmoprotectants against devastating effects of drought and salt stresses.

4.2 Biosynthesis of Osmolytes

4.2.1 Proline

The amino acid Pro plays a highly beneficial role in plants exposed to various stress conditions as an osmolyte. The vital roles of Pro have also been recognized as a metal chelator, an antioxidative

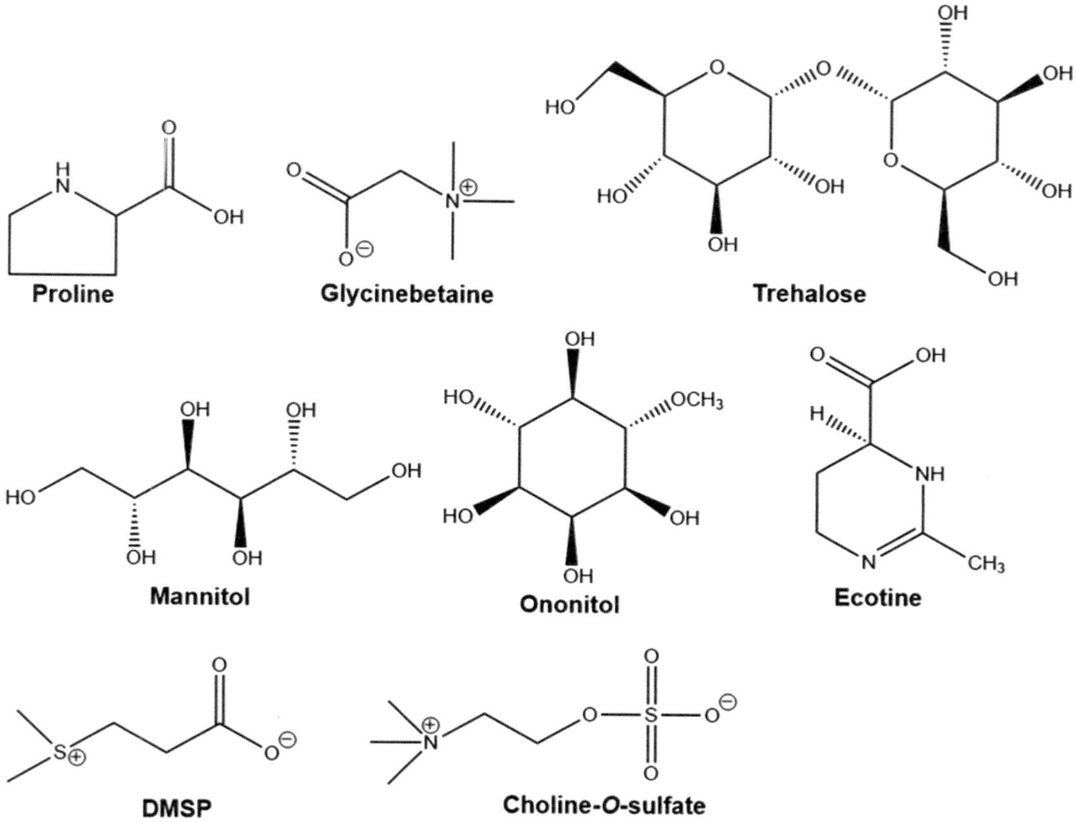

Fig. 4.1 Structure of different representative osmolytes. *DMSP* dimethylsulfoniopropionate

defense molecule, and a signaling molecule. Stressful environment results in an overproduction of Pro in plants which is highly associated with stress tolerance development imparting cellular turgor or osmotic balance, stabilizing membranes, controlling reactive oxygen species (ROS) within tolerable range, and thus preventing plant cell from oxidative damage. Moreover, the beneficial roles of Pro have been confirmed through transgenic approaches where plants overexpressing Pro biosynthesis gene accumulated higher level of Pro and showed enhanced tolerance to various environmental stresses (Hayat et al. 2012).

Accumulation of Pro could be due to de novo synthesis or decreased degradation or both (dos Reis et al. 2012). In plants, it is synthesized via two pathways, viz., glutamate (Glu) and ornithine (Orn) pathways (Fig. 4.2). However, the earlier one takes part mostly in Pro synthesis during osmotic stress (Lv et al. 2011; Witt et al. 2012). Glutamate is first converted into an intermediate glutamic semialdehyde (GSA) by the action of Δ^1-pyrroline-5-carboxylate synthetase (P5CS) which is spontaneously cyclized into pyrroline-5-carboxylate (P5C). This intermediate P5C is finally converted into Pro which is catalyzed by Δ^1-pyrroline-5-carboxylate reductase (P5CR). The potentiality of Pro synthesis is largely dependent on the activity and expression of the catalyzing enzymes. Transgenic tobacco plants overexpressing P5CS have shown increased concentration of Pro and provided better protection against abiotic stress (Kishor et al. 1995). In most of the higher plants, *P5CS* is encoded by two genes, whereas P5CR is encoded by a single gene (Verbruggen et al. 1993; Armengaud et al. 2004; Szabados and Savoure

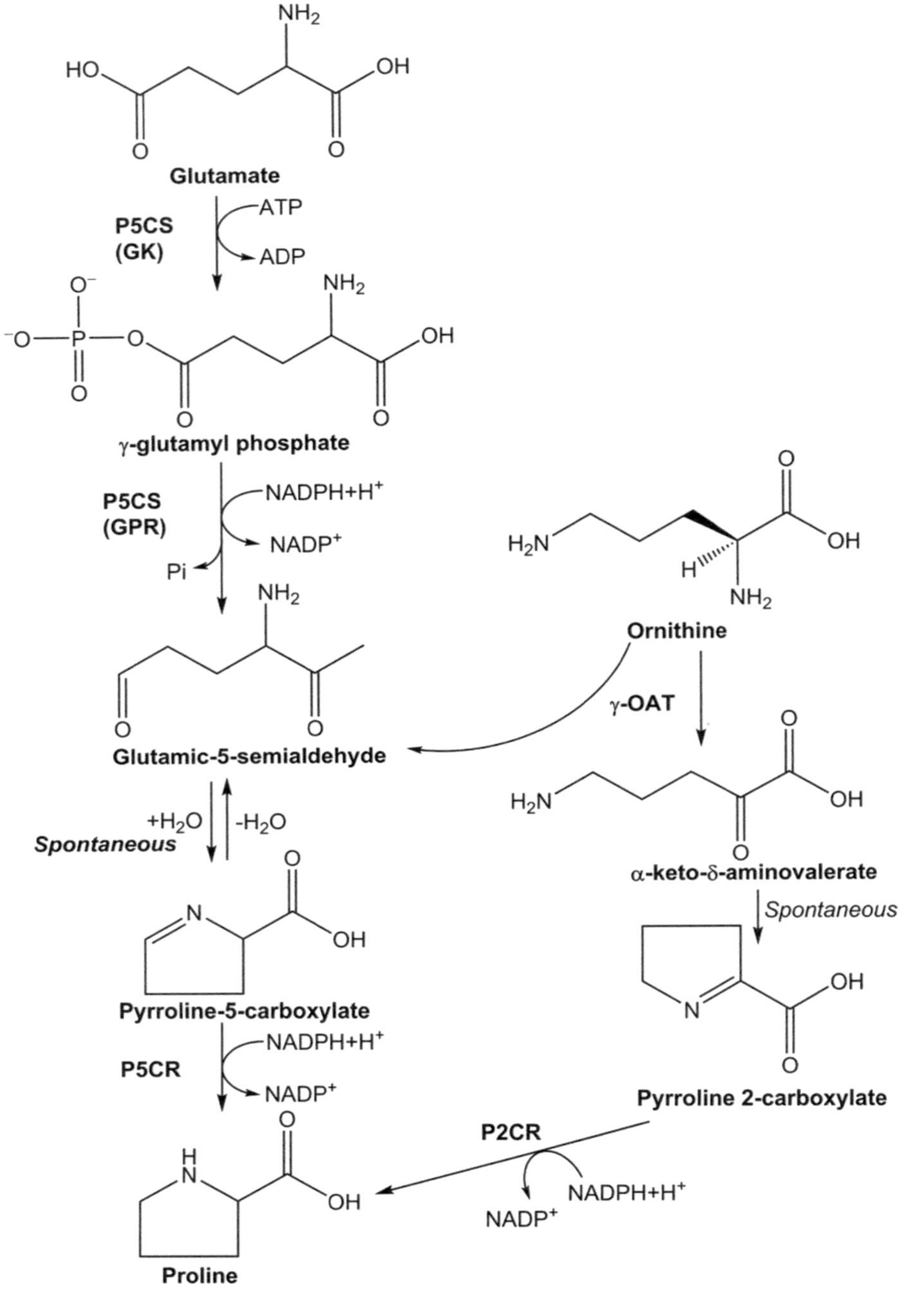

Fig. 4.2 Schematic representation of Pro biosynthesis in plants. *P5CS* Δ^1-pyrroline-5-carboxylate synthetase, *GK* γ-glutamyl kinase, *GPR* γ-glutamyl phosphate reductase, *P5CR* Δ^1-pyrroline-5-carboxylate reductase, *δ-OAT* ornithine δ-aminotransferase, *P2CR* pyrroline 2-carboxylate

2010). Proline catabolism occurred in mitochondria where two enzymes, viz., Pro dehydrogenase (ProDH) and P5C dehydrogenase (P5CDH), metabolize Pro into Glu. In Orn pathway, Orn can be transaminated either to GSA or to α-keto-δ-aminovalerate (KAV) by the activity of ornithine δ-aminotransferase (δ-OAT). This KAV is spontaneously cyclized to pyrroline 2-carboxylate

(P2C), which is finally converted into Pro by the activity of P2C reductase (P2CR; Trovato et al. 2008). The Orn pathway of Pro synthesis is common in some situation like seedling establishment and under high-N input (Delauney et al. 1993; Xue et al. 2009).

Biosynthesis, catabolism, and transport between cells and different cellular compartments determine intercellular Pro level. It is suggested that the biosynthetic enzymes (P5CS1, P5CS2, and P5CR). are localized at the cytosolic region. Mitochondrial localization is suggested for the enzymes (PDH1/ERD5, PDH2, P5CDH, and OAT) involved in Pro catabolism (Szabados and Savoure 2010).

4.2.2 Glycinebetaine

Betaine is widely distributed in higher plants, synthesized in many plant species in higher amount under environmental stress conditions (Rhodes and Samaras 1994). Betaines are quaternary ammonium compounds (QACs) which contain a carboxylic acid group. They may be generally regarded as fully N-methylated amino or imino acids. Different forms of Bet exist depending upon plant species including glycinebetaine (GB), proline betaine, β-alanine betaine, choline-O-sulfate, and 3-dimethylsulfoniopropionate (Rhodes and Samaras 1994; McNeil et al. 1999). Among different kinds of betaine, GB is the most common. Glycinebetaine, also called as original betaine (N,N,N-trimethylglycine), was first discovered from *Beta vulgaris* which is later found to be distributed in microorganisms, plants, and animals. It is one of the most abundant quaternary ammonium compounds that occurs during dehydration stress in plants (Ashraf and Foolad 2007). It is mostly found in chloroplast where it takes part in protecting the thylakoid membrane and osmotic adjustment. In contrast to existence of different GB accumulator plant species, some are considered to be non-accumulators such as *Arabidopsis*, rice, and tobacco.

Glycinebetaine is synthesized from choline in a two-step oxidation by a ferredoxin (Fd)-dependent choline monooxygenase (CMO) and a betaine aldehyde dehydrogenase (BADH) with a strong preference for NAD^+ (Fig. 4.3). In almost all biological systems, conversion of choline to GB is through a two-step dehydrogenation/oxygenation via the unstable intermediate betaine

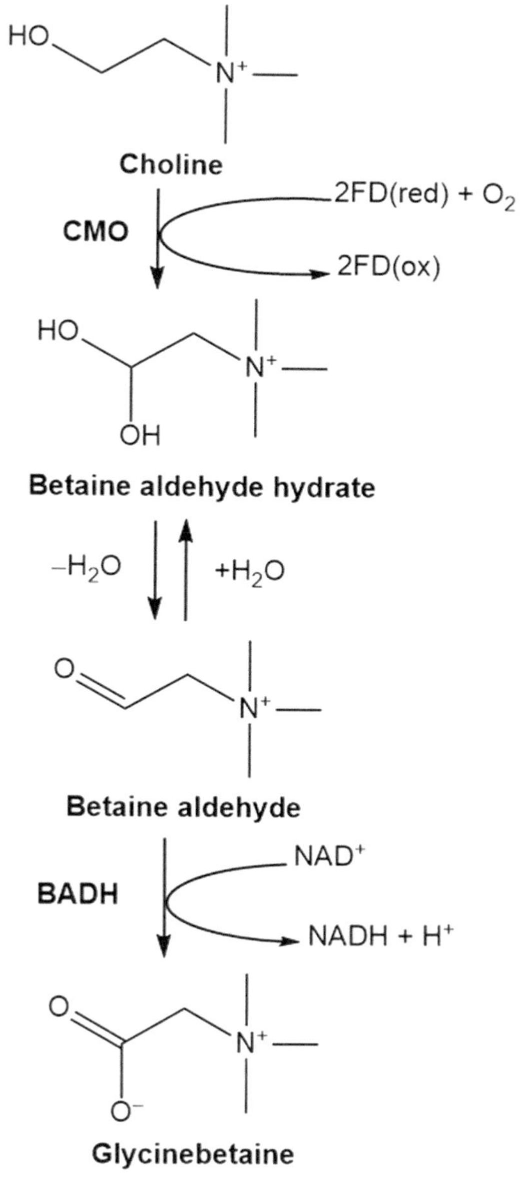

Fig. 4.3 Schematic representation of GB biosynthesis in plants. *CMO* choline monooxygenase, *BADH* betaine aldehyde dehydrogenase

aldehyde. In higher plants, GB biosynthesis occurs at chloroplasts. In the first step, choline is converted to an unstable compound betaine aldehyde by CMO. In the second step, betaine aldehyde is converted to GB by BADH (Rathinasabapathi et al. 1997). Besides this, other pathways of GB synthesis such as direct N-methylation of glycine also exists (Weretilnyk et al. 1989). Unlike higher plants, bacteria produce GB from choline using choline oxidase (COX) instead of CMO. The first enzyme is much responsible to synthesize GB, and in many plant studies, overexpression of BADH is found to be highly associated with GB synthesis and better tolerance to osmotic stress (Rathinasabapathi et al. 2001; Sakamoto and Murata 2001; Sulpice et al. 2003).

There are other biosynthesis pathways of GB in other higher plants and in some other microorganisms which are also considered as important, because genes that encode the enzymes involved in the biosynthesis of GB have been cloned from other microorganisms in transgenic approaches to enhance stress tolerance. Some evidences which are genes for CMO and BADH from higher plants (Rathinasabapathi et al. 1997), CDH and BADH from *Escherichia coli* (Landfald and Strom 1986), and choline oxidase (i.e., *codA*) from *Arthrobacter globiformis* (Deshnium et al. 1995) have been exploited in transgenic approaches.

4.2.3 Trehalose

Trehalose is a nonreducing disaccharide composed of two glucose residues (α-D-glucopyranosyl-1,1-α-D-glucopyranoside) which are bonded by α-α-(1 1) linkage. Trehalose was first reported in rye (Richards et al. 2002; Elbein et al. 2003). It has been detected in a wide range of organisms, including plants, bacteria, fungi, and invertebrates (Elbein et al. 2003). Among the disaccharides, Tre has particular properties; both of the reducing ends of the molecule form glycosidic bond. Therefore, Tre is resistant to acidic hydrolysis and is stable in solution at high temperatures, even under acidic conditions (Richards et al. 2002). Trehalose is a membrane and molecule stabilizer; it bears water replacement mechanism where Tre replaces water by establishing hydrogen bonds with membranes or macromolecules during dehydration or freezing (Crowe 2007). Trehalose has glass formation mechanism. Trehalose not only crystallizes but also solidifies into a glassy state. It is the only sugar which remains in glass-like state when completely dehydrated (Richards et al. 2002; Elbein et al. 2003). There are few previous reports regarding Tre accumulation in some desiccation-tolerant plants (Drennan et al. 1993; Bianchi et al. 1993; Albini et al. 1994). Trehalose plays multiple roles in confirming stress tolerance in plants. Recently published reports explore the roles of Tre in multiple species transcripts encoding putative Tre biosynthetic genes (Fernandez et al. 2010).

There are different Tre biosynthesis pathways existing in plants, bacteria, fungi, yeast, and algae (Goddijn and van Dun 1999; Avonce et al. 2006; Paul et al. 2008). Trehalose biosynthesis in higher plants only occurs through trehalose phosphate synthase (TPS)-trehalose phosphate phosphatase (TPP) pathway (which is also known as OtsA-OtsB pathway) (Fig. 4.4). In the first step, binding of glucose-6-phosphate to uridine diphosphoglucose (UDP-glucose) is catalyzed by TPS to produce trehalose-6-phosphate (T6P). In the second step, trehalose-6-phosphate phosphatase (TPP) catalyzes the dephosphorylation of T6P to Tre. In bacteria, OtsA and OtsB enzymes catalyze the conversion of glucose-6-phosphate and UDP-glucose to Tre. Yeasts TPS1 and TPS2 (homologues of TPS and TPP, respectively) catalyze this process (Bell et al. 1998; Paul et al. 2008). Trehalose breaks down to two glucose residues which have been found in almost all organisms synthesizing Tre (Richards et al. 2002; Elbein et al. 2003; Paul et al. 2008).

4.3 Plant Responses to Salinity and Drought

4.3.1 Salinity

Salinity severely affects plant physiological processes which primarily appears as osmotic and

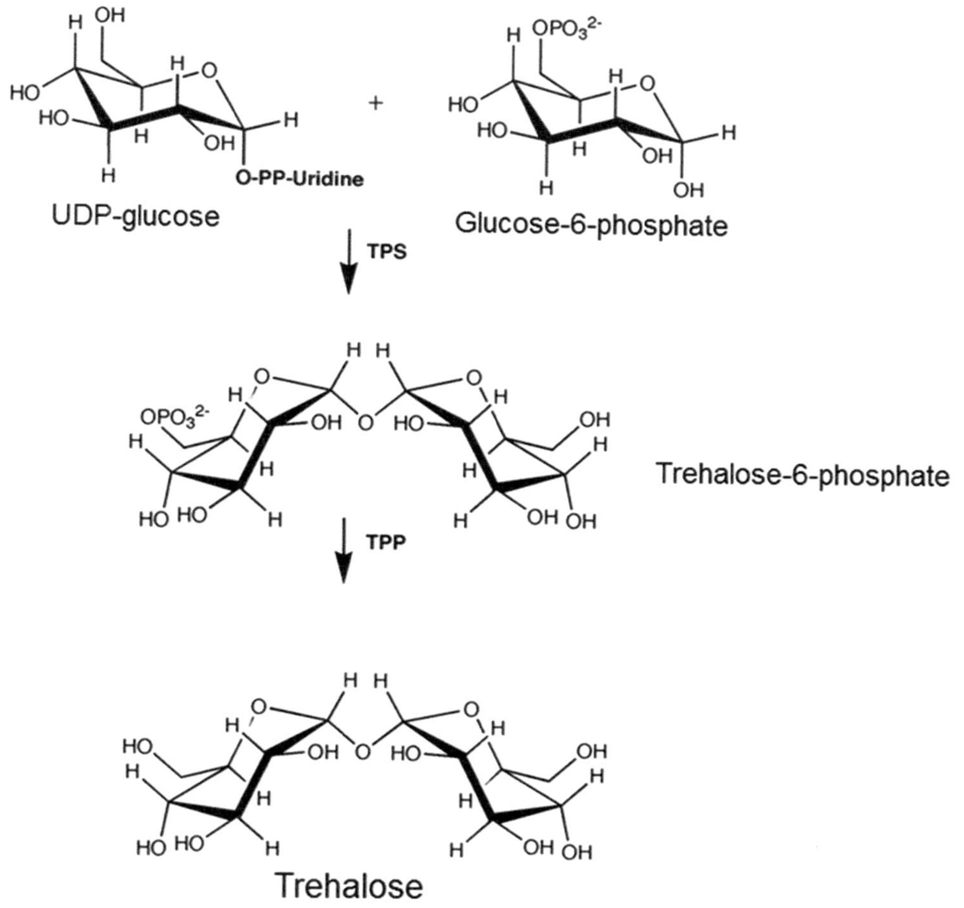

Fig. 4.4 Schematic representation of Tre biosynthesis in plants. *TPS* trehalose phosphate synthase, *TPP* trehalose phosphate phosphatase

ionic stresses causing water deficit, phytotoxicity, and nutrient imbalance (Munns 2002a). Sodium (Na^+) interferes with K^+ uptake, disturbs stomatal regulation, and causes water loss. Chlorine (Cl^-) induces chlorotic toxicity symptoms as a result of impaired chlorophyll (chl) biosynthesis (Tavakkoli et al. 2010). Romero-Aranda et al. (2001) observed that increase of salt in the root medium restricts water transportation and can lead to a decrease in leaf water potential. Salt-induced adverse effects were documented in all growth stages of different plants in numerous research findings. High salinity inhibits seed germination (Khan and Weber 2008). Salinity may alter water imbibition by seeds due to lower osmotic potential of germination media (Khan and Weber 2008) which affects germination. Salinity alters the activity of enzymes of nucleic acid metabolism (Gomes-Filho et al. 2008) and other proteins' metabolism (Yupsanis et al. 1994; Dantas et al. 2007), disturbs hormonal balance (Khan and Rizvi 1994), and reduces the utilization of seed reserves (Promila and Kumar 2000; Othman et al. 2006) which result in disturbed germination. Salt stress can damage ultrastructure of cell, tissue, and organs of germinating seedlings (Koyro 2002; Rasheed 2009). Delay in seed germination and reduced germination percentage were reported for tomato (Cuartero and Fernandez-Munoz 1999) and mustard (Bordi 2010) and mung bean (Nahar and Hasanuzzaman 2009). Reduction in germination rate, length of radicle and plumule, seedling length, and seedling vigor were reported in maize (Khodarahmpour et al. 2012). Decreased shoot and root weight, plant height, leaf number, and total biomass were reported in soybean plant due to salt stress (Dolatabadian et al. 2011). Rice plants showed remarkable reduction in plant height and tiller number and leaf area index under salt stress (Hasanuzzaman et al. 2009). Lauchli and Grattan (2007) reported that salt stress adversely affected performances of grain crops and cowpea plants at vegetative, flowering, and seed filling stage. The early reproductive stage was considered as the most sensitive stage to salt stress. The decrease in photosynthetic pigment contents including chl, carotenoid, and xanthophyll under salt stress has been reported in various studies with different plants (Saha et al. 2010; Amirjani 2011; Chutipaijit et al. 2011). Decrease in stomatal conductance and photosynthetic rate is another common effect of salt stress (Parida et al. 2004). Changes in enzyme activity, alterations in cytoplasmic structure, and negative feedback by reduced sink activity are some other reasons of reduced photosynthesis under salt stress (Iyengar and Reddy 1996). Reduced stomatal closure, reduced CO_2 availability and inhibition in carbon fixation, and excessive excitation energy in chloroplast under salt stress result in increased generation of ROS such as superoxide ($O_2^{\bullet -}$), hydrogen peroxide (H_2O_2), hydroxyl radical (OH•), and singlet oxygen ($^1O_2$) (Parida and Das 2005; Ahmad and Sharma 2008; Ahmad et al. 2010a, 2011) causing severe cellular damage through oxidation of lipids, proteins, and nucleic acids (Pastori and Foyer 2002; Apel and Hirt 2004; Ahmad et al. 2010a, b). Salt-induced overproduction of ROS has been documented in grain crop, oil seed crop, pulse crop, vegetable, and fruit plants (Gueta-Dahan et al. 1997; Dionisio-Sese and Tobita 1998; Mittova et al. 2004; Ahmad et al. 2009, 2010b). Besides different ROS, salt stress also overproduces cytotoxic methylglyoxal (MG) and causes subsequent damages to cell. Several studies showed that MG levels amplified with an increase in salt stress (Yadav et al. 2005; El-Shabrawi et al. 2010). Adverse effects of salt stress in all the growth stages result in loss of economic yield including biomass production and grain yield. The number of pods plant^{-1} and seeds pod^{-1} and seed weight and total yield were reduced in different varieties of mung bean under salt stress with the highest reduction of total yield by 77 % (Nahar and Hasanuzzaman 2009). Salt stress severely affected the yield components of rice, and as a result, yield was significantly reduced up to 50 % (Hasanuzzaman et al. 2009). Yield loss under salt stress has also been reported in wheat (Quarrie et al. 2005), in barley (Ellis et al. 2002), in tomato (Ghanem et al. 2011), and in many other crops. The damage effects of salt stress were highly depended on intensity of salt stress, and in general, long-duration stress exerted more severe effects. In contrast to salt-sensitive crops, salt-tolerant crop varieties and halophyte produce higher yield under salt stress.

4.3.2 Drought

Drought stress creates dehydration within the plants which is characterized by changes in water relations, nutrient uptake, biochemical and physiological processes, membrane structure, and subcellular organelles (Tuba et al. 1996; Sarafis 1998; Yordanov et al. 2003). Seed water imbibition is the prerequisite of seed germination, drought arrests that process and hamper seed germination process. In sunflower, with the increase of severity of drought stress, the germination process is inhibited. Drought also increased abnormal seedling and mean germination time and caused delayed seedling emergence (Kaya et al. 2006). Okra showed reduced percentage of germination, increased average time necessary for germination, reduced radical and plumule length, and reduced fresh and dry weight (FW and DW, respectively) in response to drought stress (Baghizadeh and Hajmohammadrezaei 2011). Response of drought stress is specific to plant species and cultivar. The highest percentage of germination was observed in wheat cultivars (cv. Sabalan and cv. 4041) which was 96 %. In contrast, the lowest was in cv. Sissons (55 %). Germination index, root length, and seedlings FW were also negatively affected by drought differentially in different cultivars (Ghanifathi et al. 2011). Drought stress obstructs cell expansion process and subsequently obstacles cell division and growth (Prasad and Staggenborg 2008). Pervez et al. (2009) reported reduction of plant height and shoot DW in tomato plant under drought stress. Similar adverse effects were observed in chickpea under drought stress (Mafakheri et al. 2010). Drought stress severely reduced accumulations of C, N, and P; drought reduced plant height, basal diameter, leaf number, and biomass production in *Salix paraqplesia* and *Hippophae rhamnoides* (Díaz-López et al. 2012). Reduced shoot DW, osmotic potential, leaf water potential, and stomatal conductance were observed in drought-sensitive melon genotypes (CU 40 and CU 252) (Kusvuran 2012). Reduced leaf area, death of leaf tissue, and leaf drop under drought stress are often considered as drought avoidance mechanisms which are common in post-flowering and grain filling stages (Prasad and Staggenborg 2008). Drought stress differentially affects relative water content (RWC), leaf water potential, osmotic potential, pressure potential, and transpiration rate (Kirkham 2005). In potato, early-season drought substantially reduced water use efficiency (Costa et al. 1997). In contrast higher water use efficiency was reported in wheat and tomato under drought than well-watered controls which was attributed by reduced transpiration rates under drought (Abbate et al. 2004; Subramanian et al. 2006). Drought decreases photosynthesis simply by closing stomata and limiting CO_2 availability. Drought stress reduced stomatal conductance, CO_2 exchange rate, and the ratio of internal CO_2 concentration to ambient CO_2 concentration in wheat plant. These reductions affected reproductive attributes as evidenced by reduced grain number at maturity, grain yield, and biomass production (Monneveux et al. 2006). Soil drought inhibits plant growth and development. Dry matter reduction in wheat plant was reported under drought stress (Ahmad et al. 2007). Boutraa and Sanders (2001) reported that mild drought stress inhibited relative growth rate by 25 % which was mainly due to changes in net assimilation rate and photosynthetic rate. Stomatal movement is more sensitive than RuBisCO activity which is maintained even when leaf RWC drops to 50 %, while stomata are already 75 % closed (Kaiser 1987; Flexas et al. 2006). Drought-affected plants showed more decline in the photochemical chlorophyll fluorescence quenching, photosystem II quantum yield reduction, and diminished electron transport rate. Photoinhibition and photorespiration often result from reduced CO_2 assimilation under drought stress (Dias and Bruggemann 2010). Reduction of chl a, chl b, and carotenoid contents was reported in different drought-affected plant species, such as wheat (Chakraborty and Pradhan 2012), mustard (Din et al. 2011), sissoo (Saraswathi and Paliwal 2011), mung bean (Pratap and Sharma 2010), and common bean (Abass and Mohamed 2011). Meiosis stage of pollen development is considered as the most sensitive stage exposed to drought (Yang et al. 2007; Kato et al. 2008). Anther dehiscence and pollen shedding and impaired pollen germination and fertilization are results of drought damage (Saini and Westgate 2000). Decrease in yield of grain legumes grown under drought conditions is largely due to the reduction in the number of pods plant^{-1} (Pilbeam et al. 1992; Lopez et al. 1996). In peanut, drought-induced yield reduction is attributed by reduced pod set, pod number, seed weight, and inhibited peg penetration to the soil surface due to hard soil surface resulting from desiccation by drought (Haro et al. 2008). Failure of reproductive development is a major cause of yield reduction in faba bean and quinoa under drought stress (Adolf et al. 2009). Early seed development like meiosis of pollen development and the following later phases of seed development are most sensitive. Ovary abortion and pollen sterility under drought in maize and in other small grains reduced the grain number (Andersen et al. 2002; Adolf et al. 2009). Drought stress shortens the grain filling period and enhances early senescence which is one of the causes of yield reduction under drought (Plaut et al. 2004). Drought stress during vegetative growth, flowering, and terminal stage enhances spikelet sterility in rice that leads to unfilled grains and reduced yield (Shahryari et al. 2008). In rice, drought stress at reproductive stage resulted in 93 % reduction in grain yield; in wheat, drought stress from anthesis to harvest stage resulted in 93 % reduction in yield (Kumar et al. 2009); in barley, drought at grain filling stage reduced grain yield by 52 % (Babaeian et al. 2011); in canola, seed yield was reduced by 62 % due to drought imposition in flowering stage (BirunAra et al. 2011); and in cowpea, drought at vegetative and reproductive stages reduced yield by 69 % (Ahmed and Suliman 2010). Drought stress can also lead to oxidative stress creating ROS which damages the membrane of cell and subcellular organelle and cellular components and disrupts biochemical life processes. In case of extreme ROS production under drought, this even leads to plant death (Li et al. 2010; Faize et al. 2011; Hasanuzzaman and Fujita 2011; Sorkheha et al. 2011). Drought stress produces highly cytotoxic compound MG which damages cell components. Drought-induced MG production was reported in different species of *Brassica* (Alam et al. 2014).

4.4 Mechanisms of Stress Protection by Osmolytes

The existing research findings explored various mechanisms by which osmolytes conferred stress tolerance. Most of the reports primarily emphasize the osmoregulatory roles. The other mechanisms of osmolytes are biological membrane protection, detoxification of toxic compounds such as ROS and MG, alleviation of ionic toxicity, protection of photosynthetic and mitochondrial structure, and metabolism. Moreover, the signaling role of osmolytes has also been designated as a vital stress-protective mechanism.

Naturally, plants can accumulate osmoprotectants at the levels ranging 5–50 μmol g^{-1}FW (Rhodes and Hanson 1993; Bohnert et al. 1995). Osmoprotectants are typically confined to cytosol, chloroplasts, and some other cytoplasmic compartments occupying 20 % or less of the volume of mature cells. Natural osmoprotectant concentrations in cytoplasmic compartments can reach or exceed 200 mM which is osmotically noteworthy in maintaining cell turgor and in generating driving osmotic gradient for water uptake under stress condition (Rhodes and Samaras 1994). The physicochemical basis of the protective effect by osmolytes involves the exclusion of osmoprotectant molecules from the hydration sphere of proteins (Timasheff 2008). This results in presenting proteins to the least possible surface area to the water so that native protein structures are thermodynamically favored. Salts enter into the hydration sphere and then interact with protein surfaces and cause unfolding. Thus, under drought or salt stress conditions, osmoprotectants serve not only to raise cellular osmotic pressure but also to protect cell constituents. Similar protective effects can be achieved through osmoprotectants under other stress conditions (Yancey 1994).

Proline protects plants against stress injury by maintaining osmoregulation. Proline is a molecular chaperon and it stabilizes subcellular structures including membranes and proteins. It is a free radical scavenger, an agent for buffering cellular redox potential under stress conditions. Proline can serve as a nitrogen/carbon source of cell (Verbruggen et al. 1996). It acts as a protein-compatible hydrotrope, maintains cytoplasmic pH, and alleviates cytoplasmic acidosis. Proline serves as a proton acceptor and is involved in the recycling of NADPH via its synthesis from glutamate (Hare and Cress 1997) and thus maintains appropriate $NADP^+$/NADPH ratio compatible with metabolism (Hayat et al. 2012). Trehalose is often considered as the most effective protecting disaccharide stabilizing dry membranes and accumulating under drought condition. In case of dehydration stress, lipid phase transitions and vesicle fusion destabilize the membrane structure. Even negligible quantity of Tre is able to inhibit vesicle fusion and depresses the phase transition temperature of dry lipids and thus maintains the liquid crystalline phase in the absence of water (Crowe et al. 1992). Tre replaces bound water associated with biological structures (Donnamaria et al. 1994) which provide Tre a high hydration potential. This potential gives Tre scope to stabilize dry biological membranes and proteins by hydrogen bonding of its hydroxyl groups to the polar groups of proteins and phosphate groups of membranes (Kawai et al. 1992). During dehydration or freezing, this is the vital mechanism of protection (Donnamaria et al. 1994). By special mechanism termed as vitrification, Tre protects against desiccation stress. Compared to other disaccharides, Tre forms protective glass-like structure which has low reactivity and makes it more stable due to its nonreducing character. This hygroscopic glass-like structure makes Tre extremely stable both at high temperature and at complete desiccated stress and allows Tre to maintain its structure and function following rehydration (Crowe and Crowe 2000; Richards et al. 2002). GB maintains membrane structure preventing peroxidation of membrane lipids (Chen et al. 2000) and protecting electron transport via complex II in mitochondria (Hamilton and Heckathorn 2001). Other mechanisms are stabilization of complex proteins and membranes, regulation of transcriptional and translational machinery, and intervention as a molecular chaperone in the refolding of enzymes (Sakamoto and Murata 2001).

Salt stress has major two primary events to create stressful environment: first is lowering water potential which causes loss of water and turgor pressure and, second, creating high ionic strength which results in imbalance among inorganic ions. Osmolytes alleviate salt stress in two major ways. At first, they lower the osmotic potential and maintain normal turgor pressure (Kempf and Bremer 1998). Second, osmolytes act as stabilizers of proteins and cell components against the denaturing effects of high ionic stress (Hincha and Hagemann 2004). Glycinebetaine plays a crucial role in effective protection against salt stress (Ashraf and Foolad 2007; Chen and Murata 2008) by osmotic adjustment (Gadallah 1999), protein stabilization, stabilization of RuBisCO (Mäkelä et al. 2000), photosynthetic apparatus protection (Cha-Um and Kirdmanee 2010), and reduction of ROS (Ashraf and Foolad 2007). Glycinebetaine is suggested to have positive impact on absorption and translocation of monovalent cations in salt-affected plants. GB improved surviving percentages and growing abilities of salt-treated plants which was associated with reduced Na^+ accumulation and with the maintenance of K^+ concentration (Lutts 2000).

Nounjan et al. (2012) reported that exogenous Pro reduced the Na^+/K^+ ratio in salt-affected rice plant. The similar result was obtained by exogenous Tre application in other plant species under salt stress (Zeid 2009; Nounjan et al. 2012).

Rapid breakdown of Pro upon relief of stress provides reducing agents which help in improving mitochondrial oxidative phosphorylation and generation of ATP for recovery and refurbishing stress damages (Hare and Cress 1997; Hare et al. 1998). Presenting in chloroplast, GB protects the thylakoid membrane and improves photosynthetic efficiency (Robinson and Jones 1986; Genard et al. 1991). Glycinebetaine helps to stabilize the PSII complex. Papageorgiou and Murata (1995) reported that GB stabilizes the oxygen-evolving PSII complex against salts and high-temperature stresses by protecting PSII from dissociation of the extrinsic polypeptides. Glycinebetaine accumulated in transgenic *Synechococcus* which counteracted the inhibitory effect of salt stress and helped to repair PSII by counteracting the salt-induced inhibition of the degradation and synthesis of D1 protein of PSII (Ohnishi and Murata 2006). Several other studies support that in salt-affected plants, GB provided protections by enhancing PSII photochemical performance (Hayashi et al. 1997; Sakamoto and Murata Alia 1998; Holmström et al. 2000). Yang et al. (2005) reported that exogenous application of GB improved CO_2 assimilation, decreased stomatal conductance, and improved PSII efficiency and growth of maize plants under salt stress. GB is suggested to stabilize RuBisCO (Takahashi and Murata 2006). Transgenic tobacco plants accumulating GB alleviated the negative effects of salt stress on the photosynthetic machinery and rendered salt tolerance (Yang et al. 2008). Trehalose has roles protecting the photosynthetic apparatus which renders improved photosynthesis (Zeid 2009). Ali and Ashraf (2011) reported the vital roles of Tre in enhancing gas exchange parameters contributing toward improved photosynthesis. Enhanced photosynthetic performance can be achieved by sugar-signaling mechanisms.

Trehalose metabolism interacting with sugar-signaling pathways enhances photosynthetic capacity (Paul et al. 2001; Zeid 2009).

Stress-induced generation of ROS is one of the most common incident in most plants, and prevention or protection from oxidative stress induced by ROS is one of the vital mechanisms of stress tolerances (Foyer and Noctor 2003; Ashraf and Foolad 2007; Ashraf 2009). The ROS can be scavenged by antioxidant defense system comprising of different components; chiefly, nonenzymatic antioxidant components ascorbic acid (AsA), glutathione (GSH), phenolic compounds, alkaloids, nonprotein amino acids, and α-tocopherols are commonly found in plants. The antioxidant system also has enzymes which are mainly superoxide dismutase (SOD), catalase (CAT), ascorbate peroxidase (APX), monodehydroascorbate reductase (MDHAR), dehydroascorbate reductase (DHAR), glutathione reductase (GR), glutathione peroxidase (GPX), and glutathione-*S*-transferase (GST) (Hasanuzzaman et al. 2012). Improvement of the antioxidant defense system is considered to be effective in the development of resistance and adaptive features in plants against stresses (Sharma and Dubey 2005; de Carvalho 2008; Gill and Tuteja 2010). Trehalose has role in plant oxidative defense mechanism by increasing the accumulation of antioxidative compounds under stress conditions. Exogenously applied Tre increased antioxidative secondary metabolites including AsA and tocopherol which confer oxidative stress protection (Reddy et al. 2004a, b; Posmyk et al. 2009). Research findings also proved that exogenous Tre increased the activities of enzymes of the antioxidant system and reduced oxidative stress (Ali and Ashraf 2011; Alam et al. 2014). Besides, Tre research finding proved the vital roles of other osmoprotectants such as Pro, GB, and sorbitol in imparting oxidative stress tolerance in plants which was primarily achieved by osmolyte-induced enhancement of antioxidant defense mechanism (Ahmed et al. 2010; Theerakulpisut and Gunnula 2012; Hasanuzzaman et al. 2014a). In vitro experiments with mannitol, Pro, and sorbitol have been shown to be effective scavengers of ROS (Smirnoff and Cumbes 1989).

Some of the most recent reports exposed that osmolytes can significantly reduce the toxicity of MG by reducing its content. Osmolytes reduced MG by improving the components of glyoxalase system which was one of the mechanisms of osmolyte-induced stress protection. Alam et al. (2014) reported the beneficial role of Tre in detoxification of MG by improving activities of enzymes Gly I and Gly II (glyoxalase I and glyoxalase II, respectively) and contents of GSH under drought stress in different species of *Brassica*. Similar beneficial roles were evidenced to contribute by exogenous Pro and GB application in different varieties of rice plant under salt stress (Hasanuzzaman et al. 2014a).

Although there are some reports expressing signaling functions of osmolytes, they are still lacking proper pathways. As a signaling molecule, Tre activates multiple physiological and biochemical processes and responses. This signaling behavior is involved in increasing the accumulation of different secondary metabolites including antioxidants under stress conditions which impart stress protection (Avonce et al. 2004; Zeid 2009). Studies with transgenic approaches proved that plants overexpressing genes for osmoprotectants increased accumulation of certain osmoprotectants together with induction or upregulation of other mechanisms related to stress protection. Upregulation of Pro level is associated with accumulation of H_2O_2, increased activities of antioxidative enzymes (SOD, POD, APX, CAT), and upregulated transcript levels of genes encoding antioxidant enzymes (Cu/ZnSOD, MnSOD, CytAPX, CatC) in *O. sativa* seedlings under salt stress. Further, exogenous Pro addition, increased endogenous Pro and transcript levels of *P5CS* and *.P5CR*, and transcription of genes encoding several antioxidant enzymes are indication for signaling function of Pro (Nounjan et al. 2012). Wheat and sugarcane plants overexpressing *P5CS* not only accumulated Pro but also reduced oxidative stress (Molinari et al. 2007; Vendruscolo et al. 2007). In maize and cotton, overexpression of GB synthesis gene *betA* (gene encoding choline dehydrogenase, an enzyme of GB biosynthesis) or *AhCMO* (choline monooxygenase, catalyst in GB synthesis) resulted in reduced oxidative stress (Quan et al. 2004; Lv et al. 2007; Zhang et al. 2009a). These evidences demonstrate the signaling behavior of different osmolytes. Proline accumulation in plants has been reported to be mediated by both ABA-dependent and ABA-independent signaling pathways (Chiang and Dandekar 1995; Shinozaki and Yamaguchi-Shinozaki 1997; Hare et al. 1999; Zhu 2001, 2002) subjected to environmental stresses. Moreover, ABA accumulation in plants under osmotic stress has been reported to regulate expression of *P5CS* (Xiong et al. 2001).

4.5 Role of Osmolytes in Conferring Stress Tolerance

4.5.1 Salinity

The accumulation of osmolytes including Pro, Bet, and sugars is a well-known adaptive mechanism in plants against salt stress (Parida and Das 2005; Ashraf and Foolad 2007). Moreover, use of exogenous osmolytes improved plant performance under salt stress. Several examples have been presented in Table 4.1. Roles of Pro as a compatible osmolyte or osmoprotectant and its antioxidant feature to quench ROS made itself a potent protectant of plants from the adverse abiotic stresses including salt stress (Smirnoff and Cumbes 1989; Matysik et al. 2002). Werner and Finkelstein (1995) found that an *Arabidopsis* Pro-deficient mutant could not accumulate sufficient amount of Pro and it was unable to continue growth on saline media, while the wild type accumulated Pro and continued better growth performances. Proline can induce the expression of salt stress-responsive proteins involved in plant adaptation to salt stress (Khedr et al. 2003). In *Pancratium maritimum* plants subjected to severe salt stress, the activities of antioxidant enzymes CAT and peroxidase (POD) reduced. Activities of these enzymes were significantly higher in the presence of Pro; thus, plants were protected from NaCl-induced oxidative damage (Khedr et al. 2003). Proline improves salt tolerance in tobacco plants by increasing the activities of enzymes involved in the antioxidant defense system (Hoque et al. 2008). According to Hoque et al. (2008), salt stress increased protein carbonylation and contents of GSH and oxidized glutathione (GSSG) and enhanced the activities of GST and Gly II enzymes in *Nicotiana tabacum* L. cv. BY-2. Salt stress decreased redox state of thiol and the activities of GPX and Gly I enzymes. Proline provides protective action against NaCl-induced oxidative damage by enhancing antioxidant defense and MG detoxification systems and by reducing protein carbonylation. Proline functions for stabilizing functional units of electron transport and stabilizing the membranes, proteins, and enzymes such as RuBisCO (Hamilton and Heckathorn 2001). It protects the photosynthetic apparatus (Ashraf et al. 2008) by functioning as an oxygen radical scavenger (Heuer 2003). Olive plants were adversely affected in terms of growth and physiology when subjected to salt stress. Proline supplementation alleviated the adverse effects of salt stress by maintaining plant water status and soluble sugar contents. Proline maintained better photosynthetic activity of olive plants under salt stress. The antioxidant activity was also enhanced by Pro. Thus, by helping in physiological adaptation under salinity, Pro developed salt stress tolerance in olive plants (Ahmed et al. 2010). Exposure of rice seeds to increasing concentration of NaCl (0, 100, 200, 300, and 400 mM) harshly reduced germination percentage, root and shoot length, chl content, and protein content. In contrary, rice seeds pretreated with Pro (1, 5, and 10 mM) and grown at different NaCl concentrations neutralized the adverse effect of salt stress and showed better performance. Among different doses of Pro, 10 mM was not effective in improving plant growth under higher-salinity levels of 300 and 400 mM NaCl (Deivanai et al. 2011).

According to Sobahan et al. (2009), exogenous Pro and Bet suppressed Na-induced trisodium-8-hydroxy-1,3,6-pyrenetrisulfonic acid (an apoplastic tracer) uptake and Na^+ accumulation, while Pro and Bet slightly increased

Table 4.1 Summary of the protective effects of different exogenously applied osmoprotectants under salt stress

Name of the plant	Salinity doses and duration	Doses of protectant	Protective effects	References
Olea europaea L. cv. Chemlali	100 and 200 mM NaCl, 6 months	25 and 50 mM Pro	Modulated antioxidative enzyme activities Increased photosynthetic activity and plant growth Maintained suitable plant water status under salinity conditions	Ahmed et al. (2010)
Nicotiana tabacum BY-2 cells	200 mM NaCl, 7 days	20 mM Pro	Increased FW Enhanced activities of POD and CAT	Hoque et al. (2007)
Oryza sativa L. cv. BRRI dhan49 and cv. BRRI dhan54	150 and 300 mM NaCl, 48 h	5 mM GB	Enhanced antioxidant system Reduction of oxidative stress parameters (lipid peroxidation and H_2O_2) and lipoxygenase activity Improved MG detoxification system Increased leaf chl, RWC	Hasanuzzaman et al. (2014a)
Oryza sativa L. ssp. *indica*	150 mM NaCl	50 mM GB	Increased chl and carotenoid content Increased WUE Increased seed weight and yield	Cha-Um and Kirdmanee (2010)
Nicotiana tabacum BY-2 cells	200 mM NaCl, 7 days	20 mM GB	Increased FW Increased activity of POD	Hoque et al. (2007)
Oryza sativa L. cv. Nipponbare	25 mM NaCl, 12 h	1 and 5 mM Pro and GB	Reduced Na^+ uptake in rice plants Increased K^+/Na^+ ratio	Sobahan et al. (2009)
Cucumis melo L. cv. Yuhuang and cv. Xuemei	100 mM NaCl, 5 days	0.2 mM Pro	Increased FW and DW Increased photosynthesis, quantum yield of PSII, and chl content Reduced $O_2^{\bullet-}$ level and H_2O_2 content Enhanced activities of SOD, POD, APX, CAT, and DHAR	Yan et al. (2011)
Oryza sativa L. cv. KDML105	100 mM NaCl, 6 days	10 mM Pro	Increased FW and DW Reduced Na^+/K^+ ratio Increased endogenous Pro and transcript levels of P5CS and .P5CR Upregulated transcription of genes encoding several antioxidant enzymes	Nounjan et al. (2012)
Oryza sativa L. cv. Nipponbare	150 mM NaCl, 5 days	5 mM GB	Prevented salt-induced swelling of thylakoids, disintegration of grana stacking and intergranal lamellae, and disruption of mitochondria	Rahman et al. (2002)

(continued)

Table 4.1 (continued)

Name of the plant	Salinity doses and duration	Doses of protectant	Protective effects	References
Oryza sativa L. cv. KDML105	100 mM NaCl, 6 days	10 mM Tre	Increased FW and DW Reduced Na^+/K^+ ratio and endogenous Pro	Nounjan et al. (2012)
Oryza sativa L. cv. BRRI dhan49 and cv. BRRI dhan54	150 and 300 mM NaCl, 48 h	5 mM Pro	Increased contents of AsA and GSH, ratio of GSH/GSSG, and activities of APX, MDHAR, DHAR, GR, GPX, CAT, Gly I, and Gly II Improved oxidative stress and MG toxicity tolerance	Hasanuzzaman et al. (2014a)
Zea mays cv. Giza 2	NaCl (with 0.2 MPa), 14 days	10 mM Tre	Increased plant height, root and shoot DW, and leaf RWC Increased photosynthetic pigment and nucleic acid content Increased soluble sugar and soluble protein content Decreased ion leakage Increased K^+/Na^+ ratio	Zeid (2009)
Oryza sativa L. cv. KDML105	170 mM NaCl, 24 h	5 mM and 10 mM sorbitol, 5 mM and 10 mM Tre	Enhanced growth Reduced H_2O_2 and lipid peroxidation (indicated by malondealdehyde, MDA), contents, and electrolyte leakage	Theerakulpisut and Gunnula (2012)

the K^+ content which led to a high K^+/Na^+ ratio under saline conditions. Responses of salt-sensitive (BRRI dhan49) and salt-tolerant (BRRI dhan54) rice (*Oryza sativa* L.) varieties were studied and compared under two levels of salt stress (150 and 300 mM NaCl, 48 h) and with exogenous Pro and GB application with the salt stress. Rice seedlings exposed to salt stresses significantly reduced leaf RWC and chl content; salt stress increased endogenous Pro, lipid peroxidation, and H_2O_2 levels. Sensitive and tolerant varieties showed differential responses in terms of antioxidant and glyoxalase system. Salt stress reduced AsA, GSH, and GSH/GSSG, APX, MDHAR, DHAR, GR, GPX, CAT, and Gly I activities in sensitive variety, and these parameters were increased in tolerant variety. But Gly II, GST, and SOD activities were increased in both cultivars. Exogenous Pro and GB application alleviated salt damages by improving antioxidant and glyoxalase system and improving physiological adaptation. The salt-tolerant BRRI dhan54 showed better tolerance. Among two protectants, Pro rendered better protection than GB (Hasanuzzaman et al. 2014a).

Salt-sensitive *Citrus sinensis* "Valencia Late" cell culture exhibited reduced growth rate in salt media containing >200 mM NaCl which corroborates with reduced Pro accumulation. When the same cell culture specimen was provided with exogenous Pro (5 mM), the growth performance was better (Lima-Costa et al. 2008). *Cucumis melo* seedlings subjected to 100 mM NaCl showed reduced growth, chl content, and photosynthetic performances. Seedlings were in severe oxidative stress due to distorted antioxidant metabolism under salt stress. Yan et al. (2011) demonstrated that application of exogenous 0.2 mM Pro with salt containing nutrient solution alleviated those adverse affects. Proline supplementation with salt increased FW and DW of seedlings. Exogenous Pro significantly alleviated the decrease of photosynthesis, quantum yield of PSII, and chl content. Compared with NaCl alone, exogenous Pro also reduced ROS including the $^1O_2$ and H_2O_2 content by the enhancing activities of SOD, POD, APX, CAT, and DHAR.

Salt stress disturbed antioxidant and MG detoxification systems and causes oxidative damage in *Nicotiana tabacum* L. cv. BY-2 cells grown in suspension culture. Upregulation of the antioxidant and glyoxalase systems induced by GB provides protection against NaCl-induced oxidative damage in plants. Exogenous GB application reduced protein carbonylation and enhanced glutathione redox state and activities of GPX, GST, and Gly I under salt stress (Hoque et al. 2008). Absorption and translocation of monovalent cations in salt-stressed rice plants was regulated by GB. Exogenous GB was associated with reduced sodium accumulation and maintenance of K^+ concentration in salinized plants (Lutts 2000). Rice seedlings showed ultrastructural damages such as swelling of thylakoids, disintegration of grana stacking and intergranal lamellae, and disruption of mitochondria under salt stress (150 mM NaCl). Pretreatment of GB alleviated the adverse effects of salt stress. Glycinebetaine has the capability for the production of many vacuoles in the root cells to act as storage of Na^+ which was supposed to prevent Na^+ accumulation in the shoots and its subsequent damage effects (Rahman et al. 2002). Foliar spraying of GB in salt-sensitive rice plants exposed to salt (150 mM of NaCl) stress showed improved performance by maintaining better water use efficiency (WUE) and stabilizing photosynthetic pigment which led higher CO_2 assimilation and higher photosynthesis rate. Glycinebetaine also had its effects on overall growth performance of rice plants (Cha-Um and Kirdmanee 2010).

Trehalose functions as a compatible solute which is well known. Exogenous Tre treatment improved antioxidant capacity and reduced Na^+/K^+ ratio and endogenous Pro level in salt stress-affected rice seedlings. Transcription of P5CS and .P5CR was upregulated, and the activities of SOD and POX were decreased, whereas the activity of APX is increased by Tre. Trehalose upregulated the transcription of all antioxidant enzyme genes. Trehalose could not affect the growth inhibition during salt stress (Nounjan et al. 2012). Metabolic activity of maize seedlings was altered by Tre application which helped to alleviate damage effects of salt stress. Presoaking maize seeds with Tre (10 mM) increased Hill reaction activity and photosynthetic pigment and nucleic acid content in salt-affected seedlings. Trehalose treatment reduced oxidative stress which is indicated by stabilization of the plasma membranes, reduced rate of ion leakage, and reduced lipid peroxidation of maize root cells. Tre application increased the ratio of K^+/Na^+ in the leaves of maize seedlings (Zeid 2009). Exogenous Tre (up to 5 mM) reduced Na^+ accumulation and growth inhibition of rice plant exposed to salt stress. Higher concentrations (10 mM) prevented chl loss in leaf blades and preserved root integrity (Garcia et al. 1997). Alfalfa (*Medicago sativa* L.) plants exposed to salt stress accumulated negligible quantity of Tre in roots and bacteroids which was not enough to act an osmoprotective function (Fougere et al. 1991). In other studies, *Medicago truncatula* showed downregulation of Tre under salt stress (Lopez et al. 2008).

Sorbitol (5 and 10 mM) and Tre (5 and 10 mM) showed protective roles in salt-sensitive cultivar of rice (*Oryza sativa* L. cv. KDML105) when subjected to salt stress (170 mM). Salinity markedly reduced growth of this cultivar. Salinity also resulted in oxidative stress as indicated by significant increase of H_2O_2, lipid peroxidation, and membrane electrolyte leakage. However, sorbitol and Tre supplementation ameliorated those adverse effects and improved the growth performance (Theerakulpisut and Gunnula 2012).

4.5.2 Drought

Drought-induced accumulation of osmolytes has been reported in different plant species. Research findings proved that exogenous application of osmoprotectants can enhance drought tolerance (Table 4.2).

Different levels of drought stress (−0.15, −0.49, −1.03, and −11.76 MPa) were applied on cultivars (704 and 301) of *Zea mays* with poly-

Table 4.2 Summary of the protective effects of different exogenously applied osmoprotectants under drought stress

Name of the plant	Drought doses and duration	Doses of protectant	Protective effects	References
Zea mays L. cv. Agaiti-2002 and cv. EV-1098	Reduced irrigation throughout the growing period	30 mM Pro	Increased seed sugar, oil, protein, moisture, fiber and ash, and oil oleic and linoleic acid contents Increased antioxidant contents such as phenolics, carotenoids, flavonoids, and tocopherols Enhanced oil DPPH (1,1-diphenyl-2-picryl-hydrazyl) free radical scavenging activity	Ali et al. (2013)
Zea mays L. cv. Agaiti-2002 and cv. EV-1098	Reduced water content to 60 % field capacity, 15 days	30 and 60 mM Pro	Increased growth Increased photosynthetic rate, *Ci* and *gs*, and photosynthetic pigment levels	Ali et al. (2007)
Helianthus annuus L.	Reduced water at vegetative and flowering stage	100 mM GB	Increased head diameter, number of achene, 1000-achene weight, achene yield, and oil yield	Hussain et al. (2008)
Zea mays L.	Drought stress was created reducing irrigation interval	50 ppm, 100 ppm, 150 ppm GB	Increased chl a and b contents, plant height, yield, and 1000-seed weight	Miri and Armin (2013)
Triticum aestivum L.	Drought stress at tillering, flower initiation, and milking stage	100 mM GB	Improved rate of transpiration and photosynthesis and uptake of P and Ca but reduced N, K, and Na uptake	Raza et al. (2012)
Triticum aestivum L.	Withholding irrigation at milking stage	50,100, and 150 mM GB	Improved plant water potential, plant height, spike length, number of spikelets spike^{-1}, number of grains spike^{-1}, and grain yield	Raza et al. (2014)
Oryza sativa L. cv. Pathumthani 1	Reduced (25 %) water content	100 mM GB	Increased maximum quantum yield of PSIIand photon yield of PSII, net photosynthetic rate, plant height, and yield traits such as panicle length and weight, fertility percentage, and weight of one-hundred grain	Cha-um et al. (2013)
Helianthus annuus L.	Reduced irrigation frequency	50 and 100 mM GB	Increased leaf water and turgor potentials and yield	Iqbal et al. (2008)
Lycopersicum esculentum Mill. cv. PS	Reduced irrigation frequency	5 and 10 mM GB	Increased shoot height, root length, leaf number, leaf area, shoot FW, total shoot DW, RWC, and stress tolerance index, number of flowers, fruit number, and weight of fruit	Rezaei et al. (2012)

(continued)

Table 4.2 (continued)

Name of the plant	Drought doses and duration	Doses of protectant	Protective effects	References
Brassica napus, B. campestris, and B. juncea	15 % PEG, 2 days	5 mM Tre	Increased growth and photosynthetic pigment content Improved antioxidant system and reduced oxidative stress Improved glyoxalase system and reduced MG toxicity	Alam et al. (2014)
Zea mays L.	Reduced irrigation	30 mM Tre	Increased biomass production, photosynthetic attributes, and plant-water relation parameters Increased contents of antioxidants such as tocopherols and phenolics and increased activities of POD and CAT in antioxidant system	Ali and Ashraf(2011)
Triticum aestivum L. Callus	Drought stress (50 % PEG), 5 and 10 days and thereafter 5-day recovery	50 mM Tre	Increased cell viability and biomass Increased content of AsA and GSH and SOD and GR activity	Ma et al. (2013)

ethylene glycol (PEG). Drought resulted in solute accumulation. The cultivars 704 and 301 showed increases in free Pro level 1.56–3.13 times higher than the control treatment in response to drought stress. The increase in 704 var. was higher than 301 var. Increased Pro played roles in minimizing the damage caused by dehydration (Mohammadkhan and Heidari 2008). Fifty-six days of water deficit exposure resulted in increased foliar free Pro level by 2.4-fold and GB level by 2.5-fold of drought-stressed oak compared to control (Oufir et al. 2009).

Roles of exogenous Pro as a potential osmoprotectant to ameliorate the adverse effects of drought stress were studied on maize seed, seed oil composition, and oil antioxidant activity. Water stress reduced kernel sugar, oil, protein, and moisture contents and seed macro- and microelements, but it increased seed fiber and ash contents. Water stress increased oil oleic acid content and decreased linoleic acid content. Exogenous foliar-applied Pro significantly increased seed sugar, oil, protein, moisture, fiber, and ash contents both under well-irrigated and water deficit conditions. Exogenous Pro also increased oleic and linoleic acid contents and increased antioxidant concentrations: phenolics, carotenoids, flavonoids, and tocopherols positively correlated with the enhanced oil DPPH (1,1-diphenyl-2-picryl-hydrazyl) free radical scavenging activity (Ali et al. 2013). Another experiment with exogenous Pro (30 and 60 mM as foliar spray) showed that Pro significantly improved drought (60 % field capacity, 15 days) tolerance capacity of maize which is clearly evident from improved growth, photosynthetic rate, sub-stomatal CO_2 (*Ci*), stomatal conductance (*gs*), and photosynthetic pigment levels under drought stress (compared to drought treatment without exogenous Pro application) (Ali et al. 2007).

In sunflower, the effects of exogenous GB were studied to investigate its roles in improving plant performances under water deficit condition. Three levels (0, 50, and 100 mM) of GB were applied as seed treatments or as foliar application at the vegetative or reproductive growth stage. Water stress significantly decreased leaf water contents and osmotic and turgor potentials. Seed treatment of GB was not effective in alleviating drought damage, whereas foliar application of GB at the vegetative or reproductive growth stage increased leaf water content and turgor potentials under water stress, and between the doses, foliar spray with 100 mM GB was more effective (Iqbal et al. 2008). Corn plants were subjected to drought stress by reducing irrigation intervals. Drought stress significantly reduced growth, yield components, and yield of corn. In the contrary, application of exogenous GB (50, 100, and 150 ppm) in drought-affected corn plants altered the plant physiology so that its growth, yield components, and yield were improved significantly. Among different concentrations, 150 ppm concentration while spraying before flowering had great positive effects (Miri and Armin 2013). Sunflower plant was exposed to water stress at their vegetative and flowering stage. Water stress significantly decreased the head diameter, number of achenes, 1000-achene weight, achene yield, and oil yield. Drought stress increased the free leaf Pro and achene oil contents. Exogenous GB (100 mM) resulted in further increase of free Pro level and significantly improved these yield attributes by ameliorating the harmful effects of water stress. However, GB showed more prominent effects when it was applied on the flowering stage (Hussain et al. 2008). Raza et al. (2012) reported that application of GB effectively alleviated adverse effects of drought stress in wheat plant. Drought stress was imposed at the tillering, flower initiation, and milking stages, and GB supplementation improved the rate of transpiration and photosynthesis and uptake of P and Ca, whereas GB reduced N, K, and Na uptake. In other findings, Raza et al. (2014) again reported that different doses of GB (50, 100, and 150 mM) were beneficial for improving drought tolerance of wheat plant when the stress was imposed at the milking stage. Application of GB significantly improved plant water potential and increased plant height, spike length, the number of spikelets spike^{-1}, the number of grains spike^{-1}, and grain yield. Among different doses, 100 mM GB contributed best results. Application of 100 mM GB also rendered drought tolerance in rice plant in terms of various physiological parameters; GB improved maximum quantum yield of PSII and total chl and photon yield of PSII, net photosynthetic rate, plant height, and yield traits such as panicle length and weight, fertility percentage, and one-hundred-grain weight (Cha-um et al. 2013).

The ameliorative effects of foliar-applied Tre on water stress-affected maize seedlings were investigated. Water stress significantly reduced the plant biomass production, photosynthetic attributes, and water relation parameters. Water stress generated oxidative stress and altered the activities of antioxidant enzymes and levels of nonenzymatic antioxidant components. Foliar-applied Tre (30 mM) significantly increased plant biomass production and improved photosynthetic attributes and plant-water relation parameters including leaf water potential, solute potential, turgor potential, and RWC. Trehalose application relieves plants from oxidative damage by enhancing the activities of some key antioxidant enzymes (POD and CAT) and levels of nonenzymatic compounds (tocopherols and phenolics) (Ali and Ashraf 2011). Different species *of Brassica* such as *B. napus*, *B. campestris*, and *B. juncea* were subjected to drought stress (15 % PEG) for 48 h. Drought-affected *Brassica* seedlings showed growth reduction, reduced photosynthetic pigment, elevated Pro, oxidative stress, and MG toxicity. But exogenous Tre (5 mM) application significantly reduced oxidative stress by improving antioxidant system. Methylglyoxal toxicity generated from drought stress was also reduced by exogenous Tre where Tre reduced MG content by improving glyoxalase system components. Trehalose also rendered increased photosynthetic pigment level, regulated endogenous Pro, increased leaf RWC, and improved overall growth of plant (Alam et al. 2014). During studying with wheat callus under drought stress, Ma et al. (2013) observed severe oxidative damage and reduced callus growth. Exogenous Tre (50 mM) application in drought medium efficiently altered oxidative stress. Trehalose was efficient to increase the activity of SOD and GR; Tre also increased the levels of AsA and GSH which were involved in the reduction of ROS and subsequent oxidative damage. The overall effect of Tre was better callus development under drought stress (Ma et al. 2013).

4.6 Transgenic Approaches of Osmolyte-Induced Salinity and Drought Stress Tolerance

The beneficial roles of osmolytes under salt and drought stress have been proven in many studies. However, all of the osmoprotectants do not occur in plants. One of the most important features of osmoprotectants is that their beneficial effects are generally not species specific. This feature opens a great scope so that alien osmoprotectants can be engineered into plants to protect their new host plant from the destructive damage effects (Rhodes and Hanson 1993; Bohnert et al. 1995). In Table 4.3, transgenic plants engineered to osmolyte-synthesizing or metabolic genes and their tolerance to stresses have been presented.

Transgenic plant lines of *Petunia hybrida* cv. "Mitchell" were developed transformed by D1-pyrroline-5-carboxylate synthetase genes (*AtP5CS* from *Arabidopsis thaliana* L. or *OsP5CS* from *Oryza sativa* L.), and they accumulated higher Pro, improved growth, and were more tolerant to drought stress (Yamada et al. 2005). Wheat plants transformed with the *Vigna aconitifolia P5CS* encoding enzyme involved in Pro biosynthesis were subjected to 15 days of water. A significant increase in accumulation of Pro was noticed. The transgenic wheat plants also showed enhanced tolerance to drought stress, and the tolerance was due to protection mechanisms against oxidative stress, rather than by osmotic adjustment (Vendruscolo et al. 2007). Molinari et al. (2007) also reported Pro as a component of antioxidative defense system rather than as an osmotic adjustment mediator. In their study with transgenic sugarcane transformed with a heterologous *P5CS* gene, they ascertained a 2.5-fold higher Pro level after 9 days of water stress. But no osmotic adjustment was noticed in these Pro-overproducing plants. On the other hand, the photochemical efficiency of PSII was higher (65 %), and lipid peroxidation and chl were lower which helped to increase biomass production of transgenic line. The *.P5CR* was overexpressed in *Arabidopsis* under different stress conditions including salt, PEG, ABA, and heat stresses. Under salt stress, its expression resulted in increased Pro accumulation and root growth. This transgenic line decreased MDA content under NaCl, PEG, and ABA stresses (Ma et al. 2008). Overexpression of Pro *P5CR* increased Pro accumulation and affected amino acid metabolism of soybean (*Glycine max* L. Merr. cv. Ibis) which also exhibited improved drought tolerance (Simon-Sarkadi et al. 2006).

Glycinebetaine is synthesized either by the oxidation/dehydrogenation of choline or by the *N*-methylation of glycine (Chen and Murata 2002). So, genes that encode GB-biosynthetic enzymes have been cloned from various microorganisms and plants, and transgenic plants of various species have been produced which accumulate GB at a variety of levels and exhibit enhanced tolerance to a variety of abiotic stresses. Maize inbred line *DH4866* which was transformed with the *betA* gene encoding choline dehydrogenase accumulated higher level of GB and was more tolerant to drought stress than wild-type plants. Transgenic line had enhanced GB accumulation which conferred greater protection of the integrity of the cell membrane and greater activity of enzymes. Transgenic line showed better germination and growth at young seedling stage. Grain yield of transgenic plants was significantly higher, compared to that of wild-type plants under drought stress (Quan et al. 2004). Drought tolerance capacity of homozygous transgenic cotton (*Gossypium hirsutum* L.) plants overexpressing *betA* was tested at its different growth stages. Compared to wild type, transgenic cotton accumulated significantly higher levels of GB either before or after drought stress. Transgenic plants were more tolerant to drought stress from young seedlings to flowering plants. Increased RWC and photosynthesis, better osmotic adjustment, reduced ion leakage, and lipid membrane peroxidation under drought stress were considered as vital indicators of drought tolerance (Lv et al. 2007). Transgenic cotton was developed by introducing CMO gene (*AhCMO*) cloned from *Atriplex hortensis*. Seedlings of the transgenic lines accumulated

Table 4.3 Transgenic plants engineered to osmolyte-synthesizing or metabolic genes and their tolerance to stresses

Transgenic plant	Gene	Gene sources	Tolerance response in transgenic plant	References
Triticum aestivum L. cv. CD200126	*P5CS*	*Vigna aconitifolia* L.	Drought tolerance through reduced oxidative stress	Vendruscolo et al. (2007)
Petunia hybrida cv. Mitchell	*P5CS*	*Arabidopsis thaliana* L. and *Oryza sativa* L.	Higher Pro content and better growth performance under drought stress	Yamada et al. (2005)
Saccharum spp. var. RB855156	*P5CS*	*Vigna aconitifolia L.*	Drought tolerance by Pro accumulation and improving antioxidative defense	Molinari et al. (2007)
Glycine max L. Merr. cv. Ibis	*P5CR*	*Arabidopsis thaliana*	Improved drought stress tolerance	Simon-Sarkadi et al. (2006)
Arabidopsis thaliana L.	*P5CR*	*Triticum aestivum*	Improved salt tolerance	Ma et al. (2008)
Triticum aestivum L.	*betA*	*Escherichia coli*	Increased GB and chl, with reduced Na^+/K^+ ratios and solute potential, reduced cell membrane damage, higher photosynthesis rates and vigorous growth, and higher yield under salt stress	He et al. (2010)
Zea mays L.	*betA*	*Escherichia coli*	Higher GB accumulation, integrity of the cell membrane, better seedling growth, and higher yield under drought stress	Quan et al. (2004)
Gossypium hirsutum L.	*betA*	*Escherichia coli*	Increased RWC and photosynthesis, better osmotic adjustment, reduced ion leakage, and lipid membrane peroxidation under drought stress	Lv et al. (2007)
Gossypium hirsutum L.	*AhCMO*	*Atriplex hortensis*	Higher GB synthesis, reduced osmotic potential, electrolyte leakage and MDA accumulation, improved cell membrane, and higher photosynthetic capacity under salt stress	Zhang et al. (2009)
Nicotiana tabacum L.	*BADH*	*Spinacia oleracea* L.	Higher CO_2 assimilation and seedling growth under salt stress	Yang et al. (2008)
Oryza sativa	*otsA* and *otsB*	*Escherichia coli*	Sustained higher Tre level, better plant growth, less photooxidative damage, better mineral balance under salt and drought, and low-temperature stresses	Garg et al. (2002)
Solanum tuberosum L.	*TPS1*	*Pichia angusta*	Increased Pro, inositol, and raffinose contents Increased fructose, galactose, and glucose contents Enhanced drought tolerance	Kondrák et al. (2012)

(continued)

Table 4.3 (continued)

Transgenic plant	Gene	Gene sources	Tolerance response in transgenic plant	References
Solanum tuberosum L.	*TPS1*	*Pichia angusta*	Increased yield and longer dormancy period under drought stress Increased metabolite production including sugar, proteins, osmolytes, and hormones	Juhász et al. (2014)
Triticum aestivum L. cv. Bobwhite	*mtlD* (gene for mannitol-1-phosphate dehydrogenase)	*Escherichia coli*	Enhanced drought and salt tolerance Increased shoot FW, DW, plant height, and flag leaf length, compared to non-transformed	Abebe et al. (2003)
Arabidopsis thaliana	*mt1D*	*Escherichia coli*	Increased mannitol accumulation and salt tolerance	Thomas et al. (1995)
Nicotiana tabacum	*imt1* (gene for myo-inositol *O*-methyltransferase)	*Carpobrotus edulis*	D-Ononitol accumulation and enhanced drought and salt tolerance	Sheveleva et al. (1997)
Nicotiana tabacum	*SacB* (gene for levansucrase)	*Bacillus subtilis*	Fructan accumulation and enhanced drought tolerance	Pilon-Smits et al. (1995)

26 % and 131 % higher GB than those of non-transgenic plants both under normal and salt stress (150 mM NaCl). Transgenic line showed higher salt as it reduced osmotic potential, electrolyte leakage, and MDA accumulation and improved cell membrane and higher photosynthetic capacity under salt stress condition (Zhang et al. 2009). The *betA* gene encodes choline dehydrogenase which catalyzes choline to produce betaine aldehyde and catalyze betaine aldehyde to GB. The *betA* gene was transformed in wheat and its response was examined under salt stress (200 mM NaCl). Transformed wheat plants showed increased GB and chl, with reduced Na^+/K^+ ratios and solute potential. Reduced cell membrane damage, higher photosynthesis rates, and vigorous growth were also characteristic beneficial effects in transformed wheat plants. In field trial under salt stress, the transformed plants showed higher germination rates, higher tiller production, and higher grain yields in comparison to the wild-type plants. Thus, transgenic plants showed their tolerance to salt stress (He et al. 2010). Tobacco plants were introduced with *BADH* gene for betaine aldehyde dehydrogenase having the ability to synthesize GB in chloroplasts. Salt stress (75 and 150 mM NaCl) affects different biochemical and growth parameters which were assessed in transgenic and wild type. The decrease in CO_2 assimilation was lower in transgenic compared to wild-type plants. Decreases in the activities of ribulose-1,5-bisphosphate carboxylase/oxygenase, chloroplastic fructose-1,6-bisphosphatase, fructose-1,6-bisphosphate aldolase, and phosphoribulokinase were higher in wild type compared to transgenic plants. Transgenic plants also showed significantly improved seedling growth, compared to wild type (Yang et al. 2008).

In general, Tre does not accumulate in plants in quantities high enough to act as compatible solute which can give protection or prevention under stress conditions. Evidences exist showing that transgenic plants or engineered plants overexpressing microbial Tre biosynthesis genes accumulated low levels of Tre and became more tolerant to abiotic stresses. Transgenic sugarcane plant (*Saccharum officinarum* L.) enhanced its Tre production, it was subjected to drought stress, and it exhibited increased shoot and root biomass production, compared with the non-transgenic plant (Wang et al. 2005). Rice plants overexpressing Tre synthesis genes (*otsA* and *otsB*) accumulated higher Tre and showed higher photosynthetic capacity, decreased photooxidative damage, and improved ion uptake and partitioning during salt stress. These physiological adaptations rendered enhanced salt, drought, and low-temperature stress tolerances in rice plants without any growth abnormalities (Garg et al. 2002). Responses of the drought-sensitive potato cultivar White Lady and the drought-tolerant *TPS1* transgenic variant were examined under prolonged drought stress. Potato cultivar that expresses the yeast *TPS1* gene exhibited improved drought tolerance. Transgenic potato plant showed increased fructose, galactose, and glucose contents which contributed increased accumulation of Pro, inositol, and raffinose. The transgenic potato plant showed better drought tolerance, compared to the wild type (Kondrák et al. 2012). Transgenic potato cultivar expressing *TPS1* and its wild type were exposed to periodic drought stress which resulted in an average 30 % yield loss in *TPS1* plants and 50 % yield loss in wild type. *TPS1* tubers showed longer dormancy period than wild-type tubers and exhibited alterations in the concentrations of different metabolites. *TSP1* tubers showed higher accumulation of fructose, glucose, mannose, sucrose, aspergine, phenylalanine, abscisic acid, jasmonic acid, indoleacetic acid, and salicylic acid under stress conditions either exposed in fresh or storage period (Juhász et al. 2014). Tomato plants overexpressing yeast Tre synthesis genes showed enhanced tolerance to drought, salt, and oxidative stresses (Cortina and Culianez-Macia 2005). Rice *OsTPP1* and *OsTPP2* were transiently induced by cold, salt, and drought stress as well as external ABA applications which enhanced tolerance of rice plant (Pramanik and Imai 2005; Shima et al. 2007). Transgenic *Arabidopsis* plants overexpressing *AtTPS1* were more drought tolerant and showed glucose- and ABA-insensitive phenotypes (Avonce et al. 2004). Tobacco and tomato transgened with yeast *TPS1* gene proved to be drought tolerant (Romero et al. 1997; Cortina and Culianez-Macia 2005). Despite of some existing reports regarding the beneficial effects of Tre, some research results with genetically engineered plants related to Tre metabolism exhibited altered morphology due to toxicity of high Tre concentrations (Schluepmann et al. 2003; Cortina and Culianez-Macia 2005). Transgenic plants with microbial Tre biosynthesis were reported to show developmental aberrations (López-Gómez and Lluch 2012). These research findings raise inquiry about Tre as a compatible solute (Schluepmann et al. 2003; Cortina and Culianez-Macia 2005).

Previous researches with model transgenic plants have demonstrated that cellular accumulation of mannitol can alleviate abiotic stress. Ectopic expression of the *mtlD* gene for the biosynthesis of mannitol in wheat ensured better tolerance to water stress and salinity. Wheat (*Triticum aestivum* L. cv. Bobwhite) was transformed with the *mtlD* gene of *Escherichia coli*, and its calli were exposed to −1.0 MPa of water stress and 100 mM NaCl. Compared to non-transformed, *mtlD*-transformed wheat showed less reduction in shoot FW and DW, plant height, and flag leaf length under those stress conditions (Abebe et al. 2003). *Olea europaea* mannitol carrier *OeMaT1* is important for regulation of polyol transporters either in sink or in source tissues. Mannitol transport and compartmentation by *OeMaT* are significant to allocate this source of carbon and energy which has been proven vital in salt tolerance and olive ripening. Addition of 100–500 mM NaCl to cultured cells enhanced the capacity of the polyol/H^+ symport system and the amount of *OeMaT1* transcripts and depression in mannitol dehydrogenase activity. Mannitol-grown cells remained viable for 24 h in 250 and 500 mM NaCl-containing growing media. Moreover, *OeMaT1* transcripts increased throughout maturation of olive fruits, suggesting that an *OeMaT* is involved in the accumulation of mannitol during ripening of olive (Conde et al. 2007). Conde et al. (2011) also reported that salt and drought significantly increased mannitol transport activity and *OeMaT1* expression and conferred olive trees to cope with salinity and drought by coordinating mannitol transport with intracellular metabolism. Engineered *Nicotiana tabacum* and *Populus tomentosa* synthesizing mannitol through introduction of an *Escherichia coli* mannitol-1-phosphate dehydrogenase (*mltD*, catalyzing biosynthesis of mannitol from fructose) resulted in more salt-tolerant plants (Tarczynski et al. 1993; Hu et al. 2005). *Arabidopsis* engineered by *mltD* gene enhanced seed germination under salt stress (Thomas et al. 1995).

Sheveleva et al. (1997) reported that tobacco plant transformed with *imt1* accumulated significantly higher D-ononitol and showed enhanced drought and salt tolerance. Tobacco transformed with *sacB* gene for levansucrase accumulated significantly higher fructan. Levansucrase generates fructan from fructose. The fructan-producing plants performed significantly better. Compared to wild type, plants had a 55 % more rapid growth rate, 33 % greater FW, and 59 % greater DW under drought conditions (Pilon-Smits et al. 1995).

4.7 Conclusion and Future Perspectives

Since osmotic stress is one of the common phenomena during salinity, drought, metal toxicity, and other stresses, attempts to synthesize higher amount of osmolytes would be an effective mechanism of stress tolerance to plants. Although considerable reports are available on the role of osmolytes in abiotic stress tolerance, the exact mechanisms of their protective mechanisms are yet to be revealed. The biosynthesis mechanism of some osmolytes, e.g., Tre, in higher plants is still a mystery. Therefore, further research should be focused on the exploration of those mechanisms and identification of novel genes involved in Tre biosynthesis in plants under stressful condition. Research findings proved that transgenic plants overexpressing genes for different osmoprotectants showed enhanced tolerance to abiotic stresses. But in some cases, abrasion, growth reduction, and abnormal growth were noticed in transgenic plants. To overcome these problems, extensive research should be exploited. Osmolytes, without acting as osmoprotectants, can also enhance the activities of different enzymes related to plant biochemical and physiological process. The mechanisms by which osmoprotectants regulate activities of enzymes are enhanced and are undoubtedly interesting and demand insightful studies. The signaling behavior of different osmolytes is still under dark and should be explored. Notably, findings from experimental field research studies are still scarce. Field studies of the ability of exogenous applications of osmolytes to mitigate abiotic stress are likely to prop up its extended application to crop plants practically.

References

Abass SM, Mohamed HI (2011) Alleviation of adverse effects of drought stress on common bean (*Phaseolus vulgaris* L.) by exogenous application of hydrogen peroxide. Bangladesh J Bot 41:75–83

Abbate PE, Dardanellib JL, Cantareroc MG, Maturanoc M, Melchiorid RJM, Sueroa EE (2004) Climatic and water availability effects on water-use efficiency in wheat. Crop Sci 44:474–483

Abebe T, Guenzi AC, Martin B, Cushman JC (2003) Tolerance of mannitol-accumulating transgenic wheat to water stress and salinity. Plant Physiol 131:1748–1755

Acquaah G (2007) Principles of plant genetics and breeding. Blackwell, Oxford

Adolf VI, Jacobsen SE, Liu F, Jensen CR, Andersen MN (2009) Effects of drought and salt stress during early reproductive growth stages in faba bean and quinoa. IOP Conf Series Earth Environ Sci 6:372008

Ahmad P, Sharma S (2008) Salt stress and phyto-biochemical responses of plants. Plant Soil Environ 54:89–99

Ahmad F, Rahmatullah T, Aziz M, Maqsood A, Tahir MA, Kanwal S (2007) Effect of silicon application on wheat (*Triticum aestivum* L.) growth under water deficiency stress. Emir J Food Agric 19:1–7

Ahmad P, Jaleel CA, Azooz MM, Nabi G (2009) Generation of ROS and non-enzymatic antioxidants during abiotic stress in plants. Bot Res Intern 2:11–20

Ahmad P, Jaleel CA, Salem MA, Nabi G, Sharma S (2010a) Roles of enzymatic and non-enzymatic antioxidants in plants during abiotic stress. Crit Rev Biotechnol 30:161–175

Ahmad P, Jaleel CA, Sharma S (2010b) Antioxidative defence system, lipid peroxidation, proline metabolizing enzymes and biochemical activity in two genotypes of *Morus alba* L. subjected to NaCl stress. Russ J Plant Physiol 57:509–517

Ahmad P, Nabi G, Ashraf M (2011) Cadmium-induced oxidative damage in mustard [*Brassica juncea* (L.) Czern. & Coss.] plants can be alleviated by salicylic acid. South Afr J Bot 77:36–44

Ahmed FE, Suliman ASH (2010) Effect of water stress applied at different stages of growth on seed yield and

water use efficiency of cowpea. Agric Biol J N Am 1:534–540
Ahmed CB, Rouina BB, Sensoy S, Boukhriss M, Abdullah FB (2010) Exogenous proline effects on photosynthetic performance and antioxidant defense system of young olive tree. J Agric Food Chem 58:4216–4222
Alam MM, Nahar K, Hasanuzzaman M, Fujita M (2014) Trehalose-induced drought stress tolerance: a comparative study among different *Brassica* species. Plant Omics 7:271–283
Albini FM, Murelli C, Patritti G, Rovati M, Zienna P, Vita Finzi P (1994) Low molecular weight substances from the resurrection plant *Sporobolus stapfianus*. Phytochemistry 37:137–142
Ali Q, Ashraf M (2011) Induction of drought tolerance in maize (*Zea mays* L.) due to exogenous application of trehalose: growth, photosynthesis, water relations and oxidative defence mechanism. J Agron Crop Sci 197:258–271
Ali Q, Ashraf M, Athar HUR (2007) Exogenously applied proline at different growth stages enhances growth of two maize cultivars grown under water deficit conditions. Pak J Bot 39:1133–1144
Ali Q, Anwar F, Ashraf M, Saari N, Perveen R (2013) Ameliorating effects of exogenously applied proline on seed composition, seed oil quality and oil antioxidant activity of maize (*Zea mays* L.) under drought stress. Int J Mol Sci 14:818–835
Amirjani MR (2011) Effect of salinity stress on growth, sugar content, pigments and enzyme activity of rice. Int J Bot 7:73–81
Andersen MN, Asch F, Wu Y, Jensen CR, Nñsted H, Mogensen VO, Koch KE (2002) Soluble invertase expression is an early target of drought stress during the critical, abortion-sensitive phase of young ovary development in maize. Plant Physiol 130:591–604
Apel K, Hirt H (2004) Reactive oxygen species: metabolism, oxidative stress and signal transduction. Annu Rev Plant Biol 55:373–399
Armengaud P, Thiery L, Buhot N, Grenier-De March G, Savouré A (2004) Transcriptional regulation of proline biosynthesis in *Medicago truncatula* reveals developmental and environmental specific features. Physiol Plant 120:442–450
Ashraf M (2009) Biotechnological approach of improving plant salt tolerance using antioxidants as markers. Biotechnol Adv 27:84–93
Ashraf M, Foolad MR (2007) Roles of glycine betaine and proline in improving plant abiotic stress resistance. Environ Exp Bot 59:206–216
Ashraf M, Athar HR, Harris PJC, Kwon TR (2008) Some prospective strategies for improving crop salt tolerance. Adv Agron 97:45–109
Avonce N, Leyman B, Mascorro-Gallardo JO, Van Dijck P, Thevelein JM, Iturriaga G (2004) The *Arabidopsis* trehalose-6-P synthase *AtTPS1* gene is a regulator of glucose, abscisic acid, and stress signaling. Plant Physiol 136:3649–3659
Avonce N, Mendoza-Vargas A, Morett E, Iturriaga G (2006) Insights on the evolution of trehalose biosynthesis. BMC Evol Biol 6:109–113
Babaeian M, Esmaeilian Y, Tavassoli A, Asgharzade A, Sadeghi M (2011) The effects of water stress, manure and chemical fertilizer on grain yield and grain nutrient content in barley. Sci Res Essay 6:3697–3701
Baghizadeh A, Hajmohammadrezaei M (2011) Effect of drought stress and its interaction with ascorbate and salicylic acid on okra (*Hibiscus esculents* L.) germination and seedling growth. J Stress Physiol Biochem 7:55–65
Bell W, Sun W, Hohmann S, Wera S, Reinders A, De Virgilio C, Wiemken A, Thevelein JM (1998) Composition and functional analysis of the *Saccharomyces cerevisiae* trehalose synthase complex. J Biol Chem 272:33311–33319
Bianchi G, Gamba A, Limiroli R, Pozzi N, Elster R, Salamini F, Bartels D (1993) The unusual sugar composition in leaves of the resurrection plant *Myrothamnus flabellifolia*. Physiol Plant 87:223–226
BirunAra A, Shekari F, Hassanpouraghdam MB, Khorshidi MB, Esfandyari E (2011) Effects of water deficit stress on yield, yield components and phenology of canola (*Brassica napus* L.) at different growth stages. J Food Agric Environ 9:506–509
Bohnert HJ, Nelson DE, Jensen RG (1995) Adaptations to environmental stresses. Plant Cell 7:1099–1111
Bordi A (2010) The influence of salt stress on seed germination, growth and yield of canola cultivars. Not Bot Horti Agrobot Cluj 38:128–133
Boutraa T, Sanders FE (2001) Influence of water stress on grain yield and vegetative growth of two cultivars of bean (*Phaseolus vulgaris* L.). J Agron Crop Sci 187:251–257
Chakraborty U, Pradhan B (2012) Drought stress-induced oxidative stress and antioxidative responses in four wheat (*Triticum aestivum* L.) varieties. Arch Agron Soil Sci 58:617–623
Cha-um S, Kirdmanee C (2010) Effect of glycinebetaine on proline, water use, and photosynthetic efficiencies, and growth of rice seedlings under salt stress. Turk J Agric For 34:517–527
Cha-um S, Samphumphuang T, Kirdmanee C (2013) Glycinebetaine alleviates water deficit stress in indica rice using proline accumulation, photosynthetic efficiencies, growth performances and yield attributes. Aust J Crop Sci 7:213–218
Chen THH, Murata N (2002) Enhancement of tolerance to abiotic stress by metabolic engineering of betaines and other compatible solutes. Curr Opin Plant Biol 5:250–257
Chen THH, Murata N (2008) Glycinebetaine: an effective protectant against abiotic stress in plants. Trend Plant Sci 13:499–505
Chen WP, Li PH, Chen THH (2000) Glycinebetaine increases chilling tolerance and reduces chilling-induced lipid peroxidation in *Zea mays* L. Plant Cell Environ 23:609–618

Chiang HH, Dandekar AM (1995) Regulation of proline accumulation in *Arabidopsis thaliana* (L) Heynh. during development and in response to desiccation. Plant Cell Environ 18:1280–1290

Chutipaijit S, Cha-um S, Sompornpailin K (2011) High contents of proline and anthocyanin increase protective response to salinity in *Oryza sativa* L. spp. indica. Aust J Crop Sci 5:1191–1198

Conde C, Silva P, Agasse A, Lemoine R, Delrot S, Tavares R, Gero's H (2007) Utilization and transport of mannitol in *Olea europaea* and implications for salt stress tolerance. Plant Cell Physiol 48:42–53

Conde A, Silva P, Agasse A, Conde C, Gero's H (2011) Mannitol transport and mannitol dehydrogenase activities are coordinated in *Olea europaea*under salt and osmotic stresses. Plant Cell Physiol 52:1766–1775

Cortina C, Culianez-Macia FA (2005) Tomato abiotic stress enhanced tolerance by trehalose biosynthesis. Plant Sci 169:75–82

Costa LD, Vedove GD, Gianquinto G, Giovanardi R, Peressotti A (1997) Yield, water use efficiency and nitrogen uptake in potato: influence of drought stress. Potato Res 40:19–34

Crowe JH (2007) Trehalose as a "chemical chaperone": fact and fantasy. Adv Exp Med Biol 594:143–158

Crowe JH, Crowe LM (2000) Preservation of mammalian cells—learning nature's tricks. Nature Biotechnol 18:145–146

Crowe JH, Hoekstra FA, Crowe LM (1992) Anhydrobiosis. Annu Rev Physiol 54:570–599

Cuartero J, Fernandez-Munoz R (1999) Tomato and salinity. Sci Hort 78:83–125

Dantas BF, De Sa RL, Aragao CA (2007) Germination, initial growth and cotyledon protein content of bean cultivars under salinity stress. Rev Bras de Sementes 29:106–110

de Carvalho MCH (2008) Drought stress and reactive oxygen species: production, scavenging and signaling. Plant Signal Behav 3:156–165

Deivanai S, Xavier R, Vinod V, Timalata K, Lim OF (2011) Role of exogenous proline in ameliorating salt stress at early stage in two rice cultivars. J Stress Physiol Biochem 7:157–174

Delauney AJ, Hu CA, Kishor PB, Verma DP (1993) Cloning of ornithine delta-aminotransferase cDNA from *Vigna aconitifolia* by trans-complementation in *Escherichia coli* and regulation of proline biosynthesis. J Biol Chem 268:18673–18678

Deshnium P, Los DA, Hayashi H, Mustardy L, Murata N (1995) Transformation of *Synechococcus* with a gene for choline oxidase enhances tolerance to salt stress. Plant Mol Biol 29:897–907

Dias MC, Bruggemann W (2010) Limitations of photosynthesis in *Phaseolus vulgaris* under drought stress: gas exchange, chlorophyll fluorescence and Calvin cycle enzymes. Photosynth 48:96–102

Díaz-López L, Gimenob V, Simónc L, Martínez V, Rodríguez-Ortegab WM, García-Sánchez F (2012) *Jatropha curcas* seedlings show a water conservation strategy under drought conditions based on decreasing leaf growth and stomatal conductance. Agric Water Manage 105:48–56

Din J, Khan SU, Ali I, Gurmani AR (2011) Physiological and agronomic response of canola varieties to drought stress. J Anim Plant Sci 21:78–82

Dionisio-Sese ML, Tobita S (1998) Antioxidant responses of rice seedlings to salinity stress. Plant Sci 135:1–9

Dolatabadian A, Modarressanavy SAM, Ghanati F (2011) Effect of salinity on growth, xylem structure and anatomical characteristics of soybean. Not Sci Biol 3:41–45

Donnamaria MC, Howard EI, Grigera JR (1994) Interaction of water with α, α-trehalose in solution: molecular dynamics simulation approach. J Chem Soc Faraday Trans 90:2731–2735

dos Reis SP, Lima AM, de Souza CRB (2012) Recent molecular advances on downstream plant responses to abiotic stress. Int J Mol Sci 13:8628–8647

Drennan PM, Smith MT, Goldsworth D, van Staden J (1993) The occurrence of trehalose in the leaves of the desiccation-tolerant angiosperm *Myrothamnus flabellifolius* Welw. J Plant Physiol 142:493–496

Elbein AD, Pan YT, Pastuszak I, Carroll D (2003) New insights on trehalose: a multifunctional molecule. Glycobiology 13:17–27

Ellis RP, Forster BPP, Gordon DC, Handley LL, Keith RP, Lawrence P, Meyer R, Powell W, Robinson D, Scrimgeour CM, Young G, Thomas WTB (2002) Phenotype/genotype associations of yield and salt tolerance in a barley mapping population segregating for two dwarfing genes. J Exp Bot 53:1163–1176

El-Shabrawi H, Kumar B, Kaul T, Reddy MK, Singla-Pareek SL, Sopory SK (2010) Redox homeostasis, antioxidant defense, and methylglyoxal detoxification as markers for salt tolerance in Pokkali rice. Protoplasma 245:85–96

Faize M, Burgos L, Faize L, Piqueras A, Nicolas E, Barba-Espin G et al (2011) Involvement of cytosolic ascorbate peroxidase and Cu/Zn-superoxide dismutase for improved tolerance against drought stress. J Exp Bot 62:2599–2613

Fernandez O, Béthencourt L, Quero A, Sangwan RS, Clément C (2010) Trehalose and plant stress responses: friend or foe? Trends Plant Sci 15:409–417

Flexas J, Ribas-Carbo M, Bota J, Galmes J, Henkle M, Martinez-Canellas S, Medrano H (2006) Decreased rubisco activity during water stress is not induced by decreased relative water content but related to conditions of low stomatal conductance and chloroplast CO_2 concentration. New Phytol 172:73–82

Fougere F, Le Rudulier D, Streeter JG (1991) Effects of salt stress on amino acid, organic acid, and carbohydrate composition of roots, bacteroids, and cytosol of alfalfa (*Medicago sativa* L.). Plant Physiol 96:1228–1236

Foyer CH, Noctor G (2003) Redox sensing and signalling associated with reactive oxygen in chloroplasts, peroxisomes and mitochondria. Physiol Plant 119:355–364

Gadallah MAA (1999) Effect of proline and glycinebetaine on *Vicia faba* responses to salt stress. Biol Plant 42:249–257

Garcia AB, Engler JA, Iyer S, Gerats T, Van Montague M, Caplan AB (1997) Effects of osmoprotectants upon NaCl stress in rice. Plant Physiol 115:159–169

Garg AK, Kim LK, Owens TG, Ranwalam AP, Choi YD, Kochian LV, Wu RJ (2002) Trehalose accumulation in rice plants confers high tolerance levels to different abiotic stresses. Proc Natl Acad Sci U S A 99: 15898–15903

Genard H, Le Saos J, Hillard J, Tremolieres A, Boucaud J (1991) Effect of salinity on lipid composition, glycine betaine content and photosynthetic activity in chloroplasts of *Suaeda maritime*. Plant Physiol Biochem 29:421–427

Ghanem ME, Albacete A, Smigocki AC, Frebort I, Pospisilova H et al (2011) Root-synthesised cytokinins improve shoot growth and fruit yield in salinized tomato (*Solanum lycopersicum* L.) plants. J Exp Bot 62:125–140

Ghanifathi T, Valizadeh M, Shahryari R, Shahbazi H (2011) Effect of drought stress on germination indices and seedling growth of 12 bread wheat genotypes. Adv Environ Biol 5:1034–1039

Gill SS, Tuteja N (2010) Reactive oxygen species and antioxidant machinery in abiotic stress tolerance in crop plants. Plant Physiol Biochem 48:909–930

Goddijn OJM, van Dun K (1999) Trehalose metabolism in plants. Trends Plant Sci 4:315–319

Gomes-Filho E, Machado Lima CRF, Costa JH, da Silva AC, da Guia Silva Lima M, de Lacerda CF, Prisco JT (2008) Cowpea ribonuclease: properties and effect of NaCl-salinity on its activation during seed germination and seedling establishment. Plant Cell Rep 27:147–157

Gueta-Dahan Y, Yaniv Z, Zilinskas BA, Ben-Hayyim G (1997) Salt and oxidative stress: similar and specific responses and their relation to salt tolerance in citrus. Planta 204:460–469

Hamilton EW, Heckathorn SA (2001) Mitochondrial adaptations to NaCl. Complex I is protected by antioxidants and small heat-shock proteins, whereas complex II is protected by proline and betaine. Plant Physiol 126:1266–1274

Hare PD, Cress WA (1997) Metabolic implications of stress induced proline accumulation in plants. Plant Growth Regul 21:79–102

Hare PD, Cress WA, Staden JV (1998) Dissecting the roles of osmolyte accumulation during stress. Plant Cell Environ 21:535–553

Hare PD, Cress WA, Staden JV (1999) Proline synthesis and degradation: a model for elucidating stress-related signal transduction. J Exp Bot 50:413–434

Haro R, Dardanelli J, Otegui M, Collino D (2008) Grain yield determination of peanut crops under water deficit: soil strength effects on pod set, the source sink ratio and radiation use efficiency. Field Crop Res 109:24–33

Hasanuzzaman M, Fujita M (2011) Selenium pretreatment upregulates the antioxidant defense and methylglyoxal detoxification system and confers enhanced tolerance to drought stress in rapeseed seedlings. Biol Trace Elem Res 143:1758–1776

Hasanuzzaman M, Fujita M, Islam MN, Ahamed KU, Nahar K (2009) Performance of four irrigated rice varieties under different levels of salinity stress. Int J Integ Biol 6:85–90

Hasanuzzaman M, Hossain MA, da Silva JAT, Fujita M (2012) Plant responses and tolerance to abiotic oxidative stress: antioxidant defense is a key factor. In: Bandi V, Shanker AK, Shanker C, Mandapaka M (eds) Crop Stress and its management: perspectives and strategies. Springer, Germany, pp 261–316

Hasanuzzaman M, Nahar K, Fujita M (2013a) Plant response to salt stress and role of exogenous protectants to mitigate salt-induced damages. In: Ahmad P, Azooz MM, Prasad MNV (eds) Ecophysiology and responses of plants under salt stress. Springer, New York, pp 25–87

Hasanuzzaman M, Nahar K, Fujita M, Ahmad P, Chandna R, Prasad MNV, Öztürk M (2013b) Enhancing plant productivity under salt stress – relevance of polyomics. In: Ahmad P, Azooz MM, Prasad MNV (eds) Salt stress in plants: omics, signaling and responses. Springer, Berlin, pp 113–156

Hasanuzzaman M, Alam MM, Rahman A, Hasanuzzaman M, Nahar K, Fujita M (2014a) Exogenous proline and glycine betaine mediated upregulation of antioxidant defense and glyoxalase systems provides better protection against salt-induced oxidative stress in two rice (*Oryza sativa* L.) varieties. Biomed Res Int 2014:757219

Hasanuzzaman M, Nahar K, Gill SS, Fujita M (2014b) Drought stress responses in plants, oxidative stress and antioxidant defense. In: Gill SS, Tuteja N (eds) Climate change and plant abiotic stress tolerance. Wiley, Weinheim, pp 209–249

Hayashi H, Alia Mustardy L, Deshnium P, Ida M, Murata N (1997) Transformation of *Arabidopsis thaliana* with the *codA* gene for choline oxidase: accumulation of glycinebetaine and enhanced tolerance to salt and cold stress. Plant J 12:133–142

Hayat S, Hayat Q, Alyemeni MN, Wani AS, Pichtel J, Ahmad A (2012) Role of proline under changing environments. Plant Signal Behav 7:1456–1466

He C, Yang A, Zhang W, Gao Q, Zhang J (2010) Improved salt tolerance of transgenic wheat by introducing betA gene for glycine betaine synthesis. Plant Cell Tiss Organ Cult 101:65–78

Heuer B (2003) Influence of exogenous application of proline and glycinebetaine on growth of salt stressed tomato plants. Plant Sci 165:693–699

Hincha DK, Hagemann M (2004) Stabilization of model membranes during drying by compatible solutes involved in the stress tolerance of plants and microorganisms. The Biochem J 383:277–283

Holmström KO, Somersalo S, Mandal A, Palva ET, Welin B (2000) Improved tolerance to salinity and low temperature in transgenic tobacco producing glycine betaine. J Exp Bot 51:177–185

Hoque MA, Banu MNA, Okuma E, Amako K, Nakamura K, Shimoishi Y, Murata Y (2007) Exogenous proline and glycinebetaine increase NaCl-induced ascorbate-glutathione cycle enzyme activities, and proline improves salt tolerance more than glycinebetaine in tobacco Bright Yellow-2 suspension-cultured cells. J Plant Physiol 164:1457–1468

Hoque MA, Okuma E, Nakamara Y, Shimoishi Y, Murata Y (2008) Proline and glycinebetaine enhance antioxidant defense and methylglyoxal detoxification systems and reduce NaCl-induced damage in cultured tobacco cells. J Plant Physiol 165:813–824

Hu L, Lu H, Liu Q, Chen X, Jiang X (2005) Overexpression of mtlD gene in transgenic *Populus tomentosa* improves salt tolerance through accumulation of mannitol. Tree Physiol 25:1273–1281

Hussain M, Malik MA, Farooq M, Ashraf MY, Cheema MA (2008) Improving drought tolerance by exogenous application of glycinebetaine and salicylic acid in sunflower. J Agron Crop Sci 194:193–199

Iqbal N, Ashraf M, Ashraf MY (2008) Glycinebetaine, an osmolyte of interest to improve water stress tolerance in sunflower (*Helianthus annuus* L.): water relations and yield. South Afr J Bot 74:274–281

Iyengar ERR, Reddy MP (1996) Photosynthesis in highly salt-tolerant plants. In: Pessaraki M (ed) Handbook of photosynthesis. Marcel Dekker, New York, pp 897–909

Juhász Z, Balmer D, Sós-Hegedűs A, Vallat A, Mauch-Mani B, Bánfalvi Z (2014) Effects of Drought Stress and Storage on the Metabolite and Hormone Contents of Potato Tubers Expressing the Yeast Trehalose-6-phosphate Synthase 1. Gene J Agril Sci 6:142–166

Kaiser WM (1987) Effects of water deficit on photosynthetic capacity. Physiol Plant 71:142–149

Kato Y, Kamoshita A, Yamagishi J (2008) Preflowering abortion reduces spikelet number in upland rice (*Oryza sativa* L.) under water stress. Crop Sci 48:2389–2395

Kawai H, Sakurai M, Inoue Y, Chujo R, Kobayashi S (1992) Hydration of oligosaccharides: anomalous hydration ability of trehalose. Cryobiology 29: 599–606

Kaya DM, Okcu G, Atak M, Cikili Y, Kolsarici O (2006) Seed treatments to overcome salt and drought stress during germination in sunflower (*Helianthus annuus* L.). Eur J Agron 24:291–295

Kempf B, Bremer E (1998) Uptake and synthesis of compatible solutes as microbial stress responses to high-osmolarity environments. Arch Microbiol 170: 319–330

Khan MA, Rizvi Y (1994) Effect of salinity, temperature and growth regulators on the germination and early seedling growth of *Atriplex griffithii* var. Stocksii. Can J Bot 72:475–479

Khan MA, Weber DJ (2008) Ecophysiology of high salinity tolerant plants (tasks for vegetation science), 1st edn. Springer, Amsterdam

Khedr AHA, Abbas MA, Wahid AAA, Quick WP, Abogadallah GM (2003) Proline induces the expression of salt-stress-responsive proteins and may improve the adaptation of *Pancratium maritimum* L. to salt-stress. J Exp Bot 54:2553–2562

Khodarahmpour Z, Ifar M, Motamedi M (2012) Effects of NaCl salinity on maize (*Zea mays* L.) at germination and early seedling stage. Afr J Biotechnol 11:298–304

Kirkham MB (2005) Principles of soil and plant water relations. Elsevier, Amsterdam

Kishor PBK, Hong Z, Miao GH, Hu CAA, Verma DPS (1995) Overexpression of [delta]-pyrroline-5-carboxylate synthetase increases proline production and confers osmotolerance in transgenic plants. Plant Physiol 108:1387–1394

Kondrák M, Marincs F, Antal F, Juhász Z, Bánfalvi Z (2012) Effects of yeast trehalose-6-phosphate synthase 1 on gene expression and carbohydrate contents of potato leaves under drought stress conditions. BMC Plant Biol 12:1–12

Koyro HW (2002) Ultrastructural effects of salinity in higher plants. In: Lauchli A, Luttge U (eds) Salinity: environment – plants – molecules. Kluwer, Amsterdam, pp 139–157

Kumar A, Verulkar S, Dixit S, Chauhan B, Bernier J, Venuprasad R, Zhao D, Shrivastava MN (2009) Yield and yield-attributing traits of rice (*Oryza sativa* L.) under lowland drought and suitability of early vigor as a selection criterion. Field Crops Res 114:99–107

Kusvuran S (2012) Effects of drought and salt stresses on growth, stomatal conductance, leaf water and osmotic potentials of melon genotypes (*Cucumis melo* L.). Afr J Agric Res 7:775–781

Landfald B, Strom AR (1986) Choline-glycine betaine pathway confers a high level of osmotic tolerance in *Escherichia coli*. J Bacteriol 165:849–855

Lauchli A, Grattan SR (2007) Plant growth and development under salinity stress. In: Jenks MA, Hasegawa PM, Mohan JS (eds) Advances in molecular breeding towards drought and salt tolerant crops. Springer, Berlin, pp 1–32

Li CH, Li Y, Wuyun TN, Wu GL, Jiang GM (2010) Effects of high concentration ozone on soybean growth and grain yield. Ying Yong Sheng Tai Xue Bao 21:2347–2352

Lima-Costa ME, Ferreira S, Duarte A, Ferreira AL (2008) Alleviation of salt stress using exogenous proline on a citrus cell line. Acta Hort 868:109–112

Lopez FB, Johansen C, Chauhan YS (1996) Effect of timing of drought stress on phenology, yield and yield components of a short-duration pigeon pea. J Agron Crop Sci 177:311–320

Lopez M, Tejera NA, Iribarne C, Lluch C, Herrera-Cervera JA (2008) Trehalose and trehalase in root nodules of *Medicago truncatula* and *Phaseolus vulgaris* in response to salt stress. Physiol Planta 134:575–582

López-Gómez M, Lluch C (2012) Trehalose and abiotic stress tolerance. In: Ahmad P, Prasad MNV (eds) Abiotic stress responses in plants: metabolism, productivity and sustainability. Springer, New York, pp 253–265

Lutts S (2000) Exogenous glycinebetaine reduces sodium accumulation in salt-stressed rice plants. IRRN 25:9–40

Lv S, Yang A, Zhang K, Wang L, Zhang J (2007) Increase of glycinebetaine synthesis improves drought tolerance in cotton. Mol Breed 20:233–248

Lv WT, Lin B, Zhang M, Hua XJ (2011) Proline accumulation is inhibitory to Arabidopsis seedlings during heat stress. Plant Physiol 15:1921–1933

Ma L, Zhou E, Gao L, Mao X, Zhou R, Jia J (2008) Isolation, expression analysis and chromosomal location of P5CR gene in common wheat (*Triticum aestivum* L.). S Afric J Bot 74:705–712

Ma C, Wang Z, Kong B, Lin T (2013) Exogenous trehalose differentially modulate antioxidant defense system in wheat callus during water deficit and subsequent recovery. Plant Growth Regul 70:275–285

Mafakheri A, Siosemardeh A, Bahramnejad B, Struik PC, Sohrabi Y (2010) Effect of drought stress on yield, proline and chlorophyll contents in three chickpea cultivars. Aust J Crop Sci 4:580–585

Mahajan S, Tuteja N (2005) Cold, salinity and drought stresses: an overview. Arch Biochem Biophys 444:139–158

Mäkelä P, Kärkkäinen J, Somersalo S (2000) Effect of glycine betaine on chloroplast ultrastructure, chlorophyll and protein content, and RuBPCO activity in tomato grown under drought or salinity. Biol Plant 43:471–475

Matysik J, Alia BB, Mohanty P (2002) Molecular mechanisms of quenching of reactive oxygen species by proline under stress in plants. Curr Sci 82:525–532

McNeil SD, Nuccio ML, Hanson AD (1999) Betaines and related osmoprotectants. Targets for metabolic engineering of stress resistance. Plant Physiol 120:945–949

Miri HR, Armin M (2013) The interaction effect of drought and exogenous application of glycine betaine on corn (*Zea mays* L.). European J Exp Biol 3:197–206

Mittova V, Guy M, Tal M, Volokita M (2004) Salinity upregulates the antioxidative system in root mitochondria and peroxisomes of the wild salt-tolerant tomato species *Lycopersicon pennellii*. J Exp Bot 55:1105–1113

Mohammadkhan N, Heidari R (2008) Drought-induced accumulation of soluble sugars and proline in two maize varieties. World Applied Sci J 3:448–453

Molinari HBC, Marur CJ, Daros E, de Campos MKF, de Carvalho JFRP et al (2007) Evaluation of the stress-inducible production of proline in transgenic sugarcane (*Saccharum* spp.): osmotic adjustment, chlorophyll fluorescence and oxidative stress. Physiol Plant 130:218–229

Monneveux P, Rekika D, Acevedo E, Merah O (2006) Effect of drought on leaf gas exchange, carbon isotope discrimination, transpiration efficiency and productivity in field grown durum wheat genotypes. Plant Sci 170:867–872

Munns R (2002) Comparative physiology of salt and water stress. Plant Cell Environ 25:239–250

Munns R, Tester M (2008) Mechanisms of salinity tolerance. Annu Rev Plant Biol 59:651–681

Nahar K, Hasanuzzaman M (2009) Germination, growth, nodulation and yield performance of three mungbean varieties under different levels of salinity stress. Green Farm 2:825–829

Nounjan N, Nghia PT, Theerakulpisut P (2012) Exogenous proline and trehalose promote recovery of rice seedlings from salt-stress and differentially modulate antioxidant enzymes and expression of related genes. J Plant Physiol 169:596–604

Ohnishi N, Murata N (2006) Glycinebetaine counteracts the inhibitory effects of salt stress on the degradation and synthesis of the D1 protein during photoinhibition in *Synechococcus* sp. PCC 7942. Plant Physiol 141:756–765

Othman Y, Al-Karaki G, Al-Tawaha AR, Al-Horani A (2006) Variation in germination and ion uptake in barley genotypes under salinity conditions. World J Agric Sci 2:11–15

Oufir M, Schulz N, Vallikhan PSS, Wilhelm E, Burg K, Hausman JF et al (2009) Simultaneous measurement of proline and related compounds in oak leaves by high-performance ligand-exchange chromatography and electrospray ionization mass spectrometry for environmental stress studies. J Chromatogr A 1216:1094–1099

Papageorgiou GC, Murata N (1995) The unusually strong stabilizing effects of glycinebetaine on the structure and function of the oxygen-evolving photosystem II complex. Photosynth Res 44:243–252

Parida AK, Das AB (2005) Salt tolerance and salinity effect on plants: a review. Ecotoxicol Environ Saf 60:324–349

Parida AK, Das AB, Mohanty P (2004) Investigations on the antioxidative defense responses to NaCl stress in a mangrove, *Bruguiera parviflora*: differential regulations of isoforms of some antioxidative enzymes. Plant Growth Regul 42:213–226

Pastori GM, Foyer CH (2002) Common components, networks, and pathways of cross-tolerance to stress. The central role of "redox" and abscisic acid-mediated controls. Plant Physiol 129:460–468

Paul M, Pellny T, Goddijn O (2001) Enhancing photosynthesis with sugar signals. Trends Plant Sci 6:197–200

Paul MJ, Primavesi LF, Jhurreea D, Zhang Y (2008) Trehalose metabolism and signaling. Annu Rev Plant Biol 59:417–441

Peñuelas J, Sardans J, Estiarte M, Ogaya R, Carnicer J, Coll M et al (2013) Evidence of current impact of climate change on life: a walk from genes to the biosphere. Global Change Biol 19:2303–2338

Pervez MA, Ayub CM, Khan HA, Shahid MA, Ashraf I (2009) Effect of drought stress on growth, yield and seed quality of tomato (*Lycopersicon esculentum* L.). Pak J Agric Sci 46:174–178

Pilbeam CJ, Akatse JK, Hebblethwaite PD, Wright CD (1992) Yield production in two contrasting forms of spring-sown faba beans in relation to water supply. Field Crops Res 29:273–287

Pilon-Smits EAH, Ebstamp MJM, Paul MJ, Jeuken MJW (1995) Improved performance of transgenic fructan-accumulating tobacco under drought stress. Plant Physiol 107:125–130

Pitman MG, Läuchli A (2002) Global impact of salinity and agricultural ecosystems. In: Läuchli A, Lüttge U (eds) Salinity: environment – plants – molecules. Kluwer Academic, Dordrecht, pp 3–20

Plaut Z, Butow BJ, Blumenthal CS, Wrigley CW (2004) Transport of dry matter into developing wheat kernels and its contribution to grain yield under post-anthesis water deficit and elevated temperature. Field Crops Res 86:185–198

Posmyk MM, Kontek R, Janas KM (2009) Antioxidant enzymes activity and phenolic compounds content in red cabbage seedlings exposed to copper stress. Ecotoxicol Environ Safet 72:596–602

Pramanik MHR, Imai R (2005) Functional identification of a trehalose 6-phosphate phosphatase gene that is involved in transient induction of trehalose biosynthesis during chilling stress in rice. Plant Mol Biol 58:751–762

Prasad PVV, Staggenborg SA (2008) Impacts of drought and/or heat stress on physiological, developmental, growth, and yield processes of crop plants. In: Response of crops to limited water: understanding and modeling water stress effects on plant growth processes, vol 1, Advances in agricultural systems modeling series. Soil Science Society of America, Madison, pp 301–355

Pratap V, Sharma VK (2010) Impact of osmotic stress on seed germination and seedling growth in black gram (*Phaseolus mungo*). J Environ Biol 31:721–726

Promila K, Kumar S (2000) *Vigna radiata* seed germination under salinity. Biol Plant 43:423–426

Quan R, Shang M, Zhang H, Zhao Y, Zhang J (2004) Engineering of enhanced glycine betaine synthesis improves drought tolerance in maize. Plant Biotechnol J 2:477–486

Quarrie SA, Steed A, Calestani C, Semikhodskii A, Lebreton C, Chinoy C, Steele N et al (2005) A high-density genetic map of hexaploid wheat (*Triticum aestivum* L.) from the cross Chinese Spring x SQ1 and its use to compare QTLs for grain yield across a range of environments. Theor Appl Genet 110:865–880

Rahman MS, Miyake H, Takeoka Y (2002) Effect of exogenous glycinebetaine on growth and ultrastructure of salt-stressed rice seedlings (*Oryza sativa* L.). Plant Prod Sci 5:33–44

Rasheed R (2009) Salinity and extreme temperature effects on sprouting buds of sugarcane (*Saccharum officinarum* L.): some histological and biochemical studies. Ph.D. thesis, Department of Botany, Univ of Agriculture, Faisalabad

Rathinasabapathi B, Burnet M, Russel BL, Gage DA, Liao PC, Nye GJ et al (1997) Choline monooxygenase, an unusual iron-sulfur enzyme catalyzing the first step of glycine betaine synthesis in plants: prosthetic group characterization and cDNA cloning. Proc Natl Acad Sci U S A 94:3454–3458

Rathinasabapathi B, Fouad WM, Sigua CA (2001) Beta-alanine betaine synthesis in the Plumbaginaceae. Purification and characterization of a trifunctional, *S*-adenosyl-L-methionine-dependent *N*-ethyltransferase from *Limonium latifolium* leaves. Plant Physiol 126:1241–1249

Raza MAS, Saleem MF, Ashraf MY, Ali A, Asghar HN (2012) Glycinebetaine applied under drought improved the physiological efficiency of wheat (*Triticum aestivum* L.) plant. Soil Environ 31:67–71

Raza MAS, Saleem MF, Shah GM, Khan IH, Raza A (2014) Exogenous application of glycinebetaine and potassium for improving water relations and grain yield of wheat under drought. J Soil Sci Plant Nutri 14:348–364

Reddy AR, Chaitanya KV, Vivekanandan M (2004a) Drought-induced responses of photosynthesis and antioxidant metabolism in higher plants. J Plant Physiol 161:1189–1202

Reddy AR, Chaitanya KV, Jutur PP, Sumithra K (2004b) Differential antioxidative responses to water stress among five mulberry (*Morus alba* L.) cultivars. Environ Exp Bot 52:33–42

Rezaei MA, Jokar I, Ghorbanli M, Kaviani B, Kharabian-Masouleh A (2012) Morpho-physiological improving effects of exogenous glycine betaine on tomato (*Lycopersicum esculentum* Mill.) cv. PS under drought stress conditions. Plant Omics 5:79–86

Rhodes D, Hanson AD (1993) Quaternary ammonium and tertiary sulfonium compounds in higher plants. Annu Rev Plant Physiol Plant Mol Biol 44:357–384

Rhodes D, Samaras Y (1994) Genetic control of osmoregulation in plants. In: Strange SK (ed) Cellular and molecular physiology of cell volume regulation. CRC Press, Boca Raton, pp 347–361

Richards AB, Krakiwka S, Dexter LB, Schid H, Wolterbeek APM, Waalkens-Berendsen DH et al (2002) Trehalose: a review of properties, history of use and human tolerance, and results of multiple studies. Food Chem Toxicol 40:871–898

Robinson SP, Jones GP (1986) Accumulation of glycine betaine in chloroplasts provides osmotic adjustment during salt stress. Aust J Plant Physiol 13:659–668

Rodríguez M, Canales E, Borrás-Hidalgo O (2005) Molecular aspects of abiotic stress in plants. Biotechnol Appl 22:1–10

Romero C, Belles JM, Vaya JL, Serrano R, Culianez-macia FA (1997) Expression of the yeast trehalose-6-phosphate synthase gene in transgenic tobacco plants: peiotropic phenotypes include drought tolerance. Planta 201:293–297

Romero-Aranda R, Soria T, Cuartero S (2001) Tomato plant-water uptake and plant-water relationships under saline growth conditions. Plant Sci 160:265–272

Rontein D, Basset G, Hanson AD (2002) Metabolic engineering of osmoprotectant accumulation in plants. Metab Eng 4:49–56

Saha P, Chatterjee P, Biswas AK (2010) NaCl pretreatment alleviates salt stress by enhancement of antioxidant defense system and osmolyte accumulation in mungbean (*Vigna radiata* L. Wilczek). Indian J Exp Biol 48:593–600

Saini HS, Westgate ME (2000) Reproductive development in grain crops during drought. Adv Agron 68:59–96

Sakamoto A, Murata N (2001) The use of bacterial choline oxidase, a glycinebetaine-synthesizing enzyme, to create stress-resistant transgenic plants. Plant Physiol 125:180–188

Sakamoto A, Murata Alia N (1998) Metabolic engineering of rice leading to biosynthesis of glycinebetaine and tolerance to salt and cold. Plant Mol Biol 38:1011–1019

Sarafis V (1998) Chloroplasts: a structural approach. J Plant Physiol 152:248–264

Saraswathi SG, Paliwal K (2011) Drought induced changes in growth, leaf gas exchange and biomass production in *Albizia lebbeck* and *Cassia siamea* seedlings. J Environ Biol 32:173–178

Schluepmann H, Pellny T, Van Dijken A, Smeekens S, Paul MJ (2003) Trehalose 6-phosphate is indispensable for carbohydrate utilisation and growth in *Arabidopsis thaliana*. Proc Natl Acad Sci USA 100:6849–6854

Shahryari R, Gurbanov E, Gadimov A, Hassanpanah D (2008) Tolerance of 42 bread wheat genotypes to drought stress after anthesis. Pak J Biol Sci 11:1330–1335

Sharma P, Dubey RS (2005) Drought induces oxidative stress and enhances the activities of antioxidant enzymes in growing rice seedlings. Plant Growth Regul 46:209–221

Sheveleva E, Chmara W, Bohnert HJ, Jensen RG (1997) Increased salt and drought tolerance by D-ononitol production in transgenic *Nicotiana tabacum* L. Plant Physiol 115:1211–1219

Shima S, Matsui H, Tahara S, Imai R (2007) Biochemical characterization of rice trehalose-6-phosphate phosphatases supports distinctive functions of these plant enzymes. FEBS J 274:1192–1201

Shinozaki K, Yamaguchi-Shinozaki K (1997) Gene expression and signal transduction in water-stress response. Plant Physiol 115:327–334

Simon-Sarkadi L, Kocsy G, Várhegyi Á, Galiba G, De Ronde JA (2006) Stress induced changes in the free amino acid composition in transgenic soybean plants having increased proline content. Biol Plant 50:793–796

Smirnoff N, Cumbes QJ (1989) Hydroxyl radical scavenging activity of compatible solutes. Phytochem 28:1057–1060

Sobahan MA, Arias CR, Okuma E, Shimoishi Y, Nakamura Y, Hirai Y, Mori IC, Murata Y (2009) Exogenous proline and glycinebetaine suppress apoplastic flow to reduce Na^+ uptake in rice seedlings. Biosci Biotechnol Biochem 73:2037–2042

Sorkheha K, Shirana B, Rouhia V, Khodambashia M, Sofob A (2011) Regulation of the ascorbate-glutathione cycle in wild almond during drought stress. Russ J Plant Physiol 58:76–84

Subramanian KS, Santhanakrishnan P, Balasubramanian P (2006) Responses of field grown tomato plants to arbuscular mycorrhizal fungal colonization under varying intensities of drought stress. Sci Hortic 107:245–253

Sulpice R, Tsukaya H, Nonaka H, Mustardy L, Chen TH, Murata N (2003) Enhanced formation of flower in salt-stressed *Arabidopsis* after genetic engineering of the synthesis of glycine betaine. Plant J 36:165–178

Szabados L, Savoure A (2010) Proline: a multifunctional amino acid. Trends Plant Sci 15:89–97

Takahashi S, Murata N (2006) Glycerate-3-phosphate, produced by CO_2 fixation in the Calvin cycle, is critical for the synthesis of the D1 protein of photosystem II. Biochim Biophys Acta 1757:198–205

Tarczynski MC, Jensen RG, Bohnert HJ (1993) Stress protection of transgenic tobacco by production of the osmolyte mannitol. Sci 259:508–510

Tavakkoli E, Rengasamy P, McDonald GK (2010) High concentrations of Na^+ and Cl^- ions in soil solution have simultaneous detrimental effects on growth of faba bean under salinity stress. J Exp Bot 61:4449–4459

Theerakulpisut P, Gunnula W (2012) Exogenous sorbitol and trehalose mitigated salt stress damage in salt-sensitive but not salt-tolerant rice seedlings. Asian J Crop Sci 4:165–170

Thomas JC, Sepahi M, Arendall B, Bohnert HJ (1995) Enhancement of seed germination in high salinity by engineering mannitol expression in *Arabidopsis thaliana*. Plant Cell Environ 18:801–806

Timasheff SN (2008) A physiochemical basis for selection of osmolytes by nature. In: Somero GN, Osmond CB, Bolis CL (eds) Water and Life. Springer, Berlin, pp 70–84

Trovato M, Mattioli R, Costantino P (2008) Multiple roles of proline in plant stress tolerance and development. Rendiconti Lincei 19:325–346

Tuba Z, Lichtenthaler HK, Csintalan Z, Nagy Z, Szente K (1996) Loss of chlorophylls, cessation of photosynthetic CO_2 assimilation and respiration in the poikilochlorophyllous plant *Xerophyta scabrida* during desiccation. Physiol Plant 96:383–388

Vendruscolo ECG, Schuster I, Pileggi M, Scapim CA, Molinari HBC et al (2007) Stress-induced synthesis of proline confers tolerance to water deficit in transgenic wheat. J Plant Physiol 164:1367–1376

Verbruggen N, Villarroel R, Van Montagu M (1993) Osmoregulation of a pyrroline-5-carboxylate reductase gene in *Arabidopsis thaliana*. Plant Physiol 103:771–781

Verbruggen N, Hua XJ, May M, Van Montagu M (1996) Environmental and developmental signals modulate proline homeostasis: evidence for a negative transcriptional regulator. Proc Natl Acad Sci U S A 93: 8787–8791

Wang Z, Zhang S, Yang B, Li Y (2005) Trehalose synthase gene transfer mediated by Agrobacterium tumefaciens enhances resistance to osmotic stress in sugarcane. Sugar Tech 7:49–54

Weretilnyk EA, Bednarek S, McCue KF, Rhodes D, Hanson AD (1989) Comparative biochemical and immunological studies of the glycine betaine synthesis pathway in diverse families of dicotyledons. Planta 178:342–352

Werner JE, Finkelstein RR (1995) *Arabidopsis* mutants with reduced response to NaCl and osmotic stress. Physiol Plant 93:659–666

Witt S, Galicia L, Lisec J, Cairns J, Tiessen A, Araus JL et al (2012) Metabolic and phenotypic responses of greenhouse-grown maize hybrids to experimentally controlled drought stress. Mol Plant 5:401–417

Xiong L, Ishitani M, Lee H, Zhu JK (2001) The *Arabidopsis* LOS5/ABA3 locus encodes a molybdenum cofactor sulfurase and modulates cold stress and osmotic stress responsive gene expression. Plant Cell 13:2063–2083

Xue X, Liu A, Hua X (2009) Proline accumulation a transcriptional regulation of proline biosynthesis and degradation in *Brassica napus*. BMB Rep 42:28–34

Yadav SK, Singla-Pareek SL, Ray M, Reddy MK, Sopory SK (2005) Methylglyoxal levels in plants under salinity stress are dependent on glyoxalase I and glutathione. Biochem Biophys Res Commun 337:61–67

Yamada M, Morishita H, Urano K, Shiozaki N, Yamaguchi-Shinozaki K, Shinozaki K, Yoshiba Y (2005) Effects of free proline accumulation in petunias under drought stress. J Exp Bot 56:1975–1981

Yan Z, Guo S, Shu S, Sun J, Tezuka T (2011) Effects of proline on photosynthesis, root reactive oxygen species (ROS) metabolism in two melon cultivars (*Cucumis melo* L.) under NaCl stress. Afr J Biotechnol 10:18381–18390

Yancey PH (1994) Compatible and counteracting solutes. In: Strange SK (ed) Cellular and Molecular Physiology of Cell Volume Regulation. CRC Press, Boca Raton, pp 81–109

Yang X, Liang Z, Lu C (2005) Genetic engineering of the biosynthesis of Glycinebetaine enhances photosynthesis against high temperature stress in transgenic tobacco plants. Plant Physiol 138:2299–2309

Yang JC, Zhang JH, Liu K, Wang ZQ, Liu LJ (2007) Abscisic acid and ethylene interact in rice spikelets in response to water stress during meiosis. J Plant Growth Regul 26:318–328

Yang X, Liang Z, Wen X, Lu C (2008) Genetic engineering of the biosynthesis of glycinebetaine leads to increased tolerance of photosynthesis to salt stress in transgenic tobacco plants. Plant Mol Biol 66:73–86

Yordanov I, Velikova V, Tsonev T (2003) Plant responses to drought and stress tolerance. Bulg J Plant Physiol, Special issue:187–206

Yupsanis T, Moustakas M, Domiandou K (1994) Protein phosphorylation-dephosphorylation in alfalfa seeds germinating under salt stress. J Plant Physiol 143: 234–240

Zeid IM (2009) Trehalose as osmoprotectant for maize under salinity-induced stress. Res J Agric Biol Sci 5:613–622

Zhang J (2011) China's success in increasing per capita food production. J Exp Bot 62:3707–3711

Zhang H, Dong H, Li W, Sun Y, Chen S, Kong X (2009a) Increased glycine betaine synthesis and salinity tolerance in AhCMO transgenic cotton lines. Mol Breeding 23:289–298

Zhang Y, Primavesi LF, Jhurreea D, Andralojc PJ, Mitchell RAC, Powers SJ et al (2009b) Inhibition of SNF1-related protein kinase1 activity and regulation of metabolic pathways by trehalose-6-phosphate. Plant Physiol 149:1860–1871

Zhu JK (2001) Cell signaling under salt, water and cold stresses. Curr Opin Plant Biol 4:401–406

Zhu JK (2002) Salt and drought stress signal transduction in plants. Annu Rev Plant Physiol Plant Mol Biol 53:247–273

Glycine Betaine: Role in Shifting Plants Toward Adaptation Under Extreme Environments

5

Asim Masood, Tasir S. Per, Mohd Asgher, Mehar Fatma, M. Iqbal R. Khan, Faisal Rasheed, Sofi J. Hussain, and Nafees A. Khan

Abstract

World is facing serious threat in terms of global food requirement for the eradication of hunger. The latest FAO estimates clearly indicate that about 805 million people are still chronically undernourished in the year 2012–2014, and the area under cultivation is diminishing because of various anthropogenic activities including contamination of soil by heavy metal, industrial salts, etc. Therefore, it becomes the need of the hour that certain crops will be designed that have the ability to withstand such harsh environments and help in sustainable crop development. The research in this area has already gone a step ahead, and scientist has developed crops through genetic engineering that due to certain induced traits can tolerate extreme environments. Recent advances in production of transgenic plants with higher rate of GB accumulation help them in their survival against abiotic stresses and are the promising tool for sustainable agriculture. Transgenic *Arabidopsis*, cotton, etc. have already been reported to accumulate higher levels of betaine and provide tolerance against abiotic stresses. Further, recent developments indicate that phytohormones play prominent role in the regulation of S-assimilation and also are crucial signaling molecules involved in the control of plant responses under optimal and limited environmental conditions. However, in this chapter, we try to detail out the mechanism on which GB works to protect photosynthesis and growth and its interactive effect with nutrients and phytohormones.

5.1 Introduction

Glycine betaine (GB) is a quaternary amine derivative of glycine (N, N, N-trimethylglycine) and is diversely found in various plant and animal species and in some microorganisms (Rhodes and Hanson 1993; Chen and Murata 2008). The

A. Masood (✉) • T.S. Per • M. Asgher • M. Fatma
M.I.R. Khan • F. Rasheed • S.J. Hussain • N.A. Khan
Department of Botany, Aligarh Muslim University, Aligarh 202002, UP, India
e-mail: asim.bot@gmail.com; naf9@lycos.com

N. Iqbal et al. (eds.), *Osmolytes and Plants Acclimation to Changing Environment: Emerging Omics Technologies*, DOI 10.1007/978-81-322-2616-1_5

term "betaine" was coined after the name *Beta vulgaris,* the plant from which it was first isolated (Scheibler 1869). Glycine betaine also falls in the category of compatible solutes which form a group of small organic metabolites that are easily soluble in water and are nontoxic at higher concentration. Accumulation of osmolytes is an important plant response to ensure their own survival and continued existence of their offsprings aligned with the changing environmental conditions (Serraj and Sinclair 2002). Glycine betaine as an osmolyte has been reported to accumulate in response to various abiotic stresses in several plants such as *Beta vulgaris*, *Spinacia oleracea*, *Triticum aestivum*, *Hordeum vulgare*, and *Sorghum bicolor* (Ashraf and Foolad 2007). The site of its accumulation is mainly chloroplast where it guards the thylakoid membrane and maintains photochemical efficiency (Genard et al. 1991). Normally, plant accumulates GB at very low concentration, but on exposure to abiotic stress conditions, they accumulate large amounts of GB (Storey et al. 1977). There are, however, many species such as *Oryza sativa*, *Brassica* spp., *Arabidopsis thaliana*, and *Nicotiana tabacum*, which do not accumulate GB both under normal and stress conditions (Rhodes and Hanson 1993; Shirasawa et al. 2006). Now various studies have consolidated the crystalline role of GB in abiotic stress tolerance. Higher accumulation of GB protects photosynthetic machinery under stress conditions (Park et al. 2007). Glycine betaine has an important role in antagonizing the inhibition of protein biosynthesis and, therefore, enhances the photosystem II (PSII) repair, which leads to increased stress tolerance (Chen and Murata 2008). Proteins and enzymes are also protected from denaturation and deactivation by GB, and it also maintains membrane stability and cellular osmotic adjustment and detoxifies reactive oxygen species (ROS) under stress conditions (Ashraf and Foolad 2007; Wang et al. 2007). Its accumulation is generally correlated with the extent of stress tolerance (Rhodes and Hanson 1993). It helps in enhancing the antioxidant enzyme activity and therefore provides tolerance to salt and heavy metal (HM) stress (Hoque et al. 2008; Islam et al. 2009; Hossain and Fujita 2010). In many plant species, GB-induced antioxidant defense responses have been reported, such as *Triticum aestivum* (Raza et al. 2007), *Oryza sativa* (Farooq et al. 2008), *Camellia sinensis* (Kumar and Yadav 2009), and *Vigna radiata* (Hossain and Fujita 2010). Studies of GB have focused on GB-mediated tolerance to various kinds of stresses and at various stages of the life cycle of plants (Sulpice et al. 2003; Park et al. 2004). There are two possible ways of biosynthesis of GB either by the oxidation (or dehydrogenation) of choline or by the N-methylation of glycine (Chen and Murata 2002).

In non-GB-accumulating plant species, abiotic stress tolerance has been achieved either by exogenous application of GB or increase in GB biosynthesis and accumulation by adopting transgenic approaches (Chen and Murata 2008). In many transgenic species, overproduction of GB synthesizing genes exhibited increase in the production of GB and helps in enhancing tolerance capacity under salt saline, cold, drought, or high-temperature stresses. The accumulated GB in these species, however, was much lower than what is naturally found in GB-accumulating plant species under stress conditions (Rhodes and Hanson 1993). Phytohormones synthesized in plants in very minute quantity also regulate metabolic activities involved in plant tolerance to various abiotic stresses (Masood et al. 2012; Khan et al. 2013, 2015; Khan and Khan 2014; Asgher et al. 2015). Interaction of phytohormones with GB biosynthesis and accumulation has been suggested to play an important role in plant abiotic stress tolerance (Khan et al. 2012; Khan et al. 2014). There are, however, very little reports available about the role of phytohormones in GB metabolism, which is remarkable in considering the massive research, targeted to phytohormone and GB functions in plants. Several lines of GB-accumulating transgenic plants displaying their properties suggest promising strategies for the development of stress-tolerant crop plants. Mainly in plants, the work on GB is concentrated on its promising role in relation to osmotic and drought stress (McCue and Hanson 1990). In addition to its role in osmotic and drought stress,

GB becomes visible in low-temperature stress tolerance as well. It enhances growth of bacteria at chilling temperatures (Ko et al. 1994). This chapter deals with the biosynthesis of GB under varying conditions, its role in maintenance of cell osmotic balance under stress conditions, and photosynthetic machinery protection. It also involves some recent approaches to enhance GB accumulation by transgenic approaches or phytohormone application to overcome oxidative stress-induced damage in plants.

5.2 GB Biosynthesis and Accumulation

Biosynthesis of GB occurs either from choline or glycine (Chen and Murata 2002). Majority of organisms including most of the animals, plants, and microorganisms synthesize GB by conversion of choline to GB, involving two-step oxidation reactions. In first step, choline is oxidized to betaine aldehyde, and in second, the oxidation reaction gives rise to GB. The first oxidation reaction in animals and bacteria is catalyzed by membrane-bound choline dehydrogenase (CDH), and in plants, a novel Rieske-type iron-sulfur enzyme choline monooxygenase (CMO) catalyzes this step (Rathinasabapathi et al. 1997). The second reaction is catalyzed by NAD^+-dependent betaine aldehyde dehydrogenase (BADH) (Chen and Murata 2011). In contrast, GB synthesis in the microorganisms *Arthrobacter globiformis* and a closely related strain *Arthrobacter pascens* requires only one enzyme choline oxidase, which is encoded by the *codA* gene (Chen and Murata 2002). Second biosynthetic pathway of GB involves a three-step successive methylation of glycine. This biosynthetic pathway has only been detected in extreme halophilic bacteria *Actinopolyspora halophila* and *Ectothiorhodospira halochloris*. The reactions are catalyzed by two enzymes, namely, glycine sarcosine methyltransferase (GSMT) and sarcosine dimethylglycine methyltransferase (SDMT), with *S*-adenosylmethionine as the methyl group donor (Chen and Murata 2002; Takabe et al. 2006). Firstly, methylation of glycine forms sarcosine; secondly, it yields dimethylglycine; and finally, methylation step gives GB. The first and third reactions are catalyzed by GSMT and SDMT, respectively, and the second reaction is catalyzed by both GSMT and SDMT. The two enzymes have partially overlapping substrate specificity: both glycine and sarcosine serve as substrates for GSMT, while both sarcosine and *N, N*-dimethylglycine serve as the substrates for SDMT (Chen and Murata 2011).

Physiological studies have demonstrated that the level of accumulation of GB is correlated with the degree of tolerance (Saneoka et al. 1995). Since there are many plants which do not have any GB biosynthetic mechanism, genetic engineering of GB biosynthetic pathway has been found to improve plant tolerance against abiotic stress by inducing a gene for GB accumulation in these plants. A gene for CMO from spinach was the first such gene to be isolated from a higher plant (Rathinasabapathi et al. 1997), and expressed in tobacco, *Arabidopsis*, and rice, neither of which normally accumulates GB (Nuccio et al. 1998; Hibino et al. 2002; Shirasawa et al. 2006). Enhanced tolerance to salt and/or cold stress has also been achieved in transgenic rice plant expressing the gene for choline oxidase (COD) for GB accumulation (Sakamoto et al. 1998; Mohanty et al. 2002). Improved salt tolerance was also observed in COX-transgenic rice plants, which accumulated even lower amounts of GB than the CDH- and COD-transgenic rice plants (Su et al. 2006). Similar results were reported in chloroplast-targeted CMO-transgenic rice plants (Shirasawa et al. 2006). COD-transgenic rice plants exhibited increased tolerance to drought, cold, and high temperature (Sawahel 2003; Alia et al. 1998). Similar results were reported for COD-transgenic rice and BADH-transgenic tobacco plants (Sakamoto et al. 1998; Holmström et al. 2000). Salt-stress tolerance was achieved in chloroplast-targeted BADH-transgenic carrot, which grew in the presence of up to 400 mM NaCl, a concentration at which only halophytes can survive, whereas the control plants showed severe growth retardation even at 200 mM NaCl (Kumar et al. 2004). Waditee et al. (2005) demonstrated the co-expression of GSMT and DMT in *Synechococcus* and *Arabidopsis*. The transgenic *Synechococcus* and *Arabidopsis* accumulated higher levels of betaine than previously reported for choline-oxidizing enzymes. The transgenic *Synechococcus* thrived under high salinity, up to 0.6 M NaCl. Transgenic maize plants with chloroplast-targeted CDH transgene exhibited drought tolerance and improved grain yield (Quan et al. 2004). Tolerance to drought stress by accumulation of GB was more in transgenic cotton plants than their non-transformed counterparts (Lv et al. 2007). Park et al. (2004) showed chilling stress tolerance in COD-transgenic tomato plants. The degree of tolerance in plants is also correlated to its accumulation in chloroplast and cytosol. In cytosolic-targeted COD-transgenic tomato plants, GB accumulation was five- to sixfold more than chloroplast-targeted COD-tomato plants. Despite low GB accumulation, enhanced tolerance to chilling, salt, and oxidative stresses was observed in the chloroplast-targeted COD-transgenic tomato plants (Khan et al. 2009). In the chloroplast-targeted COD-transgenic tomato leaves, most of the GB accumulated in the chloroplast (Khan et al. 2009). In chloroplast-targeted COD-transgenic *Arabidopsis*, the extent of damage to the PSII complex was much lower than that observed in control plants under elevated salt concentrations (Hayashi et al. 1997).

5.3 Transport and Translocation of GB

Modest information is available about transport of GB in plants. There is possibility of occurrence of GB transporters in cell membrane, but to date, no GB transporter has been reported. The information regarding intracellular transport of GB is also scanty. Under salt stress, the majority of GB accumulates in chloroplast in spinach (Robinson and Jones 1986). Expression of *LeProT1* gene from tomato, which is homologous to *Arabidopsis* proline transporters, in yeast mutant showed that *LeProT1* transports GB with high affinity and proline and γ-amino butyric acid (GABA) with low affinity (Schwacke et al. 1999). Glycine betaine also inhibited the ProT2-mediated transport of GABA, in *Arabidopsis thaliana*, indicating that GB had strong affinity for the transporter (Breitkreuz et al. 1999). From these results, it can be suggested that transporters of both proline and GABA might function in the transport of GB. However, all the transporters of proline and GABA are not inhibited by GB. For instance, the expression of proline transporter (HvProT), cloned from salt-stressed barley roots, was not inhibited by GB, suggesting that HvProT does not act as a transporter of GB (Ueda et al. 2001). Three betaine transporter genes were isolated from betaine-accumulating *Avicennia marina* (Waditee et al. 2004), and a homologous gene BvBet/ProT1 was isolated from sugar beet (Yang et al. 2009). The fusion protein GFP-BvBet/ProT1 showed that BvBet/ProT1 was localized at the plasma membrane. Levels of mRNA for BvBet/ ProT1 were much higher in expanded leaves than in young leaves under normal and salt-stress conditions, since betaine is transported from the expanded leaves to young leaves and the accumulated levels of BvBet/ProT1 gene transcript were higher in young leaves than in expanded leaves (Yamada et al. 2009). The formation of mutant of methanogenic archaeon *Methanosarcina mazei* Gö1 with either of the two gene clusters encoding for GB transporter *ota* or *otb* being deleted revealed completely defective GB transport in *M. mazei*. On the other hand, mutation in *otb* led to increased transcription of *ota* and thus increased transport and accumulation of glycine betaine suggesting a cross talk between the two transporters (Saum et al. 2009). A novel gene, *HvProT2*, encoding *Hordeum vulgare* GB/proline transporter, was identified in *Hordeum vulgare*. Heterologous expression in *Saccharomyces cerevisiae* mutant demonstrated its affinity highest for GB, intermediate for proline, and lowest for GABA. Transient expression of fusions of *HvProT2* and green fluorescent protein in onion epidermal cells was found to be localized at the plasma membrane. It was found to be constitutively expressed in both leaves and roots, and the expression level was higher in old leaves than young leaves and roots. Moreover, it was found to be expressed in the mestome sheath and lateral root cap cells (Fujiwara et al. 2010).

Translocation of GB occurs in plants via phloem. The fact is supported by the experiment conducted by Ladyman et al. (1980). They supplied [methyl-^{14}C]betaine to well-watered, stressed, and stressed-rewatered barley plants. After three days of incubation period and recovered it from various plant organs in the form of betaine. Heat girdling of the leaf sheath prevented export of [^{14}C]betaine from the blade. After supplying tracer [^{14}C]betaine aldehyde, an immediate precursor of the betaine, the ^{14}C which was translocated to the sheath was in the form of betaine. Thus, it appears that GB is synthesized by mature leaves during stress and behaves as an inert end product, which upon rewatering is translocated to the expanded leaves. Mäkelä et al. (1996) also concluded similar results from his study on tomato. Yamada et al. (2009) exogenously applied GB to old leaves of *Beta vulgaris* plants and found its translocation more into young leaves and roots. A large fraction of GB was found to be translocated to meristem-containing tissues such as flower buds and shoot apices in tomato plants after exogenous application of GB to mature leaves (Park et al. 2006). Different levels of GB accumulation in different plant organs indicated the active and regulated translocation of GB from the site of application and accumulation. Glycine betaine was in highest levels in actively growing tissues like flowers and shoot apices in transgenic plants of *Arabidopsis* (Sulpice et al. 2003) and tomato (Park et al. 2004, 2007) which is an indication that GB is efficiently translocated from source to sink tissues via the phloem.

5.4 Regulation of GB Metabolism and Stress Tolerance: Influence of Phytohormones and Mineral Nutrients

During the life-span of plant, they are threatened by various types of abiotic stresses such as salinity, drought, chilling, and HM, and to cope up with such harsh conditions, plants tend to develop primary and secondary metabolic functions. Glycine betaine is one of the most important osmoprotectant that helps stabilizing macromolecules and certain valuable proteins and maintains cell membrane integrity in plants. High concentration of GB present in cytoplasm stabilizes folded protein structure under stress conditions (Ashraf and Foolad 2007). Phytohormones are the special chemicals synthesized by the plants in very minute quantity and regulate various metabolic activities and also play an important role under abiotic stress tolerance (Masood et al. 2012; Khan et al. 2013, 2015; Khan and Khan 2014; Asgher et al. 2014, 2015). It has been suggested that GB and phytohormones interaction can be beneficial to plants in abiotic stress tolerance (Khan et al. 2012, 2014). However, very little is known about the role of phytohormones in GB metabolism, which is remarkable considering the massive research, targeted to phytohormone and GB functions in plants. The role of exogenous and/or endogenous GB has been reported under stressful environments extensively (Gadallah 1999; Khan et al. 2012; Noreen et al. 2013; Khan et al. 2014). Comparative study has been reported by Noreen et al. (2013) that foliar application of salicylic acid (SA) proved its potential to a far greater extent compared to GB. They showed that application of SA was more effective in improving drought stress than GB and also showed that SA may help in sustaining cotton production under drought-stressed regions. Applications of brassinosteroid (epibrassinolide (EBL), a brassinosteroid) and spermidine (a polyamine) significantly modulate GB production and enhance the activity of antioxidant enzymes in chromium (Cr)-stressed *Raphanus sativus* plants. In the same study, they also suggested that individual treatment (Cr+EBL and Cr+spermidine) was not sufficient for the production of GB as compared to additive application of both hormones (Cr+EBL+spermidine) (Choudhary et al. 2012). Different roles of abscisic acid (ABA) have been reported, but still, it is not clear how ABA coordinately regulates GB metabolism in relation to GB biosynthesis in plants (Zhang et al. 2006; Hancock et al. 2011). The relationships between ABA and betaine biosynthesis in plants under environmental stresses have contradictory reports. Some reports suggested that ABA is involved in the betaine biosynthesis in plants under environmental stresses (Jagendorf and Takabe 2001; Saneoka et al. 2001). In contrast, some reports suggested that ABA may not necessarily be associated with betaine accumulation (McCue and Hanson 1992; Maldonado et al. 1997) under stress conditions. Exogenous ABA to *Pyrus bretschneideri* significantly increased the betaine concentration in the drought-stressed plants (Gao et al. 2004a). Physiological differences in salt-tolerant cultivar and salt-sensitive cultivar of wheat were examined by Khan et al. (2012) and suggested that salt-tolerant cultivar exhibited greater GB content, which was found correlative with phytohormone ethylene. Recently, Khan et al. (2014) suggested that salt tolerance was dependent on GB and ethylene formation. They showed that SA application induces GB accumulation in mung bean plants concomitant with the suppression of ethylene formation more conspicuously under salt stress than no stress, which led to improved photosynthesis and growth under salt stress. Hussain et al. (2008) reported that application of SA maximally induced GB accumulation under water stress at the flowering stage as compared to vegetative stage. Further, they showed that water-stressed sunflower plants have reduced head diameter, number of achene, 1000-achene weight, achene yield, and oil yield; however, exogenous GB and SA application significantly improved these attributes under water stress. Jasmonic acid (JA) application to *Pyrus bretschneideri* plants under drought stress significantly increased the GB concentration. The increased betaine content was due to the increased BADH activity which shows that JA is involved in the drought-induced GB accumulation (Gao et al. 2004b).

Mineral nutrition study in relation to plant metabolism is always an interesting and challenging research field for plant physiologist and biochemist. However, the interactive role of GB and mineral nutrition is still an unrevealed area. Few studies suggested that interaction of mineral nutrients and GB metabolism can be targeted to develop the tolerant genotypes. Influence of high and low nitrogen supply in wheat plants was studied by Carillo et al. (2008). They suggested that GB protects plants from high-salinity stress on treating them with nitrogen. They further state that nitrate reductase and glutamate synthase (GOGAT) are positively correlated with the compatible solutes GB and proline. Alikhani et al. (2011) showed the importance of potassium and GB in *Aeluropus lagopoides* plants under salinity stress condition. They showed that in the absence of potassium, GB content of the root and shoot was not produced, but with the application of potassium, GB accumulation increase in the root and shoot resulted in protection of photosynthetic pigments. Gobinathan et al. (2009) suggested that application of calcium (Ca) under salt stress increased GB content. Further, they concluded that calcium appears to confer greater osmoprotection by GB accumulation in *Pennisetum typoidies*. Glycine betaine can enhance heat tolerance and accumulation of heat-shock proteins in plants. Recently, Li et al. (2014a) elaborated the mechanisms induced by GB in salt-stress tolerance. They suggested that Ca^{2+} and calmodulin (CaM) increased heat-shock transcription factors and heat-shock protein gene expression resulted in regulatory interaction of Ca^{2+}-CaM in the signal transduction pathway for induction of transcription and translation of the active heat-shock proteins. These results suggest a mechanism by which GB acted as a cofactor in the NaCl induction of a Ca^{2+}-permeable current.

5.5 Glycine Betaine in Alleviation of Abiotic Stresses in Plants

Glycine betaine is one of the organic compatible solutes in plants and accumulates in cytosol and chloroplast. It is dipolar in nature but electrically neutral molecule and highly soluble in water. Glycine betaine is widely distributed in higher plants and is synthesized in many plant species at elevated rates and is one of the major organic osmolytes produced in response to various types of environmental stresses such as drought, cold, temperature, salinity, and HMs (Yang et al. 2005; Ashraf and Foolad 2007; Pasquali et al. 2008; Hoque et al. 2008; Duman et al. 2011; Bharwana et al. 2014; Rasheed et al. 2014). Exogenously applied betaine increased the activity of enzymes of ascorbate-glutathione cycle in cultured *Nicotiana tabacum* BY-2 cells exposed to Cd stress and thereby helped in reducing the ill effects produced by Cd (Islam et al. 2009). Genetic engineering of biosynthesis of betaine from choline is a potential strategy for conferring stress tolerance in stress-sensitive plants that are incapable of synthesizing these compatible solutes, through the expression of genes responsible for the formation of BADH, CHO, COD, and CDH in plants (Sakamoto and Murata 2000). As already said, some plants are incapable to accumulate GB under stress or nonstress conditions, and this might be due to lack of functional gene involved in its biosynthesis (Rhodes and Hanson 1993; Shirasawa et al. 2006). Exogenous application of GB at 2 mM concentration significantly alleviated the metal-induced toxicity by inducing antioxidant enzyme activities, reducing oxidative stress, and enhancing growth, biomass, and photosynthesis (Bharwana et al. 2014). *Arabidopsis* plants expressing N-methyltransferase genes accumulated betaine to a high level under stress conditions. Betaine levels were higher than those produced by choline-oxidizing enzymes; from the research, it was assumed that the expression of genes responsible for glycine N-methyltransferase provide abiotic stress tolerance in crop plants (Waditee et al. 2005). Hoque et al. (2008) while studying in *Nicotiana tabacum* showed that 20 mM GB provides a protective action against NaCl-induced oxidative damage by enhancing antioxidant defense. Glycine betaine increased antioxidants' activity by decreasing Cd-induced oxidative stress in *Lemna gibba* and protects photosynthetic activity of plants (Duman et al. 2011). It has also been well established that application of GB induces antioxidants' machinery against salinity in *Nicotiana tabacum* and *Vigna radiata* (Hoque et al. 2008; Hossain and Fujita 2010); HMs in *Nicotiana tabacum*, *Lemna gibba*, and *Gossypium* (Islam et al. 2009; Duman et al. 2011; Bharwana et al. 2014); cold in *Camellia sinensis* (Kumar and Yadav 2009); and drought stress in *Triticum aestivum* and *Triticum durum* (Ma et al. 2006; Gupta and Thind 2015).

In addition to other roles, GB is involved in inhibiting ROS, protection of photosynthetic machinery, activation of some stress-related genes, and membrane protection (Chen and Murata 2008). Production of betaine in plants enhanced tolerance and improved the capacity of PSII to recover from photoinhibition caused by salt and low-temperature stress by protecting protein complexes involved in the photosynthetic process (Holmström et al. 2000). Exogenous application of GB reverses adverse effect of salt stress and upregulated photosynthetic capacity by increasing endogenous level of GB, as it has a role in stomatal and non-stomatal factors (Nawaz and Ashraf 2010). The results indicate that overexpression of *AhCMO* in *Gossypium hirsutum* enhanced salt-stress tolerance, which is of great value in cotton production in the saline fields, due to elevated accumulation of GB, which provided greater protection of the cell membrane and photosynthetic capacity than in non-transgenic plant (Zhang et al. 2009). Accumulation of GB in transgenic apple expressing stress regulator gene, *Osmyb4*, was linked to enhanced drought and cold tolerance, due to the ability of this gene to activate several pathways usually induced in response to stress conditions by accumulating osmolyte (Pasquali et al. 2008). Glycine betaine is in particular effective in protecting highly complex proteins, such as the PSII complex, against Cd stress in *Arabidopsis* (Yang and Lu 2005). Bharwana et al. (2014) studied in *Gossypium* that exogenous application of GB significantly alleviates lead toxicity by improving the growth, biomass, and photosynthetic capacity and increased antioxidant enzyme activities and lowering oxidative stress. Drought tolerance of *Sesbania aculeate* was found to be associated with a high activity of nitrate reductase in leaves and nodules and high accumulation of GB content in nodules by avoiding drought stress without interfering protein synthesis (Ashraf and Iram 2005). Accumulating GB by overexpression of BADH in *Ipomoea batatas* improves abiotic stress tolerance providing a feasible approach to improve stable yield production (Fan et al. 2012). Increased in GB as osmotic adjustment resulted in higher photosynthetic capacity in salt-tolerant genotype maintained at a higher level than those in the salt-sensitive genotype of rice resulted in higher growth performance when supplemented to extreme pH and salt stress (Cha-um et al. 2009). Rasheed et al. (2014) while studying in *Triticum aestivum* cultivars, namely, Millat-2011 and Punjab-2011, found that Millat-2011 exhibited better tolerance to Cd toxicity in terms of lesser damage to photosynthetic pigments (Chlorophyll (Chl) a, Chl b) under the application of GB, and reduced oxidative stress was observed. Increased Chl contents in Cd-stressed plants might be due to accumulation of GB. In a study by Li et al. (2014b), a BADH gene from spinach was introduced into tomato through Agrobacterium-mediated transformation, and transgenic tomato lines expressing BADH gene exhibited higher abilities for GB accumulation resulted in increased tolerance to heat-enhanced photoinhibition and improve D1 protein content, which accelerated the repair of PSII. It has been well established that application of GB improves photosynthesis and growth against drought stress in *Triticum durum* (Gupta and Thind 2015), salt stress in *Brassica napus* (Athar et al. 2015), and HMs stress *Gossypium* (Islam et al. 2009; Bharwana et al. 2014). The *codA* gene for choline oxidase allows *Synechococcus* cells to accumulate GB, and this accumulation alleviates adverse effect of heat stress on the repair of PSII by accelerating the synthesis of the D1 protein resulting in improved photosynthesis and growth (Allakhverdiev et al. 2007). Introduction of BADH gene for BADH into *Nicotiana tabacum* from *Spinacia oleracea* resulted in enhanced tolerance of growth and photosynthesis to high temperatures in transgenic plants (Yang et al. 2005).

5.6 Recent Advances in GB-Induced Photosynthetic Protection under Stressful Environment

As discussed above, GB is one of the major organic osmolytes and widely distributed in higher plants and is synthesized in many plant species in response to various types of environmental stresses (Rhodes and Hanson 1993). Glycine betaine accumulates in many plants under stress and positively affects photosynthetic apparatus. Long ago, PSII has been considered the most important component of the photosynthetic apparatus. However, it appears that PSII damage indirectly affected CO_2 fixation (Weis 1981) and that CO_2 fixation was most responsive to environmental stresses via a direct effect on Rubisco activase and caused the inhibition of activation of Rubisco (Feller et al. 1998; Salvucci and Crafts-Brandner 2004). In particular, GB was effective in protecting highly complex proteins, such as the PSII complex inactivation. Several studies on the different disciplines as physiology, biochemistry, biophysics, and genetics of plants have suggested that GB plays an important role in plants under various types of environmental stresses. Glycine betaine maintained photosynthetic capacity by increasing stomatal conductance and maintaining Rubisco activity and chloroplast ultrastructure under drought stress (Nomura et al. 1998). The transformation of a GB synthesis gene into plants led to overaccumulation of glycine betaine and protected the photosynthetic apparatus from stress-induced damage (Sakamoto and Murata 2002). Glycine betaine increased the various types of stress tolerance such as salt tolerance (Holmstrom et al. 2000; Park et al. 2004), cold tolerance (Alia et al. 1998; Chen et al. 2000), and freezing tolerance of plants (Sakamoto and Murata 2000). Glycine betaine improved the nonphotochemical quenching (NPQ) by protecting violaxanthin de-epoxidase (VDE) from photodamage, ever since VDE was a key enzyme in the xanthophyll cycle and was localized in the thylakoid membrane (Havaux et al. 2000). The application of GB in salt-stressed wheat plants suggested that the increase in photosynthesis due to increase in stomatal conductance caused higher CO_2 diffusion inside the leaf, thus supporting higher photosynthetic rate (Taiz and Zeiger 2002). Chen and Murata (2002) have showed that GB involved in inhibiting ROS accumulation and protecting photosynthetic machinery leads to the activation of some stress-related genes for the membrane protection. Moreover, in the study by Yang et al. (2005), they observed the role of GB in vivo in protecting photosynthesis from high-temperature stress using transgenic tobacco. They suggested that the accumulation of GB in vivo by introducing the BADH gene for betaine aldehyde dehydrogenase into tobacco resulted in the enhanced tolerance of growth and photosynthesis to high temperatures in transgenic plants. Glycine betaine induced increase in photosynthetic capacity in plants due to stomatal or non-stomatal limitations for major controlling factors of photosynthetic rate (Athar and Ashraf 2005; Dubey 2005). Glycine betaine also stabilized the activity of repair proteins under high concentrations of salt and hence protected the photosynthetic machinery (Bartels and Sunkar 2005). The study of Raza et al. (2006) showed that the exogenous application of GB mitigated the adverse effects of salinity on photosynthetic capacity of wheat, as moderate GB accumulator. They suggested that GB improved the net CO_2 assimilation rate in wheat cultivars. The study of Allakhverdiev et al. (2007) reported that GB alleviated the inhibitory effect of moderate heat stress on the repair of PSII during photoinhibition. Wang et al. (2008) reported that GB improved the efficiency of PSII photochemistry and alleviated photoinhibition under stress state. However, pretreatment of GB having concentration of 100–150 mM had been reported to maintain the CO_2 assimilation rate and net photosynthetic rate in rice plants under drought stress (Farooq et al. 2008). Similarly, GB leads to improved water-use efficiency and pigment stabilization, as well as to increased water oxidation in PSII (represented by Fv/Fm and ΦPSII) and improved net photosynthetic rate (Cha-um and Kirdmanee 2010). In contrast, the findings of Wang et al. (2010) showed that the overaccumulation by GB in plants increased NPQ and suggested that xanthophyll cycle is involved in the improvement of protective mechanisms against photoinhibition. Glycine betaine has been reported to protect photosynthetic machinery, stabilize the structure of Rubisco, and act as oxygen radical scavenger under drought stress (Cha-um et al. 2013). Exogenous application of GB increased the tolerance of barley leaves to the supplied thermal stress at 45 °C by exerting a protective effect on the oxygen evolving complex and by increasing connectivity between PSII antennae for making a greater stability of the system PSII (Oukarroum et al. 2012) A significant decline in photosynthetic pigments of water-stressed plant directly affected the light harvesting capacity and electron transport system in PSII, indicated by chlorophyll fluorescence parameters as reported by Cha-um et al. (2013). The report of Bharwana et al. (2014) showed that the GB increased lead tolerance by enhancing the Chl synthesis, photosynthetic activities, and antioxidant enzyme activities and by lowering the electrolytic leakage, malondialdehyde content, and H_2O_2 levels. The findings of Khan et al. (2014) showed that the effects of SA on GB synthesis and ethylene formation induced antioxidant system that protected photosynthetic machinery from oxidative stress induced by salt stress and alleviated the adverse effects of salt stress on photosynthesis and growth of mung bean. Gupta and Thind (2015) observed that the photosynthesis performance in wheat was improved by the application of GB under field drought stress and suggested the role of nonenzymatic antioxidants in sustaining photosynthetic efficiency and yield stability.

5.7 Conclusion and Future Perspectives

Based on the above literature, a hypothetical scheme (Fig. 5.1) is designed to explain the possible mechanism in which GB participates and plays an important role in inducing stress resistance in plants. The figure also proposed the interactive effect of GB in relation to nutrients or phytohormones, working together in conferring abiotic stress tolerance. Recent advances in production of transgenic plants with higher rate of GB accumulation helps them in their survival against abiotic stresses are the promising tool for sustainable agriculture. The role of GB biosynthesis and accumulation at various physiological and metabolic levels has been explored, but it is yet to be completely comprehended and

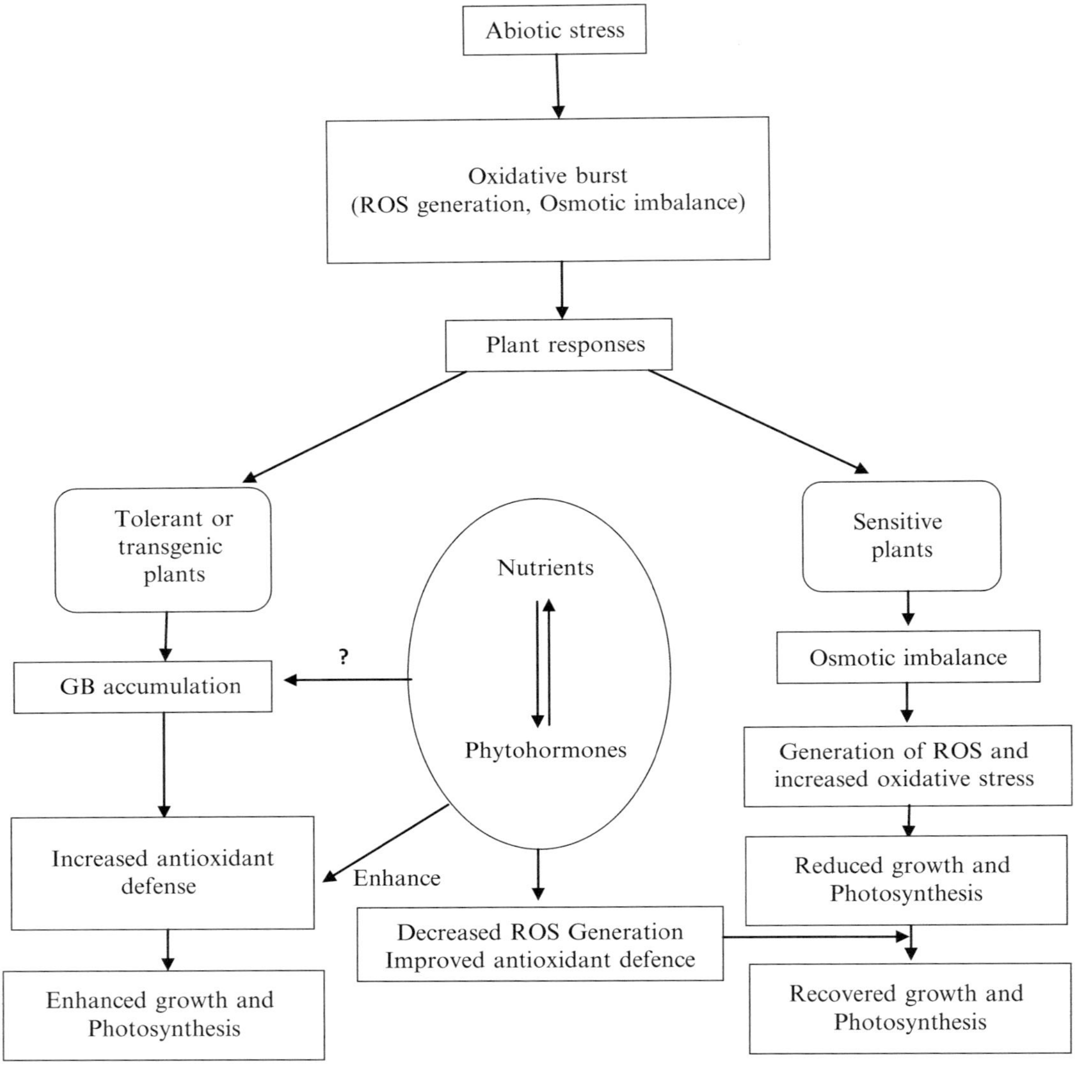

Fig. 5.1 Showing possible mechanism in which GB participates and plays an important role in inducing stress resistance in tolerant/transgenic and sensitive plant species. The research on interaction between GB and nutrients in relation to phytohormones still needed to be explored

exploited. Transgenic studies are instrumental in investigating molecular mechanisms of the protective action of GB against abiotic stress, its protection of photosynthetic machinery particularly PSII. Although GB accumulation is reported to be influenced by various phytohormones, there is still a large part which needs to be explored about the interactive role of GB and nutrient and their cross talk with different phytohormones. The protective action of GB can be implied on agricultural crops to enhance tolerance to abiotic stress. This is another field which needs further study, and this will provide another alternative besides transgenics to improve tolerance to abiotic stress by GB accumulation. There is also possibility to enhance induction of genes by GB so as to produce ROS-scavenging enzymes and dip the ROS levels in cells.

References

Alia HH, Chen THH, Murata N (1998) Transformation with a gene for choline oxidase enhances the cold tolerance of *Arabidopsis* during germination and early growth. Plant Cell Environ 21:232–239

Alikhani F, Saboora A, Razavi K (2011) Changes in osmolytes contents, lipid peroxidation and photosynthetic pigment of *Aeluropus lagopoides* under potassium deficiency and salinity. J Stress Physiol Biochem 7:5–19

Allakhverdiev SI, Los DA, Mohanty P, Nishiyama Y, Murata N (2007) Glycinebetaine alleviates the inhibitory effect of moderate heat stress on the repair of photosystem II during photoinhibition. Biochim et Biophys Acta 1767:1363–1371

Asgher M, Khan NA, Khan MIR, Fatma F, Masood A (2014) Ethylene production is associated with alleviation of cadmium-induced oxidative stress by sulfur in mustard types differing in ethylene sensitivity. Ecotoxicol Environ Saf 106:54–61

Asgher M, Khan MIR, Anjum NA, Khan NA (2015) Minimising toxicity of cadmium in plants-role of plant growth regulators. Protoplasma 252:399–413

Ashraf M, Foolad M (2007) Roles of glycine betaine and proline in improving plant abiotic stress resistance. Environ Exp Bot 59:206–216

Athar HR, Ashraf M (2005) Photosynthesis under drought stress. In: Pessarakli M (ed) Handbook of photosynthesis. CRC Press, New York, pp 793–804

Athar H, Zafar ZU, Ashraf M (2015) Glycinebetaine improved photosynthesis in canola under salt stress: evaluation of chlorophyll fluorescence parameters as potential indicators. J Agr Crop Sci. doi:10.1111/jac.12120

Bartels D, Sunkar R (2005) Drought and salt tolerance in plants. Crit Rev Plant Sci 24:23–58

Bharwana SA, Ali S, Farooq MA, Iqbal N, Hameed A, Abbas F, Ahmad MSA (2014) Glycine betaine-induced lead toxicity tolerance related to elevated photosynthesis, antioxidant enzymes suppressed lead uptake and oxidative stress in cotton. Turk J Bot 38:281–292

Breitkreuz KE, Shelp BJ, Fischer WN, Schwacke R, Rentsch D (1999) Identification and characterization of GABA, proline and quaternary ammonium compound transporters from *Arabidopsis thaliana*. FEBS Lett 450:280–284

Carillo P, Mastrolonardo G, Nacca F, Parisi D, Verlotta A, Fuggi A (2008) Nitrogen metabolism in durum wheat under salinity: accumulation of proline and glycine betaine. Funct Plant Biol 35:412–426

Cha-um S, Kirdmanee C (2010) Effect of glycinebetaine on proline, water use, and photosynthetic efficiencies, and growth of rice seedlings under salt stress. Turk J Agric For 34:517–527

Cha-um S, Supaibulwattana K, Kirdmanee C (2009) Comparative effects of salt stress and extreme pH stress combined on glycinebetaine accumulation, photosynthetic abilities and growth characters of two rice genotypes. Rice Sci 16:274–282

Cha-um S, Samphumphuang T, Kirdmanee C (2013) Glycinebetaine alleviates water deficit stress in indica rice using proline accumulation, photosynthetic efficiencies, growth performances and yield attributes. Aust J Crop Sci 7:213–218

Chen THH, Murata N (2002) Enhancement of tolerance to abiotic stress by metabolic engineering of betaines and other compatible solutes. Curr Opin Plant Biol 5:250–257

Chen THH, Murata N (2008) Glycinebetaine: an effective protectant against abiotic stress in plants. Trends Plant Sci 13:499–505

Chen THH, Murata N (2011) Glycinebetaine protects plants against abiotic stress: mechanisms and biotechnological applications. Plant Cell Environ 34:1–20

Chen WP, Li PH, Chen THH (2000) Glycinebetaine increases chilling tolerance and reduces chilling-induced lipid peroxidation in *Zea mays* L. Plant Cell Environ 23:609–618

Choudhary SP, Kanwar M, Bhardwaj R, Yu J-Q, Tran L-SP (2012) Chromium stress mitigation by polyamine-brassinosteroid application involves phytohormonal and physiological strategies in *Raphanus sativus* L. PLoS One 7:e33210

Dubey RS (2005) Photosynthesis in plants under stressful conditions. In: Pessarakli M (ed) Hand book of photosynthesis. CRC Press/Taylor and Francis Group, New York, pp 717–737

Duman F, Aksoy A, Aydin Z, Temizgul R (2011) Effects of exogenous glycinebetaine and trehalose on Cadmium accumulation and biological responses of an aquatic plant (*Lemna gibba* L). Water Air Soil Pollut 217:545–556

Fan W, Zhang M, Zhang H, Zhang P (2012) Improved tolerance to various abiotic stresses in transgenic sweet potato (*Ipomoea batatas*) expressing spinach betaine aldehyde dehydrogenase. PLoS One 7:e37344

Farooq M, Basra SMA, Wahid A, Cheema ZA, Cheema MA, Khaliq A (2008) Physiological role of exogenously applied glycinebetaine to improve drought tolerance in fine grain aromatic rice (*Oryza sativa* L.). J Agron Crop Sci 194:325–333

Feller U, Crafts-Brandner SJ, Salvucci ME (1998) Moderately high temperatures inhibit ribulose-1, 5-bisphosphate carboxylase/oxygenase (Rubisco) activase-mediated activation of Rubisco. Plant Physiol 116:539–546

Fujiwara T, Mitsuya S, Miyake H, Hattori T, Takabe T (2010) Characterization of a novel glycinebetaine/proline transporter gene expressed in the mestome sheath and lateral root cap cells in barley. Planta 232:133–143

Gadallah M (1999) Effects of proline and glycinebetaine on *Vicia faba* responses to salt stress. Biol Plant 42:249–257

Gao XP, Pan QH, Li MJ, Zhang LY, Wang XF, Shen YY, Zhang DP (2004a) Abscisic acid is involved in the water stress-induced betaine accumulation in pear leaves. Plant Cell Physiol 45:742–750

Gao XP, Wang XF, Lu YF, Zhang LY, Shen YY, Liang Z, Zhang DP (2004b) Jasmonic acid is involved in the water-stress-induced betaine accumulation in pear leaves. Plant Cell Environ 27:497–507

Génard H, Le Saos J, Billard J-P, Trémolières A, Boucaud J (1991) Effect of salinity on lipid composition, glycine betaine content and photosynthetic activity in chloroplasts of Suaeda maritime. Plant Physiol Biochem 29:421–427

Gobinathan P, Murali PV, Panneerselvam R (2009) Interactive effects of calcium chloride on salinity-induced proline metabolism in *Pennisetum typoidies*. Adv Biochem Res 3:168–173

Gupta N, Thind S (2015) Improving photosynthetic performance of bread wheat under field drought stress by foliar applied *Glycine Betaine*. J Agric Sci Technol 17:75–86

Hancock JT, Neill SJ, Wilson ID (2011) Nitric oxide and ABA in the control of plant function. Plant Sci 181:555–559

Havaux M, Bonfils JP, Lütz C, Niyogi KK (2000) Photodamage of the photosynthetic apparatus and its dependence on the leaf developmental stage in the npq1 Arabidopsis mutant deficient in the xanthophyll cycle enzyme violaxanthin de-epoxidase. Plant Physiol 124:273–284

Hayashi H, Alia Mustardy L, Deshnium P, Ida M, Murata N (1997) Transformation of *Arabidopsis thaliana* with the *codA* gene for choline oxidase; accumulation of glycinebetaine and enhanced tolerance to salt and cold stress. Plant J 12:133–142

Hibino T, Waditee R, Araki E, Ishikawa H, Aoki K, Tanaka Y, Takabe T (2002) Functional characterization of choline monooxygenase, an enzyme for betaine synthesis in plants. J Biol Chem 277:41352–41360

Holmström KO, Somersalo S, Mandal A, Palva TE, Welin B (2000) Improved tolerance to salinity and low temperature in transgenic tobacco producing glycine betaine. J Exp Bot 51:177–185

Hoque MA, Banu MNA, Nakamura Y, Shimoishi Y, Murata Y (2008) Proline and glycinebetaine enhance antioxidant defense and methylglyoxal detoxification systems and reduce NaCl-induced damage in cultured tobacco cells. J Plant Physiol 165:813–824

Hossain MA, Fujita M (2010) Evidence for a role of exogenous glycinebetaine and proline in antioxidant defense and methylglyoxal detoxification systems in mung bean seedlings under salt stress. Physiol Mol Biol Plants 16:19–29

Hussain M, Malik MA, Farooq M, Ashraf MY, Cheema MA (2008) Improving drought tolerance by exogenous application of glycinebetaine and salicylic acid in sunflower. J Agron Crop Sci 194:193–199

Islam MM, Hoque MA, Okuma E, Banu MNA, Shimoishi Y, Nakamura Y, Murata Y (2009) Exogenous proline and glycinebetaine increase antioxidant enzyme activities and confer tolerance to cadmium stress in cultured tobacco cells. J Plant Physiol 166:1587–1597

Jagendorf AT, Takabe T (2001) Inducers of glycinebetaine synthesis in barley. Plant Physiol 127:1827–1835

Khan MIR, Khan NA (2014) Ethylene reverses photosynthetic inhibition by nickel and zinc in mustard through changes in PS II activity, photosynthetic nitrogen use efficiency, and antioxidant metabolism. Protoplasma 251:1007–1019

Khan MS, Yu X, Kikuchi A, Asahina M, Watanabe KN (2009) Genetic engineering of glycine betaine biosynthesis to enhance abiotic stress tolerance in plants. Plant Biotechnol 26:125–134

Khan MIR, Iqbal N, Masood A, Khan NA (2012) Variation in salt tolerance of wheat cultivars: role of glycinebetaine and ethylene. Pedosphere 22:746–754

Khan MIR, Iqbal N, Masood A, Per TS, Khan NA (2013) Salicylic acid alleviates adverse effects of heat stress on photosynthesis through changes in proline production and ethylene formation. Plant Signal Behav 8:e26374

Khan MIR, Asgher M, Khan NA (2014) Alleviation of salt-induced photosynthesis and growth inhibition by salicylic acid involves glycinebetaine and ethylene in mungbean (*Vigna radiata* L.). Plant Physiol Biochem 80:67–74

Khan MIR, Nazir F, Asgher M, Per TS, Khan NA (2015) Selenium and sulfur influence ethylene formation and alleviate cadmium-induced oxidative stress by improving proline and glutathione production in wheat. J Plant Physiol 173:9–18

Ko R, Smith LT, Smith GM (1994) Glycine betaine confers enhanced osmotolerance and cryotolerance on *Listeria monocytogenes*. J Bacteriol 176:426–431

Kumar V, Yadav SK (2009) Proline and betaine provide protection to antioxidant and methylglyoxal detoxification systems during cold stress in *Camellia sinensis* (L.) *O. Kuntze*. Acta Physiol Plant 31:261–269

Kumar S, Dhingra A, Daniell H (2004) Plastid-expressed betaine aldehyde dehydrogenase gene in carrot cultured cells, roots, and leaves confers enhanced salt tolerance. Plant Physiol 136:2843–2854

Ladyman JAR, Hitz WD, Hanson AD (1980) Translocation and metabolism of glycine betaine in barley plants in relation to water stress. Planta 150:191–196

Li M, Guo S, Xu Y, Meng Q, Li G, Yang X (2014a) Glycine betaine-mediated potentiation of HSP gene expression involves calcium signaling pathways in tobacco exposed to NaCl stress. Physiol Plant 150:63–75

Li M, Li Z, Li S, Guo S, Meng Q, Li G, Yang X (2014b) Genetic engineering of glycine betaine biosynthesis reduces heat-enhanced photoinhibition by enhancing antioxidative defense and alleviating lipid peroxidation in tomato. Plant Mol Biol Rep 32:42–51

Lv S, Yang A, Zhang K, Wang L, Zhang J (2007) Increase of glycinebetaine synthesis improves drought tolerance in cotton. Mol Breed 20:233–248

Ma QQ, Wang W, Li YH, Li DQ, Zou Q (2006) Alleviation of photoinhibition in drought-stressed wheat (*Triticum aestivum*) by foliar-applied glycinebetaine. J Plant Physiol 163:165–175

Mäkelä P, Peltonen-Sainio P, Jokinen K, Pehu E, Setälä H, Hinkkanen R, Somersalo S (1996) Uptake and translocation of foliar-applied glycinebetaine in crop plants. Plant Sci 121:221–230

Maldonado CA, Zuniga GE, Corcuera LJ, Alberdi M (1997) Effect of water stress on frost resistance of oat leaves. Environ Exp Bot 38:99–107

Masood A, Iqbal N, Khan NA (2012) Role of ethylene in alleviation of cadmium-induced photosynthetic capacity inhibition by sulphur in mustard. Plant Cell Environ 35:524–533

McCue KF, Hanson AD (1990) Drought and salt tolerance: towards understanding and application. Trends Biotechnol 8:358–362

McCue KF, Hanson AD (1992) Salt-inducible betaine aldehyde dehydrogenase from sugar beet: cDNA cloning and expression. Plant Mol Biol 18:1–11

Mohanty A, Kathuria H, Ferjani A, Sakamoto A, Mohanty P, Murata N, Tyagi A (2002) Transgenics of an elite indica rice variety Pusa Basmati 1 harbouring the codA gene are highly tolerant to salt stress. Theor Appl Genet 106:51–57

Nawaz K, Ashraf M (2010) Exogenous application of glycinebetaine modulates activities of antioxidants in maize plants subjected to salt stress. J Agron Crop Sci 196:28–37

Nomura M, Hibino T, Takabe T, Sugiyama T, Yokota A, Miyake H, Takabe T (1998) Transgenically produced glycinebetaine protects ribulose 1,5-bisphosphate carboxylase/oxygenase from inactivation in *Synechococcus sp.* PCC7942 under salt stress. Plant Cell Physiol 39:425–432

Noreen S, Athar HR, Ashraf M (2013) Interactive effects of watering regimes and exogenously applied osmoprotectants on earliness indices and leaf area index in cotton (*Gossypium hirsutum* L.) crop. Pak J Bot 45:1873–1881

Nuccio ML, Russell BL, Nolte KD, Rathinasabapathi B, Gage DA, Hanson AD (1998) The endogenous choline supply limits glycine betaine synthesis in transgenic tobacco expressing choline monooxygenase. Plant J 16:487–496

Oukarroum A, El Madidi S, Strasser RJ (2012) Exogenous glycine betaine and proline play a protective role in heat-stressed barley leaves (*Hordeum vulgare* L.): a chlorophyll a fluorescence study. Plant Biosyst 146:1037–1043

Park EJ, Jeknić Z, Sakamoto A, DeNoma J, Yuwansiri R, Murata N, Chen THH (2004) Genetic engineering of glycinebetaine synthesis in tomato protects seeds, plants, and flowers from chilling damage. Plant J 40:474–487

Park EJ, Jeknić Z, Chen THH (2006) Exogenous application of glycinebetaine increases chilling tolerance in tomato plants. Plant Cell Physiol 47:706–714

Park EJ, Jeknić Z, Pino MT, Murata N, Chen THH (2007) Glycinebetaine accumulation in chloroplasts is more effective than that in cytosol in protecting transgenic tomato plants against abiotic stress. Plant Cell Environ 30:994–1005

Pasquali G, Biricolti S, Locatell F, Baldoni E, Mattana M (2008) Osmyb4 expression improves adaptive responses to drought and cold stress in transgenic apples. Plant Cell Rep 27:1677–1686

Quan R, Shang M, Zhang H, Zhao Y, Zhang J (2004) Engineering of enhanced glycine betaine synthesis improves drought tolerance in maize. Plant Biotechnol J 2:477–486

Rasheed R, Ashraf MA, Hussain I, Haider MZ, Kanwal U, Iqbal M (2014) Exogenous proline and glycinebetaine mitigate cadmium stress in two genetically different spring wheat (*Triticum aestivum* L.) cultivars. Brazil J Bot 37:399–406

Rathinasabapathi B, Burnet M, Russel BL, Gage DA, Liao P-C, Nye GJ, Scott P, Golbeck JH, Hanson AD (1997) Choline monooxygenase, an unusual iron-sulfur enzyme catalyzing the first step of glycine betaine synthesis in plants: prosthetic group characterization and cDNA cloning. Proc Natl Acad Sci U S A 94:3454–3458

Raza SH, Athar H, Ashraf M (2006) Influence of exogenously applied glycinebetaine on the photosynthetic capacity of two differently adapted wheat cultivars under salt stress. Pak J Bot 38:341–351

Raza SH, Athar HR, Ashraf M, Hameed A (2007) Glycinebetaine-induced modulation of antioxidant enzymes activities and ion accumulation in two wheat cultivars differing in salt tolerance. Environ Exp Bot 60:368–376

Rhodes D, Hanson AD (1993) Quaternary ammonium and tertiary sulfonium compounds in higher plants. Annu Rev Plant Physiol Plant Mol Biol 44:357–384

Robinson SP, Jones GP (1986) Accumulation of glycine betaine in chloroplasts provides osmotic adjustment during salt stress. Aust J Plant Physiol 13:659–668

Sakamoto A, Murata N (2000) Genetic engineering of glycinebetaine synthesis in plants: current status and implication for enhancement of stress tolerance. J Exp Bot 51:81–88

Sakamoto A, Murata N (2002) The role of glycine betaine in the protection of plants from stress: clues from transgenic plants. Plant Cell Environ 25:163–171

Sakamoto A, Alia, Murata N (1998) Metabolic engineering of rice leading to biosynthesis of glycinebetaine and tolerance to salt and cold. Plant Mol Biol 38:1011–1019

Salvucci ME, Crafts-Brandner SJ (2004) Relationship between the heat tolerance of photosynthesis and the thermal stability of Rubisco activase in plants from contrasting thermal environments. Plant Physiol 134:1460–1470

Saneoka H, Nagasaka C, Hahn DT, Yang WJ, Premachandra GS, Joly RJ, Rhodes D (1995) Salt tolerance of glycinebetaine-deficient and -containing maize lines. Plant Physiol 107:631–638

Saneoka H, Ishiguro S, Moghaieb RE (2001) Effect of salinity and abscisic acid on accumulation of glycinebetaine and betaine aldehyde dehydrogenase mRNA in Sorghum leaves (*Sorghum bicolor*). J Plant Physiol 158:853–859

Saum R, Mingote A, Santos H, Müller V (2009) Genetic analysis of the role of the ABC transporter Ota and Otb in glycine betaine transport in *Methanosarcina mazei* Gö1. Arch Microbiol 191:291–301

Sawahel W (2003) Improved performance of transgenic glycinebetaine-accumulating rice plants under drought stress. Biol Plant 47:39–44

Scheibler C (1869) Ueber das betain, eine im safte der zuckerrüben (*Beta vulgaris*) vorkommende pflanzenbase. Ber Dtsch Chem Ges 2:292–295

Schwacke R, Grallath S, Breitkreuz KE, Stransky E, Stransky H, Frommer WB, Rentsch D (1999) LeProT1, a transporter for proline, glycine betaine, and gamma-amino butyric acid in tomato pollen. Plant Cell 11:377–392

Serraj R, Sinclair TR (2002) Osmolyte accumulation: can it really help increase crop yield under drought conditions? Plant Cell Environ 25:333–341

Shirasawa K, Takabe T, Takabe T, Kishitani S (2006) Accumulation of glycinebetaine in rice plants that overexpress choline monooxygenase from spinach and evaluation of their tolerance to abiotic stress. Ann Bot 98:565–571

Storey R, Ahmad N, Wyn Jones RG (1977) Taxonomic and ecological aspects of the distribution of glycinebetaine and related compounds in plants. Oecologia 27:319–332

Su J, Hirji R, Zhang L, He C, Selvaraj G, Wu R (2006) Evaluation of the stress-inducible production of choline oxidase in transgenic rice as a strategy for producing the stress-protectant glycine betaine. J Exp Bot 57:1129–1135

Sulpice R, Tsukaya H, Nonaka H, Mustardy L, Chen THH, Murata N (2003) Enhanced formation of flowers in salt-stressed Arabidopsis after genetic engineering of the synthesis of glycinebetaine. Plant J 36:165–176

Takabe T, Rai V, Hibino T (2006) Metabolic engineering of glycinebetaine. In: Rai A, Takabe T (eds) Abiotic stress tolerance in plants: toward the improvement of global environment and food. Springer, Dordrecht, pp 137–151

Ueda A, Shi W, Sanmiya K, Shono M, Takabe T (2001) Functional analysis of salt inducible proline transporter of barley roots. Plant Cell Physiol 42:1282–1289

Waditee R, HibinoT TY, Nakamura T, Incharoensakdi A, Hayakawa S, Suzuki S, Futsuhara Y, Kawamitsu Y, Takabe T (2002) Functional characterization of betaine/proline transporters in betaine-accumulating mangrove. J Biol Chem 277:8373–18382

Waditee R, Bhuiyan MNH, Rai V, Aoki K, Tanaka Y, Hibino T, Suzuki S, Takano J, Jagendorf AT, Takabe T, Takabe T (2005) Genes for direct methylation of glycine provide high levels of glycinebetaine and abiotic-stress tolerance in Synechococcus and Arabidopsis. Proc Natl Acad Sci USA 102:1318–1323

Wang S, Yao Q, Tao J, Qiao Y, Zhang Z (2007) Co-ordinate expression of glycine betaine synthesis genes linked by the FMDV 2A region in a single open reading frame in *Pichia pastoris*. Appl Microbiol Biotechnol 77:891–899

Wang C, Ma XL, Hui Z, Wang W (2008) Glycine betaine improves thylakoid membrane function of tobacco leaves under low-temperature stress. Photosynthetica 46:400–409

Wang GP, Li F, Zhang J, Zhao MR, Hui Z, Wang W (2010) Overaccumulation of glycine betaine enhances tolerance of the photosynthetic apparatus to drought and heat stress in wheat. Photosynthetica 48:30–41

Weis E (1981) Reversible heat-inactivation of the Calvin cycle: a possible mechanism of the temperature regulation of photosynthesis. Planta 151:33–39

Yamada N, Promden W, Yamane K, Tamagake H, Hibino T, Tanaka Y, Takabe T (2009) Preferential accumulation of betaine uncoupled to choline monooxygenase in young leaves of sugar beet: importance of long-distance translocation of betaine under normal and salt-stressed conditions. J Plant Physiol 166:2058–2070

Yang XH, Lu CM (2005) Photosynthesis is improved by exogenous glycinebetaine in salt-stressed maize plants. Physiol Plant 124:343–352

Yang X, Liang Z, Lu C (2005) Genetic engineering of the biosynthesis of glycinebetaine enhances photosynthesis against high temperature stress in transgenic tobacco plants. Plant Physiol 138:2299–2309

Zhang JH, Jia WS, Yang JC, Ismail AM (2006) Role of ABA in integrating plant responses to drought and salt stresses. Field Crop Res 97:111–119

Zhang H, Dong H, Li W, Sun Y, Chen S, Kong X (2009) Increased glycine betaine synthesis and salinity tolerance in AhCMO transgenic cotton lines. Mol Breed 23:289–298

Metabolic Engineering of Compatible Solute Trehalose for Abiotic Stress Tolerance in Plants

6

Saroj Kumar Sah, Gurwinder Kaur, and Shabir H. Wani

Abstract

It is estimated that by 2050 the world population will reach 9.1 billion, but the production of agricultural products is the same. In order to feed the whole population, global agricultural production should be increased by 60–110 %, and to feed the additional 2.3 billion population, 70 % more food should be grown to fulfill the demand. Due to abiotic stresses, agricultural production is lowered, so now abiotic stresses are a foremost area of concern to fulfill the required food demand. The major abiotic stresses which threaten the food security worldwide are high salinity, drought, submerge tolerance, and cold. To produce stress-tolerant crops, genetic engineering of stress-signaling pathway is one of the main goals of agricultural research. In recent years, biotechnologist is trying to develop a new abiotic stress-tolerant variety by engineering a trehalose metabolism in crops which can have a substantial impact on worldwide food production. Trehalose, a nonreducing disaccharide, has tremendous effects in abiotic stress tolerance and metabolic regulation in a wide range of organisms. Trehalose-6-phosphate synthase (TPS and trehalose-6-phosphate phosphatase (TPP are two key enzymes which help in the biosynthesis of plants. Trehalose is an uncommon sugar present in bacteria, fungi, and desiccation-tolerant higher plants and has exceptional capacities to protect biomolecules by stabilizing dry biological membrane and proteins from

S.K. Sah
Department of Biochemistry, Molecular Biology, Entomology and Plant Pathology, Mississippi State University, Starkville, MS 39762, USA

G. Kaur
School of Agricultural Biotechnology, Punjab Agricultural University, Ludhiana 141 001, India

S.H. Wani (✉)
Division of Genetics and Plant Breeding, SKUAST-K, Shalimar, Srinagar 190025, JK, India
e-mail: shabirhussainwani@gmail.com

N. Iqbal et al. (eds.), *Osmolytes and Plants Acclimation to Changing Environment: Emerging Omics Technologies*, DOI 10.1007/978-81-322-2616-1_6

environmental stress. It has a multiple function and some of them are species specific. Many research groups showed that there is a linkage between trehalose and abiotic stress by conducting different experiments. They introduced trehalose biosynthetic genes to develop stress tolerance line in important crops like rice, tomato, and potato. Here, in this review, we discuss occurrence, characters, chemical and biological characteristics, uses, pathways, and successful examples in crop plants.

6.1 Introduction

The worldwide human population is growing tremendously but agricultural production is same. Therefore, to fulfill the demands of food supply, crop production should be doubled especially major crops like wheat, maize, and rice. Subsequently, the pressure to increase crop production area is continuously increasing, and due to this pressure, water uses in irrigation lead to intensify water pollution, shortage of waters, exhaustion of aquifers, saltwater infiltration, and competition with other household usages. Water is one of the foremost limiting factors for crop production. Due to this reason, plant researchers are more focused on developing drought-tolerant crops. Water stress in response to plants at different levels has been studied by many researchers. According to Almeida et al. (2007), desiccation stress is an extremely intricate process in response to molecular levels in studying the molecular mechanisms and is a difficult process because it involves many genes for cellular, biochemical, and molecular mechanisms. There are three categories in studying stress-related genes: first, mainly genes involved in signaling cascades and transcriptional controls as, for example, transcription factors, phospholipase, and MAP kinase; second, genes involved in the protection of membranes and proteins like osmoprotectants, heat-shock proteins and chaperones, late embryogenesis abundant, and free radical scavengers; and third, genes responsible in water and ion uptake and transport, for example, aquaporins (Almeida et al. 2007). Abiotic stresses like cold, heat, and drought cause cellular damage and secondary stresses as in osmotic and oxidative stresses. Preliminary stress signals prompt transcription factors and downstream signaling processes that further activate stress-responsive mechanisms to regenerate homeostasis by restoring and protecting membrane proteins. Stress tolerance is achieved due to these cellular events. Desiccation tolerance implicates a chain of molecular mechanisms. Among them, genetic engineering is the leading technology in developing water stress tolerance. Using osmoprotectant (i.e., glycine betaine, proline, ectoine, mannitol, sorbitol, fructan, trehalose, etc.) accumulation to develop stress tolerance is one of the major approaches of genetic engineering techniques (Almeida et al. 2007; Cushman and Bohnert 2000). Nonreducing disaccharides (i.e., trehalose and sucrose) provide a soluble energy source as stable molecules, and in stress conditions, they work as a protectant compound in all organisms excluding vertebrates. Trehalose is formed through 1–1 alpha bond linkage between two glucose moieties, i.e., a-D-glucopyranosyl-1 and 1-a-D-glucopyranoside, which is considered to be better than nonreducing disaccharides such as proline (Garcia et al. 1997). Trehalose is a nonreducing disaccharide of glucose found in many organisms like bacteria, fungi, nematodes, and crustaceans (Elbein 1974). According to Wingler (2002), trehalose prevents proteins from denaturation and fusion of membranes during stresses like desiccation and heat stress by stabilizing membrane proteins, and its accumulation has been reported in renaissance plants like *Selaginella lepidophylla* under extreme desiccation conditions (Scott 2000). To develop salt and drought tolerance of crop plants, genetic engineering of trehalose is one of the most important approaches; trehalose also shows a chief role in plant carbohydrate metabolism (Penna 2003;

Eastmond and Graham 2003). Despite of being found in very minute amount in plants, it has a tremendous role in plant cell metabolism for abiotic stress tolerance. In most of the plants trehalose acts as signaling molecule rather than directly involved in the mitigation of abiotic stress.

There are many roles which have been anticipated by different research for both trehalose and trehalose-6-phosphate synthase (TPS). For example, it helps to regulate embryo maturation, plant growth and development, vegetative growth and effect on flowering for development of seedlings, carbohydrate metabolism, glucose, abscisic acid, and stress signaling. Microarray analysis revealed that it also affects expression level of genes during abiotic stress (Schluepmann et al. 2004; Bae et al. 2005; Almeida et al. 2007).

6.2 Trehalose Synthesis Pathway

In living organism, there are five reported pathways of trehalose biosynthesis, among them some possess more than one pathway; however, others use more than one pathway depending upon the type of stress (Barraza et al. 2013; Paul et al. 2008).

1. TPS–TPP (OtsA–OtsB) pathway
2. TreY–TreZ pathway
3. TreP pathway
4. TreS pathway
5. TreT pathway

6.2.1 TPS–TPP (OtsA–OtsB) Pathway

It is the most common pathway for trehalose biosynthesis and it has two steps, mainly present in both prokaryotes and eukaryotes. In plants, trehalose synthesis is different than both prokaryotes and eukaryotes. In this trehalose is synthesized by catalyzing trehalose 6-phosphate synthase (TPS) which provides the intermediate products, i.e., Trehalose-6-Phosphate from glucose-6-phosphate and uridine diphosphate glucose, and after that, trehalose-6-phosphate phosphatase (TPP) catalyzes the dephosphorylation of trehalose-6-phosphate to trehalose (Iordachescu and Imai 2011; Paul et al. 2008).

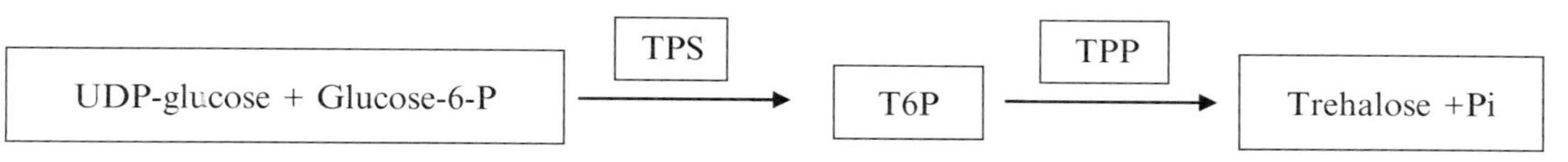

6.2.2 TreY–TreZ Pathway

This pathway was first discovered in *Arthrobacter* sp. (Maruta et al. 1995). Basically maltodextrins are converted into trehalose by two-step pathway. In the beginning, by intramolecular transglycosylation, maltooligosyl-trehalose synthase (TreY) catalyzes the conversion of maltopentaose into maltooligosyl trehalose; after that, maltooligosyltrehalose trehalohydrolase (TreZ) hydrolyses the maltooligosyl trehalose and releasing free trehalose (Maruta et al. 1995). Various researchers showed that TreY pathway is common in many bacterial species like Rhizobium, *Bradyrhizobium japonicum,* and *Corynebacterium* whereas this pathway is absent in main bacterial species like *Bacillus subtilis and Escherichia* (Maruta et al. 1996; Sugawara et al. 2010; Tzvetkov et al. 2003).

6.2.3 TreP Pathway

TreP is the second trehalose synthesis pathway found in both prokaryotes and eukaryotes. It is a potential reversible reaction that converts glucose-1-phosphate and glucose to trehalose, catalyzed by trehalose phosphorylase (TreP). This pathway was first discovered in *Euglena gracilis* (Belocopitow and Marechal 1970) and later it was found in mushrooms and bacteria (Paul et al. 2008).

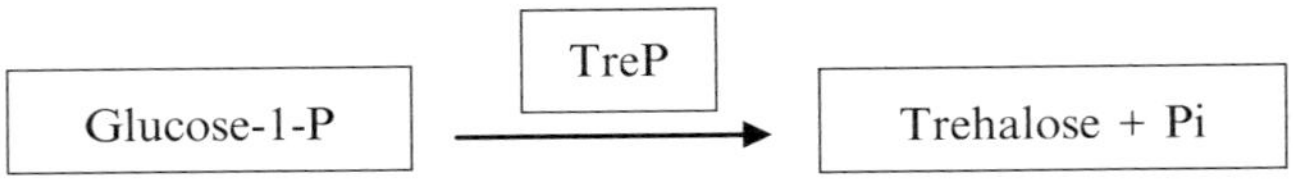

6.2.4 TreS Pathway

In this pathway, trehalose is synthesized by conversion of maltose to trehalose by trehalose synthase. This pathway is called reversible transglycosylation; TreS pathway helps during osmotic stress in *Pseudomonas syringae* (Freeman et al. 2010), while TreS functions in trehalose catabolism in *B. japonicum* (Sugawara et al. 2010). Till now, trehalose synthase has been found in bacteria only (Paul et al. 2008). First of all, trehalose synthase was cloned from *Pimelobacter* sp. R48 by Nishimoto and his coworkers in 1995.

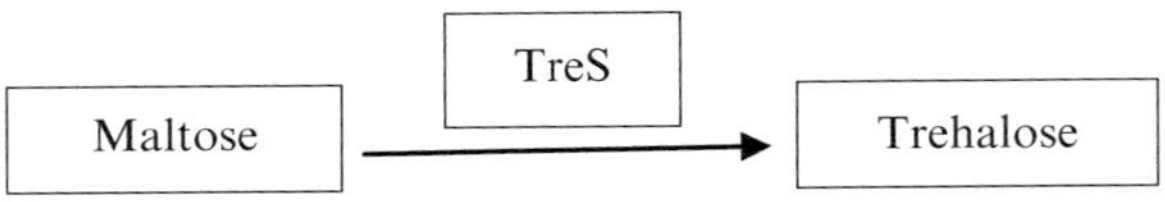

6.2.5 TreT Pathway

In this pathway, trehalose is synthesize by reversible formation of ADP-glucose and glucose by catalyzing trehalose glycosyltransferase and first of all it was reported in hyperthermophilic archaea (*Thermococcus litoralis*) (Qu et al. 2004). Trehalose glycosyltransferase is also found in bacteria and archea (Paul et al. 2008).

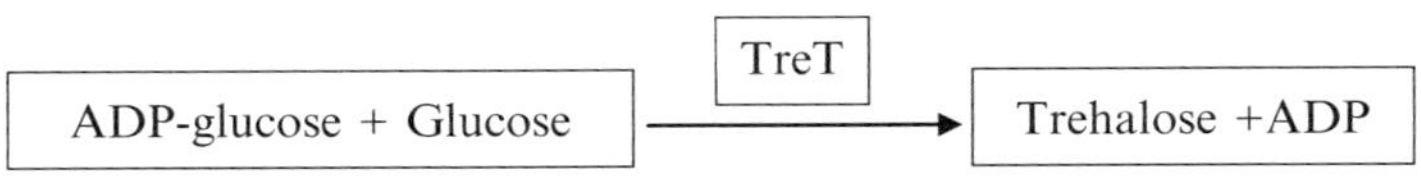

6.3 Discovery and Manifestation of Trehalose in Diverse Organisms and Plants

Wiggers (1832) discovered it first in rye ergot and later in 1858 Mitscherlich isolated trehalose from mushroom and named it as mycose, whereas Berthelot isolated novel sugar called trehalique from trehala-manna. Trehalose is combined with glycolipids and act like main structural parts in mycobacteria. According to Elbein (1974), during flight insects used trehalose from blood in the form of energy source. According to Drennan et al. 1993, trehalose was known to be accumulated in high concentrations for survival under complete dehydration in anhydrobiotic organisms, and during drought conditions, trehalose preserves the membranes (Crowe et al. 1984). Likewise, in yeast it acts as osmotic, heat and desiccation tolerant, whereas in *E. coli*, by stabilizing cell membranes and preventing protein denaturation, it protects against cold stress and it also may act like free radical scavenger (Hottiger et al. 1987; Hounsa et al. 1998; Benaroudj et al. 2001. In the very beginning, trehalose was only extracted from trehala manna that is why it used to be called as rare sugar which was extracted from a resurrection plant. After that, in 2002, Richards and his coworkers developed a technique to isolate trehalose from yeast. In early period of time, trehalose was only extracted from some limited plants like *Selaginella* species and *Myrothamnus flabellifolia* (Bianchi et al. 1993); after that, it was found that it is also present in manna-like exudates on flowers. Veluthambib et al. (1982) reported that trehalose has toxic effect on feeding cell wall and sucrose and starch metabolism.

Till 1996, due to lack of detection tools, it was assumed that trehalose has minimal relevance and it has limited functions and majority of the function has been replaced with sucrose. After 1997, trehalose became popular when Goddijn and his coworkers in 1997, by using *E. coli* genes, engineered the trehalose pathways into plants which produced phenotypes consistent with communication. Shortly after, Blazquez et al. and Vogel et al. (1998) reported functional genes encoding TPS and TPP in *Arabidopsis thaliana*, and in 2001, Leyman and his coworkers published full genomic sequence database for *Arabidopsis thaliana* which established the authenticity of many genes for trehalose synthesis. After that many more research were conducted on trehalose and found very interesting features of trehalose. Schluepmann et al. (2003) reported trehalose and T6P in *A. thaliana,* and in 2002, Eastmond et al. demonstrated the essential trehalose pathway gene (*AtTPS1*) encoding T6P for a plant. After that, trehalose become a method of choice for researcher and lots of work have been conducted to alter the pathway, metabolism, and development of trehalose (Ramon et al. 2007) which further leads to embryo development, structure of inflorescence, leaf structure, process of cell division and mechanism of cell wall synthesis, seedling development, carbon utilization and metabolism, as well as photosynthesis process. Many work has also been conducted on abiotic stresses specially drought. (Eastmond et al. 2002; Pellny et al. 2004; Gomez et al. 2006; Satoh-Nagasawa et al. 2006; Schluepmann et al. 2003; Kolbe et al. 2005; Garg et al. 2002; Almeida et al. 2005, 2007; Karim et al. 2007; Miranda et al. 2007; Pilon-Smits et al. 1998).

6.4 Role of Trehalose during Oxidative Stress

Any nonliving factors that affect negatively living organisms are reunited under the general term "abiotic stress" and its effects can be and are mitigated by a variety of defense mechanisms developed by the different biological systems in existence. Examples of abiotic stress are desiccation, salinity, and high and low temperature. There are two general mechanisms used to counteract abiotic stress: avoidance and adaptation. In the case of avoidance, organisms migrate to deeper soil layers where temperatures are within tolerable range. Adaptation to stress is based on activation of stress defense gene pathways, which results in the production of heat-shock proteins, LEA proteins, redox regulating proteins, different compatible solutes, and cytochrome P450s (Roelofs et al. 2008).

6.4.1 Desiccation Stress

During drought conditions, most of the anhydrobiotic organisms collected high concentrations of trehalose. In drought conditions, many disaccharides accumulated to protect the membranes. In dehydration conditions, membrane must be stabilized because of lipid phase transitions and vehicle fusion; for this scenario, trehalose is one of the best effective methods to stabilize dry membranes (Crowe and Crowe 1990; Crowe et al. 1992). According to Strom and Kaasen (1993), in desiccation conditions, the chance of water loss is 99 %. In this type of situation, trehalose has some beneficial aspects. In the absence of water, trehalose repressed vesicle fusion completely and depressed the phase transition temperature of dry lipids which maintained them in liquid crystalline phase, and in this case, the biological structure which are associated with bound water will be replaced by freezing trehalose molecules (Donnamaria et al. 1994). Kawai and his coworkers (1992) reported the mode of action of trehalose, and according to them, trehalose might stabilize dry biological membranes and proteins due to hydrogen bonding and polar groups of phosphate groups. On the other hand, trehalose protects phosphate groups of membranes and it also helps to protect from desiccation stress by virtue of vitrification.

Trehalose is a nonreducing sugar and hygroscopic in nature; because of this, it is more stable than other disaccharides and it has a low reactivity that is why it forms a protective glass-like structure. Trehalose glass has great stability because trehalose dihydrate on the outer surface of the glass with a little amount of water may give a structure which encloses the inner glass. It is also stable in both high temperature and during complete desiccations. So, it hold biomolecules in natural conditions from an unhealthy environment; for example, it hold biomolecules in a form that allow them to return to their native conditions and functions following rehydrations (Crowe and Crowe 2000; Richards et al. 2002).

6.4.2 Salt Stress

Due to salt stress, generally organisms are affected in two ways. Firstly, organisms lose water and turgor pressure due to low water potential, and secondly in the living cells, this creates a continuous flux of inorganic ions due to high ionic strength of their contiguous environment, basically, osmolytes, compatible solutes which help in living organisms to maintain their turgor pressure and cell volumes. Generally, there are two methods to improve the salt stress by compatible solutes: first, by maintaining the normal turgor pressure of the cells which can be achieved by lowering the osmotic potential of the cytoplasm (Kempf and Bremer 1998) and second by stabilizing proteins and cell components (Hincha and Hagemann 2004). There are many types of compatible solutes like sugars, trehalose, polyols, free amino acids, etc. Among them, trehalose is the best for prokaryotes as well as eukaryotes and it is acting like a stress protectant.

6.4.3 Temperature Stress

Due to low temperature stress, biological membranes are affected while decreasing their fluidity. Both low and high temperature stresses cause protein denaturation and aggregation. Trehalose protects protein from degradation and aggregation, and it also helps to stabilize biological membranes due to hydrogen bond and phosphate group formation.

6.5 Uses of Trehalose in Relation to Abiotic Stresses

Abiotic stresses are major environmental constraint which limits crop production. The development of stress-tolerant crops through bioengineering of stress-signaling pathways is one of the major goals of agricultural research. Osmotic adjustment is an effective component of such manipulations which results in osmoprotectant (compatible solutes) accumulation in plant systems (Blum 1988). Compatible solutes protect plants from stress: (a) detoxifying radical oxygen species and (b) stabilizing the quaternary structures of proteins to maintain their function. Trehalose which acts as a storage carbohydrate possesses the unique feature of reversible water-absorption capacity to protect biological molecules from stress induced by desiccation and appears to be superior to other sugars conferring protection (Rontein et al. 2002). Abiotic stresses such as heat, cold, or water stress were found to be possessing high concentrations of trehalose accumulation in yeast (Goddijn and van Dun 1999). In the plant kingdom, most species do not accumulate detectable levels of trehalose, except for the highly desiccation-tolerant "resurrection" plants. Low or undetectable levels of trehalose in transgenic plants could be attributed to specific trehalase activity. Trehalase activity results in trehalose degradation, while the presence of a trehalase inhibitor (validamycin A) causes severalfold accumulation of trehalose. Hence, it should be possible to increase trehalose accumulation by downregulating plant trehalase activity or by expressing the trehalose biosynthetic genes under tissue- or stress-specific regulation.

Several reports are there which suggest the presence of trehalose in higher plants and a few desiccation-tolerant plants (Bianchi et al. 1993;

Drennan et al. 1993; Albini et al. 1994). The *Arabidopsis* genome was found to contain eleven putative TPS and ten putative TPP genes through whole genome sequencing approach, whereas rice genome has nine TPSs and nine TPPs. Transgenic *Arabidopsis* plants overexpressing AtTPS1 (Avonce et al. 2004) resulted in glucose and ABA-insensitive and drought stress-tolerant phenotypes. The alteration of gene regulation involved in ABA and glucose signaling during seedling vegetative growth may account for the insensitivity, suggesting AtTPS1 and/or trehalose-6-phosphate as a major player in gene regulation and signaling during seedling development. Arabidopsis csp-1 mutant, with a point mutation in the synthase domain of another Arabidopsis TPS, AtTPS6, was also drought tolerant (Chary et al. 2008).

The accumulation of low trehalose levels in transgenic plants engineered for overexpressing the microbial trehalose biosynthesis genes shows tolerance to abiotic stresses, especially to drought stress. Holmstrom et al. (1996) first demonstrated the *Agrobacterium*-mediated transformation of tobacco using TPS gene (*TPS1*) from *S. cerevisiae* under the control of the *ats1A* gene (Rubisco small unit) promoter from *Arabidopsis*. The transgenic leaves were detached and air-dried; they were found to be having higher trehalose content and less water loss than the leaves from wild-type plants and remained fresh even after 24 h. Similarly when plantlets were air-dried, the transgenic lines were able to withstand desiccation better than wild-type plants. But trehalose accumulation was insufficient for osmotic adjustment, and desiccation tolerance could be achieved by stabilization of cellular structures and macromolecules by accumulation of trehalose. This experiment first demonstrated the association of water-deficit tolerance of intact plants and detached leaves with trehalose-6-phosphate synthase gene transformation.

Goddijn et al. (1997) did the genetic transformation of tobacco and potato plants with *otsA* and *otsB* genes from *E. coli* under the control of CaMV35S promoter. Low trehalose accumulation was obtained in transgenic tobacco, whereas in potato no traces of trehalose were detected. But the accumulation of trehalose in transgenic tobacco plants showed negative effects on plant growth such as aberrant roots, lancet-shaped leaves, and stunted growth. A potent trehalase inhibitor was incorporated to culture medium and to hydroponically growing transgenic lines of tobacco and potato to improve trehalose accumulation in both species. High levels of trehalose accumulation were recorded in both species. This was the first report which demonstrated that trehalose accumulation is dependent on trehalase activity in higher plants.

Dai et al. (2001) reported the *Agrobacterium-mediated* transformation of tobacco using *otsA* gene from *E. coli* under the control of the CaMV35S promoter. The transgenic plants so produced were water-deficit tolerant, but they also showed altered phenotypes like stunted growth. Pellny et al. (2004) developed the transgenic tobacco plants with *otsA* and *otsB* genes which showed higher photosynthetic capacities per unit leaf area and per leaf dry weight. This alteration in photosynthetic activity was found to be associated with trehalose-6-phosphate accumulation rather than trehalose.

Drought tolerance was also exhibited by tobacco and tomato plants which were transformed with yeast TPS1 gene under the control of 35S promoter (Romero et al. 1997; Cortina and Culianez-Macia 2005). The transgenic plants produced from these studies lead to growth aberrations like stunted growth in tobacco plants and abnormal root development in tomato plants. These growth abnormalities were most probably due to trehalose-6-phosphate accumulation, which has been proved to be an essential player in plant development (Eastmond et al. 2002) and act as an inhibitor of SnRK1 (a hexokinase involved in transcriptional regulation of metabolism, growth, and development in plants) (Zhang et al. 2009; Paul et al. 2010). The transgenic tobacco plants possessing drought tolerance without any growth abnormalities were obtained by targeting the TPS1 gene expression in chloroplast or by using bifunctional fusion yeast trehalose synthesis genes (Karim et al. 2007).

In tomato, the overexpression of yeast *TPS1* gene resulted in an increased chlorophyll and starch levels in normal conditions which lead to a significant advantage under salt, drought, and oxidative stress (Cortina and Culiariez-Macia 2005). Similarly, when *ScTPS1* was overexpressed in tobacco, an advantage under drought stress conditions was reported instead of an abnormal leaf phenotype (Romero et al. 1997). A slightly improved drought response was reported in potato using a drought-inducible promoter fused to *ScTPS1* which leads to longer water retention potential, maintenance of stomatal conductance, and net photosynthesis (Stiller et al. 2008).

Garcia et al. (1997) reported low level of trehalose accumulation in rice roots following three days of salt stress. External applications of trehalose (up to 5 mM) reduced Na+accumulation and growth inhibition, and higher concentrations (10 mM) prevented chlorophyll loss in leaf blades and preserved root integrity. Similarly, expression level of trehalose was found to be too low to account for an osmoprotectant role in roots and bacteroids of Alfalfa (*Medicago sativa* L.) plants under salt stress (Fougere et al. 1991). The trehalase activity was reported to be downregulated in nodules of *Medicago truncatula* under salt stress which allowed low quantities of trehalose accumulation for efficient contribution to osmoprotection (Lopez et al. 2008). Overexpression of yeast trehalose synthesis genes in tomato leads to the production of plants which were also tolerant to salt and oxidative stresses in addition to drought stress (Cortina and Culianez-Macia 2005). Salt, drought, and low temperature stress-tolerant rice plants without any growth abnormality were developed by overexpressing bacterial fused trehalose synthesis genes under the control of tissue-specific or stress-dependent promoters (Garg et al. 2002).

Trehalose was also transiently induced with chilling stress, and its accumulation was coincided with the phase change of glucose and fructose levels (Pramanik and Imai 2005). In *Arabidopsis*, AtTPS5 has a role in thermotolerance. AtTPS5 interacted with a transcriptional activator, *viz.*, MBF1c, a key regulator of thermotolerance (Suzuki et al. 2008).

The symbiotic relationship between rhizobia and legumes has an important impact on legume yields. In addition, it helps in nitrogen fixation that remained in the soil for future crops. Therefore, application of rhizobia to legume seeds prior to planting in the field is important to encourage the legume rhizobia symbiosis formation. However, survival percentage of rhizobia is very low (>5 %), because of rapid desiccation (Roughley et al. 1993). The increase in trehalose concentration in the bacterial cells with the application of external trehalose (3 mmol l-1) to *Bradyrhizobium japonicum* strain USDA 110, the survival rate of bacteria under desiccation increased by twofold to fourfold (Streeter 2003).

The availability of trehalose encoding genes from plants such as SlTPS1 from *S. lepidophylla* (Zentella et al. 1999) and the AtTPS1 from *A. thaliana* (Blazquez et al. 1998; Leyman et al. 2001) provides the possibility of generating the transgenic plants with better public acceptance as compared to plants with transgene from microorganisms. The AtTPS1 gene has been used for transformation of *Arabidopsis* to increase the accumulation of TPS and to study its physiological role (Schluepmann et al. 2004; Van Dijken et al. 2004; Avonce et al. 2004, 2005). The transgenic *Arabidopsis* plants showed higher trehalose level and obtained desiccation tolerance. This gene has been proposed as a selection marker for plant transformation (Leyman et al. 2004).

Debast et al. (2011) developed the transgenic potato plants with altered T6P levels to investigate the role of T6P signaling in their tubers. Transgenic potato lines with increased T6P levels displayed reduced starch content, decreased ATP contents, and increased respiration rate diagnostic for high metabolic activity. However, potato lines with lower T6P showed the accumulation of soluble carbohydrates, hexose phosphates, and ATP, but there was no change in starch content and a strongly reduced tuber yield was also observed. Further, the carbon partitioning between starch and soluble carbohydrates did not get altered when transgenic tubers were given the glucose feeding. Transcriptional profiling of transgenic lines in B33-TPP tubers revealed that target genes of SnRK1 which were involved in the promotion of cell proliferation and growth were downregulated while those involved in inhibiting cell cycle progression were upregulated. It also revealed that T6P-accumulation in tubers strongly delayed sprouting, while those with reduced T6P sprouted earlier than the wild type. Early sprouting of B33-TPP tubers correlated with a reduced abscisic acid content.

Martins et al. (2013) investigated the role of trehalose-6-phosphate as a sugar-signaling metabolite in *Arabidopsis* leaves which regulates the accumulation and turnover of transitory starch. The Tre6P levels increase up to 11-fold during daytime by ethanol-induced overexpression of trehalose-6-phosphate synthase. A transient increase in the rate of starch accumulation occurs in the middle of the day, but it was not linked to reductive activation of ADP-glucose pyrophosphorylase. A twofold to threefold increase in Tre6P led to significant inhibition of starch degradation during the night. The absence of maltose and maltotriose accumulation indicated that Tre6P affects an early step in the pathway of starch degradation in the chloroplasts. A higher orthophosphate content was found in starch granules isolated from induced plants than granules from noninduced control plants, consistent either with disruption of the phosphorylation-dephosphorylation cycle that is essential for efficient starch breakdown or with inhibition of starch hydrolysis by b-amylase. Nonaqueous fractionation of leaves showed that Tre6P is predominantly located in the cytosol, with estimated in vivo Tre6P concentrations of 4–7 mM in the cytosol, 0.2–0.5 mM in the chloroplasts, and 0.05 mM in the vacuole. It is proposed that Tre6P is a component in a signaling pathway that mediates the feedback regulation of starch breakdown by sucrose, potentially linking starch turnover to demand for sucrose by growing sink organs at night.

Henry et al. (2014) elucidated the impacts of the diurnal cycle and disruption of the day/night cycle on trehalose pathway gene expression and sugar metabolism in maize as characterized by the maize trehalose pathway genes. The maize genome encodes 14 trehalose-6-phosphate synthase (TPS) genes, 11 trehalose-6-phosphate phosphatase (TPP) genes, and one trehalase gene. Transcript abundance of most of these genes was affected by the day/night cycle and extended dark stress, as were sucrose, hexose sugars, starch, and trehalose-6-phosphate (T6P) levels. After extended darkness, T6P levels inversely followed the class II TPS and sucrose non-fermenting-related protein kinase 1 (SnRK1) target gene expression. Most significantly, T6P no longer tracked the sucrose levels after extended darkness.

A summarized view of stress-tolerant transgenic plants with trehalose biosynthetic gene expression has been shown below (Table 6.1).

6.6 Conclusion and Future Perspectives

Trehalose is a very important osmoprotectant found in minute amounts in plants, though it has a major role in plant cell metabolism associated with abiotic stress tolerance by using genetic engineering tools in crop plants where genes that have been used were isolated from yeast and bacterial species. An alternative strategy exists for trehalose accumulation through blocking of trehalase. In this strategy trehalose accumulation becomes higher while using trehalase inhibitors. Till date, field trial of genetically engineered trehalose-accumulating plants has not been performed, but the cloning of trehalase genes from several plants has already been achieved; therefore, in the future, transgenic research evaluating this strategy will be expected. By the help of genetically engineered plants for trehalose accumulation, different and new variety of ornamentals can be developed by changing size and flower. One of the important characters of trehalose accumulation is carbohydrate metabolism. In the context of sugar metabolism, only few things have been identified for, so attention should be focus to get more promise research on this area. Some research showed genetic engineering of overexpression genes, sometime, may lead to downstream pathways; therefore, as a pre-

Table 6.1 Trehalose biosynthetic gene expression in transgenic plants

Species transformed	Promoter	Trehalose biosynthetic genes	Tolerance	Trehalose level	Changes in phenotype	Reference
Tobacco	Rubisco	TPS1	Drought	800–3200 μg/g DW	No changes	Holmstrom et al. (1996)
Tobacco	35S	otsA	Drought	0–110 μg/g FW	Aberrant root growth, lancet-shaped leaves, stunted plants	Goddijn et al. (1997)
–	–	otsA–otsB	–	0–60 μg/g FW	Same as above, but not as pronounced, bleached interveinal tissue	Pilon-Smits et al. (1998)
Potato	Patatin	otsA	NA	3–20 μg/g FW	No changes	Goddijn et al. (1997)
–	–	otsA–otsB	NA	NA	–	–
Tobacco	35S	TPS1	Drought	<170 μg/g FW	Loss of apical dominance, inhibited growth, change in leave shape like lancet structure, partial sterility	Romero et al. (1997)
Tobacco	16SrRNA	TPS1	Drought	360–440 μg/g FW	No changes	Lee et al. (2003)
Tobacco	35S	TP	Drought	6.3 μmol/g FW	No changes	Han et al. (2005)
Tobacco	Rubisco	TPS1	Drought	~8 μg/g FW	No changes	Karim et al. (2007)
–	–	TPS1–TPS2	–	~16 μg/g FW	No changes	–
–	AtRAB18	TPS1	Drought	1–2 μg/g FW	No changes	–
–	–	TPS1–TPS2	–	4 μg/g FW	No changes	–
Arabidopsis	Rubisco	TPS1	Drought	NA	No changes	–
Rice	ABRC1-rice actin1 (ABA-inducible)	otsA–otsB	Drought, salt, cold	~48 μg/g FW	No changes	Garg et al. (2002)
–	Rice rbcS	–	–	~55 μg/g FW	No changes	–
Rice	Ubiquitin	otsA–otsB	Drought, salt, cold	310–1 036 μg/g	No changes	Jang et al. (2003)
Tomato	35S	TPS1	Drought, salt	150 μg/g FW	Short internodes and thick shoots, the color of leaves are rigid dark green, aberrant root development	Cortina and Culianez-Macia (2005)
Arabidopsis	rd29A (stress inducible)	PS1–TPS2	Drought, cold, salt, heat	8.2–16.7 μg/g FW	No changes	Miranda et al. (2007)

(continued)

Table 6.1 (continued)

Species transformed	Promoter	Trehalose biosynthetic genes	Tolerance	Trehalose level	Changes in phenotype	Reference
–	35S	–	–	8.5–38.4 μg/g FW	The color of leaves are green and smaller in size, partial sterility, glucose insensitive	–
Rice	35S	OsTPS1	Cold, high salinity, and drought	–	No changes	Li et al. (2011)
Potato	35S	TPS1	Drought	–	Dwarfism	Yeo et al. (2000)
Tobacco	35S	otsA	–	–	Altered phenotype (stunted growth); while detaching leaf transgenic plants showed less water	Dai et al. (2001)
Tobacco	35S	otsA–otsB	–	–	Altered photosynthesis	Pellny et al. (2004)
Tobacco	35S rd29A	TRE (trehalase gene)	–	–	In transgenic plants, trehalose activity is lowered	Gomez-Escobedo et al. (2004)
Tobacco	35S	TPS1	–	–	Tolerance to osmotic stress; wild-type plants are bigger than transgenic; lancet-shaped leaves are not present	Almeida et al. (2005)
Tobacco	35S	TPS1	–	–	Plants grow in media supplemented with glucose (glucose-insensitive phenotypes)	Leyman et al. (2006)

Modified from Almeida et al. (2007)

requisite, specific promoter is required to switch off and switch on. So, produced transgenic plants can be beneficial for identifying unknown pathways. It can be used as primary research material in near future to identify those unknown pathways.

References

Albini FM, Murelli C, Patritti G, Rovati M, Zienna P, Vita Finzi P (1994) Low-molecular weight substances from the resurrection plant *Sporobolus stapfianus*. Phytochemistry 37:137–142

Almeida AM, Villalobos E, Araújo SS, Leyman B, van Dijck P, Alfaro-Cardoso L, Fevereiro PS, Torné JM, Santos DM (2005) Transformation of tobacco with an *Arabidopsis thaliana* gene involved in trehalose biosynthesis increases tolerance to several abiotic stresses. Euphytica 146:165–176

Almeida AM, Santos M, Villalobos E, Araujo SS, van Dijck P, Leyman B, Cardoso LA, Santos D, Fevereiro PS, Torne JM (2007) Immunogold localization of trehalose-6- phosphate synthase in leaf segments of wild-type and transgenic tobacco plants expressing the AtTPS1 gene from *Arabidopsis thaliana*. Protoplasma 230:41–49

Avonce N, Leyman B, Mascorro-Gallardo JO, Van Dijck P, Thevelein JM, Iturriaga G (2004) The Arabidopsis trehalose-6-P synthase AtTPS1 gene is a regulator of glucose, abscisic acid, and stress signaling. Plant Physiol 136:3649–3659

Avonce N, Leyman B, Thevelein J, Iturriaga G (2005) Trehalose metabolism and glucose sensing in plants. Biochem Soc Trans 33:276–279

Bae HH, Herman E, Bailey B, Bae HJ, Sicher R (2005) Exogenous trehalose alters *Arabidopsis* transcripts involved in cell wall modification, abiotic stress, nitrogen metabolism, and plant defense. Physiol Plant 125:114–126

Barraza A, Estrada-Navarrete G, Rodriguez-Alegria ME, Lopez-Munguia A, Merino E, Quinto C, Sanchez F (2013) Down-regulation of PvTRE1 enhances nodule biomass and bacteriod number in the common bean. New Phytol 197:194–206

Belocopitow E, Marechal LR (1970) Trehalose phosphorylase from *Euglena gracilis*. Biochim Biophys Acta 198:151–154

Benaroudj N, Lee DH, Goldberg AL (2001) Trehalose accumulation during cellular stress protects cells and cellular proteins from damage by oxygen radicals. J Biol Chem 276:24261–24267

Bianchi G, Gamba A, Limiroli R, Pozzi N, Elster R, Salamini F, Bartels D (1993) The unusual sugar composition in leaves of the resurrection plant *Myrothamnus flabellifolia*. Physiol Plant 87(2):223–226

Blazquez MA, Santos E, Flores CL, Martinez-Zapater JM, Salinas J, Gancedo C (1998) Isolation and molecular characterization of the Arabidopsis TPS1 gene, encoding trehalose-6-phosphate synthase. Plant J 13:685–690

Blum A (1988) Plant breeding for stress environments. CRC Press, Boca Raton

Chary SN, Hicks GR, Choi YG, Carter D, Raikhel NV (2008) Trehalose-6-phosphate synthase/phosphatase regulates cell shape and plant architecture in *Arabidopsis*. Plant Physiol 146:97–107

Cortina C, Culiáñez-Macià FA (2005) Tomato abiotic stress enhanced tolerance by trehalose biosynthesis. Plant Sci 169(1):75–82

Crowe JH, Crowe LM (1990) Lyotropic effects of water on phospholipids. In: Franks F (ed) Water science reviews. Cambridge University Press, Cambridge, UK, pp 1–23

Crowe JH, Crowe LM (2000) Preservation of mammalian cells – learning nature's tricks. Nat Biotechnol 18:145–146

Crowe JH, Crowe LM, Chapman D (1984) Preservation of membranes in anhydrobiotic organisms: the role of trehalose. Science 223:701–703

Crowe JH, Hoekstra FA, Crowe LM (1992) Anhydrobiosis. Annu Rev Physiol 54:579–599

Cushman JC, Bohnert HJ (2000) Genomic approaches to plant stress tolerance. Curr Opin Plant Biol 3:117–124

Dai X, Wang Y, Zhou J (2001) Expression of otsA gene in tobacco and improvement stress tolerance. Wei Sheng Wu Xue Bao 41:427–431

Debast S, Nunes-Nesi A, Hajirezaei MR, Hofmann J, Sonnewald U, Fernie AR, Börnke F (2011) Altering trehalose-6-phosphate content in transgenic potato tubers affects tuber growth and alters responsiveness to hormones during sprouting. Plant Physiol 156:1754–1771

Donnamaria MC, Howard EI, Grigera JR (1994) Interaction of water with a, a-trehalose in solution: molecular dynamics simulation approach. J Chem Soc Faraday Trans 90:2731–2735

Drennan PM, Smith MT, Goldsworthy D, van Staden J (1993) The occurrence of trehalose in the leaves of the desiccation-tolerant angiosperm *Myrothamnus flabellifolius* Welw. J Plant Physiol 142:493–496

Eastmond PJ, Graham IA (2003) Trehalose metabolism: a regulatory role for trehalose-6-phosphate? Curr Opin Plant Biol 6:231–235

Eastmond PJ, van Dijken AJ, Spielman M, Kerr A, Tissier AF, Dickinson HG, Jones JD, Smeekens SC, Graham IA (2002) Trehalose-6-phosphate synthase 1, which catalyses the first step in trehalose synthesis, is essential for *Arabidopsis* embryo maturation. Plant J 29:225–235

Elbein AD (1974) The metabolism of a, a- trehalose. Adv Carbohydr Chem Biochem 30:227–256

Fougere F, Le Rudulier D, Streeter JG (1991) Effects of salt stress on amino acid, organic acid, and carbohydrate composition of roots, bacteroids, and cytosol of alfalfa (*Medicago sativa* L.). Plant Physiol 96:1228–1236

Freeman BC, Chen C, Beattie GA (2010) Identification of the trehalose biosynthetic loci of *Pseudomonas syringae* and their contribution to fitness in the phyllosphere. Environ Microbiol 12:1486–1497

Garcia AB, Engler JA, Iyer S, Gerats T, Van Montagu M, Caplan AB (1997) Effects of osmoprotectants upon NaCl stress in rice. Plant Physiol 115:50–69

Garg AK, Kim JK, Ranwala AP, Choi YD, Kochian LV, Wu RJ (2002) Trehalose accumulation in rice plants confers high tolerance levels to different abiotic stresses. Proc Natl Acad Sci U S A 99:15898–15903

Goddijn OJM, van Dun K (1999) Trehalose metabolism in plants. Trends Plant Sci 4:315–319

Goddijn OJ, Verwoerd TC, Voogd E, Krutwagen RW, de Graaf PT, van Kun K, Poels J, Ponstein AS, Damm B,

Pen J (1997) Inhibition of trehalase activity enhances trehalose accumulation in transgenic plants. Plant Physiol 113:181–190

Gomez LD, Baud S, Gilday A, Li Y, Graham IA (2006) Delayed embryo development in the ARABIDOPSIS TREHALOSE-6-PHOSPHATE SYNTHASE 1 mutant is associated with altered cell wall structure, decreased cell division and starch accumulation. Plant J 46:69–84

Gomez-Escobedo IA, Cabrera-Ponce JL, Herrera-Estrella LR, Hernández-Luna C, Montes de Oca-Luna R (2004) Mejora del crecimiento de plantas de tabaco mediante la inhibicion del gen de la trehalasa. Ciencia UANL VII:483–489

Han SE, Park SR, Kwon HB, Yi BY, Lee GB, Byun MO (2005) Genetic engineering of drought-resistant tobacco plants by introducing the trehalose phosphorylase (TP) gene from *Pleurotus sajor-caju*. Plant Cell Tiss Org Cult 82:151–158

Henry C, Bledsoe SW, Siekman A, Kollman A, Waters BM, Feil R, Stitt M, Lagrimini LM (2014) The trehalose pathway in maize: conservation and gene regulation in response to the diurnal cycle and extended darkness. J Exp Bot 65:5959–5973

Hincha DK, Hagemann M (2004) Stabilization of model membranes during drying by compatible solutes involved in the stress tolerance of plants and microorganisms. Biochem J 383:277–283

Holmstrom KO, Mantyla E, Welin B, Mandal A, Palva ET, Tunnela OE, Londesborough J (1996) Drought tolerance in tobacco. Nature 379:683–684

Hottiger T, Boller T, Wiemken A (1987) Rapid changes of heat and desiccation tolerance correlated with changes of trehalose content in *Saccharomyces cerevisiae* cells subjected to temperature shifts. FEBS LETT 220:113–115

Hounsa CG, Brandt EV, Thevelein J, Hohmann S, Prior BA (1998) Role of trehalose in survival of Saccharomyces cerevisiae under osmotic stress. Microbiology 144:671–680

Iordachescu M, Imai R (2011) Trehalose and abiotic stress in biological systems, abiotic stress and plant mechanisms and adaptations. In: Shanker A (ed) In Tech. doi:10.5772/22208

Jang IC, Oh SJ, Seo JS, Choi WB, Song SI, Kim CH, Kim YS, Seo HS, Choi YD, Nahm BH, Kim JK (2003) Expression of a bifunctional fusion of the *Escherichia coli* genes for trehalose-6-phosphate synthase and trehalose-6-phosphate phosphatase in transgenic rice plants increases trehalose accumulation and abiotic stress tolerance without stunting growth. Plant Physiol 131:516–524

Karim S, Aronsson H, Ericson H, Pirhonem M, Leyman B, Welin B, Mantyla E, Palva ET, Van Dijck P, Holmstrom KO (2007) Improved drought tolerance without undesired side effects in transgenic plants producing trehalose. Plant Mol Biol 64:371–386

Kawai H, Sakurai M, Inoue Y, Chujo R, Kobayashi S (1992) Hydration of oligosaccharides: anomalous hydration ability of trehalose. Cryobiology 29:599–606

Kempf B, Bremer E (1998) Uptake and synthesis of compatible solutes as microbial stress responses to high-osmolarity environments. Arch Microbiol 170:319–330

Kolbe A, Tiessen A, Schluepmann H, Paul M, Ulrich S, Geigenberger P (2005) Trehalose 6-phosphate regulates starch synthesis via posttranslational redox activation of ADP-glucose pyrophosphorylase. Proc Natl Acad Sci U S A 102:11118–11123

Lee SB, Kwon HB, Kwon SJ, Park SC, Jeong MJ, Han SE, Byun MO, Daniell H (2003) Accumulation of trehalose within transgenic chloroplasts confers drought tolerance. Mol Breed 11:1–13

Leyman B, Van Dijck P, Thevelein JM (2001) An unexpected plethora of trehalose biosynthesis genes in *Arabidopsis thaliana*. Trends Plant Sci 6:510–513

Leyman B, Avonce N, Ramon M, van Dijck P, Thevelein JM, Iturriaga G (2004) New selection marker for plant transformation. Methods Mol Biol 267:385–396

Leyman B, Avonce N, Ramon M, Van Dijck P, Iturriaga G, Thevelein JM (2006) Trehalose-6-phosphate synthase as an intrinsic selection marker for plant transformation. J Biotechnol 301:309–317

Li H, Su H, Kim SB, Chang YK, Hong S, Seo YG, Kim CJ (2011) Enhanced production of trehalose in *Escherichia coli* by homologous expression of otsBA in the presence of the trehalase inhibitor, validamycin A, at high osmolarity. J Biosci Bioeng 113:224–232

Lopez M, Tejera NA, Iribarne C, Lluch C, Herrera-Cervera JA (2008) Trehalose and trehalase in root nodules of Medicago truncatula and *Phaseolus vulgaris* in response to salt stress. Physiol Plant 134:575–582

Martins MC, Hejazi M, Fettke J, Steup M, Feil R, Krause U et al (2013) Feedback inhibition of starch degradation in *Arabidopsis* leaves mediated by trehalose-6-phosphate. Plant Physiol 163:1142–1163

Maruta K, Nakada T, Kubota M, Chaen H, Sugimoto T, Kurimoto M, Tsujisaka Y (1995) Formation of trehalose from maltooligosaccharides by a novel enzymatic system. Biosci Biotechnol Biochem 59:1829–1834

Maruta K, Hattori K, Nakada T, Kubota M, Sugimoto T, Kurimoto M (1996) Cloning and sequencing of trehalose biosynthesis genes from Rhizobium sp. M-11. Biosci Biotechnol Biochem 60:717–720

Miranda JA, Avonce N, Suarez R, Thevelein JM, Dijck P, Iturriaga G (2007) A bifunctional TPS-TPP enzyme from yeast confers tolerance to multiple and extreme abiotic-stress conditions in transgenic *Arabidopsis*. Planta 226:1411–1421

Paul MJ, Primavesi LF, Jhurreea D, Zhang YH (2008) Trehalose metabolism and signaling. Annu Rev Plant Biol 59:417–441

Paul MJ, Jhurreea D, Zhang Y, Primavesi LF, Delatte T, Schluepmann H, Wingler A (2010) Upregulation of biosynthetic processes associated with growth by trehalose-6-phosphate. Plant Signal Behav 5:368–392

Pellny TK, Ghannoum O, Conroy JP, Schluepmann H, Smeekens S, Andralojc J, Krause KP, Goddijn O, Paul MJ (2004) Genetic modification of photosynthesis with *E. coli* genes for trehalose synthesis. Plant Biotechnol J 2:71–82

Penna S (2003) Building stress tolerance through overproducing trehalose in transgenic plants. Trends Plant Sci 8:355–357

Pilon-Smits EAH, Terry N, Sears T, Kim H, Zayed A, Hwang S, van Dun K, Voogd E, Verwoerd TC, Krutwagen RWHH, Goddijn OJM (1998) Trehalose producing transgenic tobacco plants show improved growth performance under drought stress. J Plant Physiol 152:525–532

Pramanik MHR, Imai R (2005) Functional identification of a trehalose-6- phosphatase gene that is involved in transient induction of trehalose biosynthesis during chilling stress in rice. Plant Mol Biol 58:751–762

Qu Q, Lee SJ, Boos W (2004) TreT, a novel trehalose glycosyltransferring synthase of the hyperthermophilic archaeon *Thermococcus litoralis*. J Biol Chem 279:47890–47897

Ramon M, Rolland F, Thevelein JM, Van Dijck P, Leyman B (2007) ABI4 mediates the effects of exogenous trehalose on Arabidopsis growth and starch breakdown. Plant Mol Biol 63:195–206

Richards AB, Krakiwka S, Dexter LB, Schid H, Wolterbeek APM, Waalkens-Berendsen DH, Shigoyuki A, Murimoto M (2002) Trehalose: a review of properties, history of use and human tolerance, and results of multiple studies. Food Chem Toxicol 40:871–898

Roelofs D, Aarts MGM, Schat H, van Straalen M (2008) Functional ecological genomics to demonstrate general and specific responses to abiotic stress. Funct Ecol 22:8–18

Romero C, Belles JM, Vaya JL, Serrano R, Culianez-Macia FA (1997) Expression of the yeast trehalose-6-phosphate synthase gene in transgenic tobacco plants: pleiotropic phenotypes include drought tolerance. Planta 201:293–297

Rontein D, Basset G, Hanson AD (2002) Metabolic engineering of osmoprotectant accumulation in plants. Metabol Eng 4:49–56

Roughley RJ, Gemell LG, Thompson JA, Brockwell J (1993) The number of *Bradyrhizobium sp.* (Lupinus) applied to seed and its effect on rhizosphere colonization, nodulation and yield of lupin. Soil Biol Biochem 25:1453–1458

Satoh-Nagasawa N, Nagasawa N, Malcomber S, Sakai H, Jackson D (2006) A trehalose metabolic enzyme controls inflorescence architecture in maize. Nature 441:227–230

Schluepmann H, Pellny T, van Dijken A, Smeekens S, Paul M (2003) Trehalose 6-phosphate is indispensable for carbohydrate utilization and growth in *Arabidopsis thaliana*. Proc Natl Acad Sci U S A 100:6849–6854

Schluepmann H, Pellny T, van Dijken A, Smeekens S, Paul M (2004) Trehalose 6-phosphate is indispensable for carbohydrate utilization and growth in *Arabidopsis thaliana*. Proc Natl Acad Sci USA 100:6849–6854

Scott P (2000) Resurrection plants and the secret of eternal leaf. Ann Bot 85:159–166

Stiller I, Dulai S, Kondrak M, Tarnai R, Szabo L, Toldi O (2008) Effect of drought on water content and photosynthetic parameters in potato plants expressing the trehalose-6-phosphate synthase gene of *Saccharomyces cerevisiae*. Planta 227:299–308

Streeter JG (2003) Effect of trehalose on survival of *Bradyrhizobium japonicum* during desiccation. J Appl Microbiol 95:484–491

Strom AR, Kaasen I (1993) Trehalose metabolism in *Escherichia coli*: stress protection and stress regulation of gene expression. Mol Microbiol 8:205–210

Sugawara M, Cytryn EJ, Sadowsky MJ (2010) Functional role of *Bradyrhizobium japonicum* trehalose biosynthesis and metabolism genes during physiological stress and nodulation. Appl Environ Microbiol 76:1071–1081

Suzuki N, Bajad S, Shuman J, Shulaev V, Mittler R (2008) The transcriptional coactivator MBF1c is a key regulator of thermotolerance in *Arabidopsis thaliana*. J Biol Chem 283:9269–9275

Tzvetkov M, Klopprogge C, Zelder O, Liebl W (2003) Genetic dissection of trehalose biosynthesis in *Corynebacterium glutamicum*: inactivation of trehalose production leads to impaired growth and an altered cell wall lipid composition. Microbiology 149:1659–1673

Van Dijken A, Schluepmann H, Smeekens SCM (2004) *Arabidopsis* trehalose-6-phosphate synthase 1 is essential for normal vegetative growth and transition to flowering. Plant Physiol 135:969–977

Veluthambi K, Mahadevan S, Maheshwari R (1982) Trehalose toxicity in *Cuscuta reflexa*: sucrose content decreases in shoot tips upon trehalose feeding. Plant Physiol 69:1247–1251

Vogel G, Aeschbacher RA, MuÈller J, Boller T, Wiemken A (1998) Trehalose-6-phosphate phosphatases from *Arabidopsis thaliana*: identification by functional complementation of the yeast tps2 mutant. Plant J 13:673–683

Wiggers HA (1832) Annalen der Chemie u. Pharmazie 1:173

Wingler A (2002) The function of trehalose biosynthesis in plants. Phytochemistry 60:437–440

Yeo ET, Kwon HB, Han SE, Lee JT, Ryu JC, Byu MO (2000) Genetic engineering of drought resistant potato plants by introduction of the trehalose-6-phosphate synthase (TPS1) gene form *saccharomyces cerevisiae*. Mol Cells 10:263–268

Zentella R, Moscorro-Gallardo JO, Van Dijck P, Folch-Mallol J, Bonini B, Van Vaeck C, Gaxiola R, Covarrubias A, Nieto-Sotelo J, Thevelein J, Iturriaga GA (1999) A *Selaginella lepidophylla* trehalose-6-phosphate synthase complements growth and stress tolerance defects in a yeast tps1 mutant. Plant Physiol 119:1473–1482

Zhang Y, Primavesi LF, Jhurreea D, Andralojc PJ, Mitchell RAC, Powers SJ, Schluepmann HJ, Delatte T, Wingler A, Paul MJ (2009) Inhibition of SNF1-related protein kinase1 activity and regulation of metabolic pathways by trehalose-6-phosphate. Plant Physiol 149:1860–1871

Polyamines: Osmoprotectants in Plant Abiotic Stress Adaptation

7

Atreyee Sengupta, Mayukh Chakraborty, Jayita Saha, Bhaskar Gupta, and Kamala Gupta

Abstract

Environmental stress is one of the major threats affecting the living world. The most crucial function of plant is to induce different regulatory self-defence pathways in response to stress. Abiotic stresses such as drought, high temperature and salinity cause rapid depletion of cellular water leading to loss of crops and agricultural productivity worldwide. On exposure to these prevalent stresses, plant accumulates several polyhydroxylic compounds and zwitterionic alkylamines commonly known as osmolytes or compatible solutes. These are low molecular weight water soluble compounds providing stress tolerance when accumulated without hindering cellular mechanisms. Genes involved in the biosynthetic pathways of different osmolytes have been identified from various sources. Genetic engineering utilising these endogenous genes has resulted in development of transgenic varieties with better adaptability towards stress. Polyamines are ubiquitous biogenic amines that have been implicated in diverse cellular functions. The protective role of plant polyamines as osmolytes is still controversial and needs further investigation. However, there are some reports that suggest functional similarities between polyamines and other osmolytes commonly found in plant defence mechanism. Functions include protection of macromolecules, cellular pH maintenance, ROS scavenging, stabilisation of native

A. Sengupta • M. Chakraborty • J. Saha
Department of Biological Sciences,
Presidency University, 86/1 College Street,
Kolkata 700073, India

B. Gupta (✉)
Department of Biological Sciences, Presidency University, 86/1 College Street, Kolkata 700073, India

Department of Zoology, Government General Degree College, Singur, Hooghly District, West Bengal 712409, India
e-mail: bhaskar.dbs@presiuniv.ac.in; bhaskarzoology@gmail.com

K. Gupta (✉)
Department of Biological Sciences,
Presidency University, 86/1 College Street,
Kolkata 700073, India

Department of Botany, Government General Degree College, Singur, Hooghly District, West Bengal 712409, India
e-mail: kamala.dbs@presiuniv.ac.in; kamalagupta@gmail.com

N. Iqbal et al. (eds.), *Osmolytes and Plants Acclimation to Changing Environment: Emerging Omics Technologies*, DOI 10.1007/978-81-322-2616-1_7

protein structure, etc. Apart from these direct functions, exogenous application of polyamines results in the elevation of endogenous level of different osmolytes. Thus, polyamines play an indirect role in plant abiotic stress tolerance by participating in osmolyte synthesis in response to stress. In this review, we have dissected the role of polyamines as osmoprotectants.

7.1 Introduction

Distribution and productivity of crop plants are severely restricted by environmental stresses, and these have become an issue of growing concern worldwide. During the period of stress for continued growth and survival, plants have evolved several protective measures for acclimatisation. These mechanisms comprise a wide range of responses at the molecular, cellular and whole plant level (Gupta and Huang 2014). A plant regularly faces adverse growth conditions such as drought, salinity, chilling, freezing and high temperature which results into delayed growth and development and reduced productivity and in some cases might also lead to plant death. In order to ensure their survivability, plants have implemented several strategies which enable them to grow and prosper even during the most disastrous climatic phases. One such mechanism which helps plant to sustain them during the period of stress is by accumulating compatible solutes or osmolytes. Osmolytes are low molecular weight, highly soluble compounds that are non-toxic even in high concentrations, and their accumulation enables individuals to survive under severe extremities. Osmolytes belong to different classes of chemically diverse molecules. These comprise sugars, polyols, methylamines, amino acids and their derivatives. They are electrically neutral molecules, do not inhibit enzyme activity and increase the cell's ability to preserve osmotic balance during stress condition, thus maintaining cellular structure and function. Osmolytes are also potent protein stabilisers, protecting the membrane from deleterious effect of stress. They are widely distributed throughout varied taxonomic groups ranging from algae to higher plants (Geeta 2014).

Apart from these neutrally charged molecules that are well-known osmoprotectants and are widely distributed in different plants belonging to different taxa, there is another group of nitrogenous compounds, which have been found to accumulate in response to stress in a variety of plants and are designated as polyamines. Three polyamines, namely, putrescine (diamine), spermidine (triamine) and spermine (tetramine) are found to be involved in various physiological functions in all organisms from bacteria to plants. They are characterised as low molecular weight polycationic aliphatic nitrogenous compounds. Leeuwenhoek (1678), a Dutch microbiologist, first reported the presence of polyamine in human spermatozoa more than three centuries ago. The role of polyamines as compatible solute is well supported by their peculiar functions in regulating cellular processes such as cell division and elongation, root and shoot growth, flower and fruit development, replication, transcription, translation, membrane and cell wall stabilisation, chromatin organisation, ribosome biogenesis, adaptation against biotic and abiotic stresses, and programmed cell death (Evans and Malmberg 1989; Galston et al. 1997; Igarashi and Kashiwagi 2000; Thomas and Thomas 2001; Bais and Ravishankar 2002; Paschalidis and Roubelakis-Angelakis 2005a, b; Liu et al. 2006; Kusano et al. 2007a, b; Zhao and Yang 2008). Intricate role of polyamines in abiotic stress alleviation has also been well documented (Bouchereau et al. 1999; Groppa et al. 2003; Kasukabe et al. 2004; Imai et al. 2004; Nayyar et al. 2005; Liu et al. 2007; Yamaguchi et al. 2007; Yang et al. 2007; Rider et al. 2007; Wang et al. 2007; Groppa and Benavides 2008; Duan et al. 2008; Cuevas et al. 2008; Kuznetsov and Shevyakova 2010; Shevyakova et al. 2011; Gupta et al. 2013; Dey et al. 2013). However, the specific functions played by polyamines in stress alleviation remain un-elucidated. The focus of this chapter is to discuss the role of polyamines as osmoprotectants and as modulators of other standard osmolytes synthesised within the cell in response to abiotic stress.

7.2 Plant Response to Abiotic Stresses

Abiotic stresses such as salinity, high temperature, chemical toxicity, drought and oxidative stress cause serious damage to the biotic health throughout the world. They are responsible for the loss of 50 % crop worldwide (Boyer 1982; Bray et al. 2000). Abiotic stress induces a series of biochemical, physiological and morphological alteration that in turn affects the growth and productivity of the plant. The effects of these abiotic stresses are often interconnected and produce somewhat similar results. As a consequence, these diverse stresses activate similar signalling cascades and cellular response such as production of stress-responsive proteins, upregulation of antioxidant enzyme system, accumulation of osmoprotectants and many more (Shinozaki and Yamaguchi-Shinozaki 2000; Cushman and Bohnert 2000; Zhu 2001) which eventually enables the plant to survive the stressful condition.

7.2.1 Physiological and Biochemical Mechanism

The complexities of plant response when exposed to diverse abiotic stresses involve modulations of expression of several genes, alteration of different biochemical pathways, epigenetic regulations and post-translational modifications of various stress regulatory enzymes and proteins. Plants primarily exhibit four major lines of defence: (i) production of enzymatic and non-enzymatic antioxidants, (ii) regulation of transcription factors, (iii) membrane transport, and (iv) accumulation and synthesis of osmolytes. Primary stresses such as drought, cold, heat, salinity and heavy metal pollution often produce a similar after-effect leading to secondary stress conditions, the oxidative and osmotic stress. Main feature of oxidative stress is generation of reactive oxygen species (ROS) by reduction of molecular oxygen. ROS is normally generated during different cellular activities; however, their level rises when the plant is exposed to stressful condition. In order to protect themselves from the deleterious effect of ROS, plants adopt several defence mechanisms to combat and scavenge ROS (Greenway and Munns 1980; Asada 1994; Mittler 2002). These include production of various antioxidant enzyme systems such as superoxide dismutase (SOD), catalase (CAT), guaiacol peroxidase (GPX), ascorbate peroxidase (APX), monodehydroascorbate reductase (MDHAR), dehydroascorbate reductase (DHAR) and glutathione reductase (GR) that function in scavenging ROS. Apart from these enzymes, there are several other non-enzymatic compounds that function similarly and serve as a potent antioxidant, thus protecting the plant from oxidative damage. These include ascorbate (AsA), glutathione (GSH), carotenoids, tocopherol and phenolics (Kamal-Eldin and Appelqvist 1996; Ahmad et al. 2010; Srivastava and Dubey 2011; Rawia Eid et al. 2011) (Fig. 7.1). The second line of defence mechanism adopted by plant involves regulation of different transcription factors as they are the major modulators that control gene expression during stress. Among different transcription factors, bZIP, WRKY, AP2, NAC and C2H2 zinc finger families are the major stress-responsive members (Sun et al. 2010; Mizoi et al. 2011; Song et al. 2011; Hu et al. 2013). Several transcription factors have been identified in *Arabidopsis*, rice and wheat that have been demonstrated to play significant role in abiotic stress responses and are capable of controlling the expression of target gene by binding to the cis-acting element in the promoter of that specific gene. The third line of defence is provided by different membrane-bound carrier proteins, channel proteins, antiporters and symporters that efficiently keep the ion concentration within cytosol low during the period of stress, thereby regulating ion uptake and transport within the cell (Sairam and Tyagi 2004). The fourth and the final line of defence, which we shall also be discussing about more elaborately, is synthesis and accumulation of compatible solutes or osmolytes in response to diverse abiotic stresses. These are small group of organic compounds whose concentration is maintained within the cell either by irreversible synthesis of the compound or by combination of

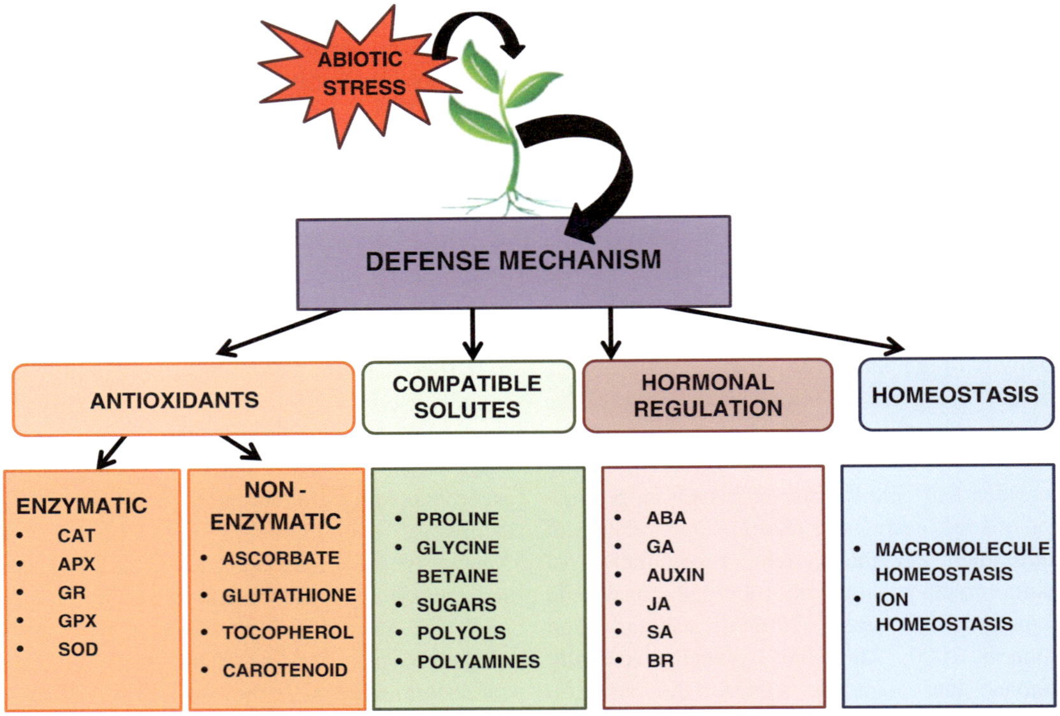

Fig. 7.1 Plant defence mechanism induced in response to abiotic stress [catalase (CAT), ascorbate peroxidase (APX), glutathione reductase (GR), guaiacol peroxidise (GPX), superoxide dismutase (SOD), abscisic acid (ABA), gibberellic acid (GA), jasmonic acid (JA), salicylic acid (SA), brassinosteroids (BR)]

synthesis and degradation. They mainly include proline (Hoque et al. 2007; Nounjan et al. 2012), glycine betaine (GB) (Ashraf and Foolad 2007), sugar (Kerepesi and Galiba 2000), polyols (Ashraf and Foolad 2007; Ford 1984) and polyamines (Kuznetsov and Shevyakova 2007; Gill and Tuteja 2010; Hussain et al. 2011; Shu et al. 2012; Alet et al. 2012; Gupta et al. 2014; Saha et al. 2015).

7.3 Osmolytes: Biochemical and Molecular Basis of Their Accumulation

Theconceptof compatible solute was first introduced by Brown (1976). They are characterised as low molecular weight water soluble compounds, mostly uncharged and polar in nature, which do not interfere with the cellular functioning even at high concentration. Accumulation of osmolytes favours acclimation of organisms to adverse environmental conditions. Genes involved in the biosynthesis of compatible solutes and their biochemical pathways have been thoroughly studied and identified from varied organisms (Klähn and Hagemann 2011). Some of them have been cited in Table 7.1. Precise role of compatible solutes is still un-elucidated. However, as their accumulation is proportional to the external osmolarity, the two major functions of these osmolytes are to protect the structure and to maintain osmotic balance within the cell via continuous water influx (Hasegawa et al. 2000).

7.3.1 Proline

Proline is an imino acid characterised by an exceptional conformational rigidity and is an essential compound functioning efficiently in primary metabolism. It protects the cell from damage by acting either as an osmotic agent or radical scavenger. In plants glutamate is the main source

Table 7.1 Expression analysis of different genes involved in osmolyte metabolic pathway during abiotic stress in different species

Gene	Functions	Species	Abiotic stress	Expression level	References
P5CS	1. Codes for enzyme pyrolline-5-carboxylate synthetase 2. Rate-limiting enzyme; catalyses the first step of proline biosynthetic pathway	*Arabidopsis*, *Medicago* and tobacco	Drought, salinity and low temperature	Upregulated	Funck et al. (2008), Armengaud et al. (2004), Ribarits et al. (2007)
PDH	1. Codes for enzyme proline dehydrogenase 2. Catalyses the first step involved in proline catabolism converting proline into pyrroline-5-carboxylate	*Arabidopsis*	Drought, salinity and low temperature	Downregulated	Funck et al. (2008)
TPS	1. Codes for enzyme trehalose-6-phosphate synthase 2. Catalyses the first step in trehalose biosynthetic pathway	Rice and *Arabidopsis*	Salinity	Upregulated	Leyman et al. (2001, 2006), Avonce et al. (2006)
TPP	1. Codes for trehalose-6-phosphate phosphatase 2. Enzyme responsible for conversion of trehalose-6-phosphate to trehalose	Rice and *Arabidopsis*	Salinity	Upregulated	Leyman et al. (2006), Avonce et al. (2006)
TRE1	1. Codes for enzyme trehalase which catalyses conversion of trehalose back to glucose	*Arabidopsis*	Drought	Upregulated	Van-Houtte et al. (2013)
GolS	1. Codes for galactinol synthase 2. Catalyses the first step in RFO biosynthesis	*Arabidopsis*	High temperature, salinity and drought	Upregulated	Panikulangara et al. (2004)

of proline biosynthesis (Fig. 7.2). Plant species mostly have two genes that code for pyrroline-5-carboxylate synthetase (P5CS) and one gene that codes for P5C reductase (P5CR). The enzymes involved in glutamate-proline biosynthetic pathways are localised in the cytoplasm and the chloroplast. In majority of plants proline is synthesised from glutamate; however, an alternative pathway exists where proline is synthesised from ornithine initially in the mitochondria (Fig. 7.2). Another striking feature of proline metabolism is repression of proline catabolism during stress, which is reactivated with the withdrawal of the stress. In this process proline is sequentially oxidised to P5C and then to glutamate by the action of enzymes proline dehydrogenase (PDH) and pyrroline-5-carboxylate dehydrogenase (P5CDH), respectively (Armengaud 2004; Lehmann et al. 2010; Szabados and Savoure 2010). Proline accumulation is triggered by desiccation, either naturally or during abiotic stress, and is degraded when rehydrated.

Accumulation of proline is directly proportional with the activity of the enzyme P5CS, which is a rate-limiting enzyme (Delauney and Verma 1993; Szekely et al. 2008) and is encoded by two homologous genes *P5CS1* and *P5CS2*. Studies have revealed upregulated expression of *P5CS1* transcript in root and shoot in response to abiotic stress compared to its homologous counterpart *P5CS2* mostly found to be upregulated in dividing cells (Strizhov et al. 1997; Szekely et al. 2008). Transcription of the gene *P5CS1* in the vegetative tissues of the plants is inducible by imposition of abiotic stresses such as drought, salinity or low temperature as well as exogenous

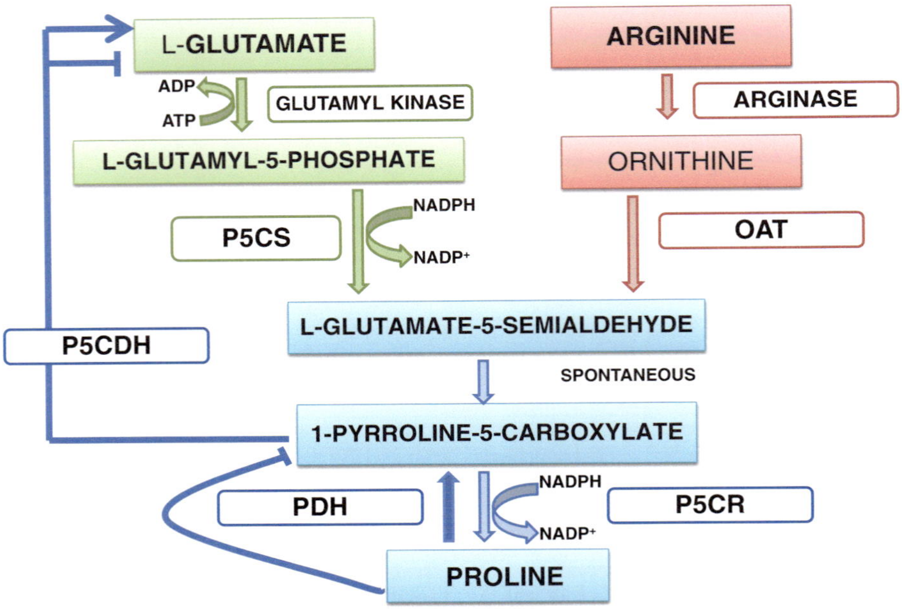

Fig. 7.2 Proline metabolic pathway during the period of abiotic stress in various plant species. [Pyrroline-5-carboxylate synthetase (P5CS), pyrroline-5-carboxylate reductase (P5CR), ornithine-delta-aminotransferase (OAT), proline dehydrogenase (PDH), pyrolline-5-carboxylate dehydrogenase (P5CDH)]

application of ABA (Strizhov et al. 1997; Ginzberg et al. 1998; Abraham et al. 2003). Genetic basis for natural variation in proline concentration in *Arabidopsis thaliana* can be accounted for the differential pattern of alternate splicing of the gene P5CS1 (Kesari et al. 2012). Accumulation of proline is also controlled epigenetically. Zhang et al. (2013) revealed the interconnection between DNA methylation and proline accumulation in rice during osmotic stress.

7.3.2 Glycine Betaine

The most abundant form of betaine available in the environment is glycine betaine. Betaines are quaternary ammonium compounds containing fully methylated nitrogen atoms. Other than glycine betaines there are several other betaine derivatives that are also prevalent in plants which include proline betaine, alanine betaine, choline-o-sulfate and 3-dimethylsulfoniopropionate. In higher plants, the level of glycine betaines was found to increase in response to various types of environmental stresses (Rhodes and Hanson 1993). The protective role of betaine was first studied in an experiment in which they are supplied to bacteria whose growth was inhibited by high salt concentration and was found that betaines play a protective role during salt stress (Le Rudulier et al. 1984). Glycine betaine is widely distributed in marine algae and at least ten flowering plant families, including Chenopodiaceae, Amaranthaceae, Compositae and Malvaceae (Blunden and Gordon 1986; Rhodes and Hanson 1993; Gage and Rathinasabapathi 1999). In biological systems, glycine betaine is synthesised either from choline or from glycine following two distinct biochemical pathways (Fig. 7.3). The plants such as spinach, maize and barley that are capable of producing glycine betaine (GB) are referred as GB accumulators, and they accumulate GB in their leaves when subjected to drought, salinity and cold stress, thereby protecting the plant from harmful effects of stress (Storey et al. 1977; Nyyssola et al. 2000). Genes that code for the enzymes involved in the glycine betaine biosynthetic pathway have been identified and cloned. Overexpression of these genes in plants confers tolerance to abiotic stresses. These

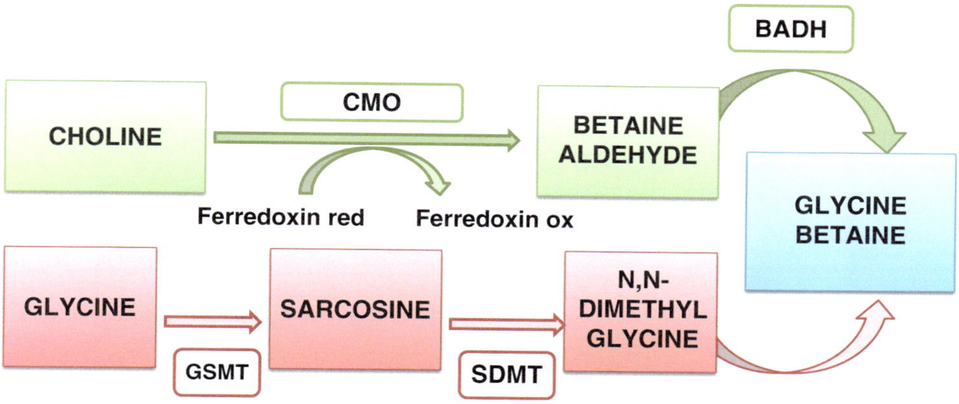

Fig. 7.3 Glycine betaine biosynthetic pathways. [Ferredoxin-dependant choline monooxygenase (CMO), NAD+-dependant betaine aldehyde dehydrogenase (BADH), glycine sarcosine methyltransferase (GSMT) and sarcosine dimethylglycine methyltransferase (SDMT)]

genes are also transferred to plants that do not accumulate GB such as *Arabidopsis thaliana*, *Brassica napus*, *Nicotiana tabacum* and *Oryza sativa*, thereby rendering them stress tolerant (McCue and Hanson 1990).

7.3.3 Sugars

Besides causing accumulation of specialised compounds, plants also accumulate specific sugar molecule to combat abiotic stress which includes trehalose, fructans and raffinose family of oligosaccharides (Drennan et al. 1993; Nakayama et al. 2007; Dos Santos et al. 2013). These sugar molecules are reported to be upregulated when the plants are subjected to abiotic stress.

7.3.3.1 Trehalose

Trehalose is non-reducing disaccharide composed of two glucose molecules bound together by an alpha-alpha linkage. It was discovered from rye in the year 1832; later it has been detected in a wide range of organism including bacteria, fungi, invertebrates and plants and is often designated as a stress alleviator (Elbein 1974; Richards et al. 2002). Initially, it was thought that trehalose was restricted to the resurrection plants such as *Selaginella lepidophylla* or *Myrothamnus flabellifolius*, where it was identified in a detectable amount (10 mg per gm fresh weight) (Goddijn and van Dun 1999). However, later experiments prove their existence in other plants such as *Arabidopsis*, tobacco and rice as well. Among the disaccharides available in the nature, trehalose exhibits a specific property. Both reducing ends of trehalose molecules are involved in glycosidic bond; as a result it is resistant to acidic hydrolysis; thus, these compounds are stable in high acidic solution even under high temperature (Richards et al. 2002). Trehalose biosynthetic pathway is a two-step process (Fig. 7.4a), and genes coding for enzymes trehalose-6-phosphate synthase [TPS (*AtTPS1*)] and trehalose-6-phosphate phosphatase [TPP (*AtTPPA* and *AtTPPB*)] were first characterised in *A. thaliana* in the late 1990s. Further in silico analysis led to the identification of ten homologues of *AtTPS1*, which can roughly be divided into two classes. Class I genes (*AtTPS1-AtTPS4*) encode for protein that has a TPS domain closely related to yeast, while, on other hand, class II genes encode protein with a TPP domain that exhibits a strong homology with *AtTPPA* and *AtTPPB*. *Oryza sativa* was reported to have nine homologues of *OsTPS* and *OsTPP* (Leyman et al. 2001, 2006; Blázquez et al. 1998; Avonce et al. 2006).

7.3.3.2 Raffinose Family Oligosaccharides

Raffinose family oligosaccharides or RFOs such as raffinose, stachyose and galactinol are ubiquitous in plants. They have been found to

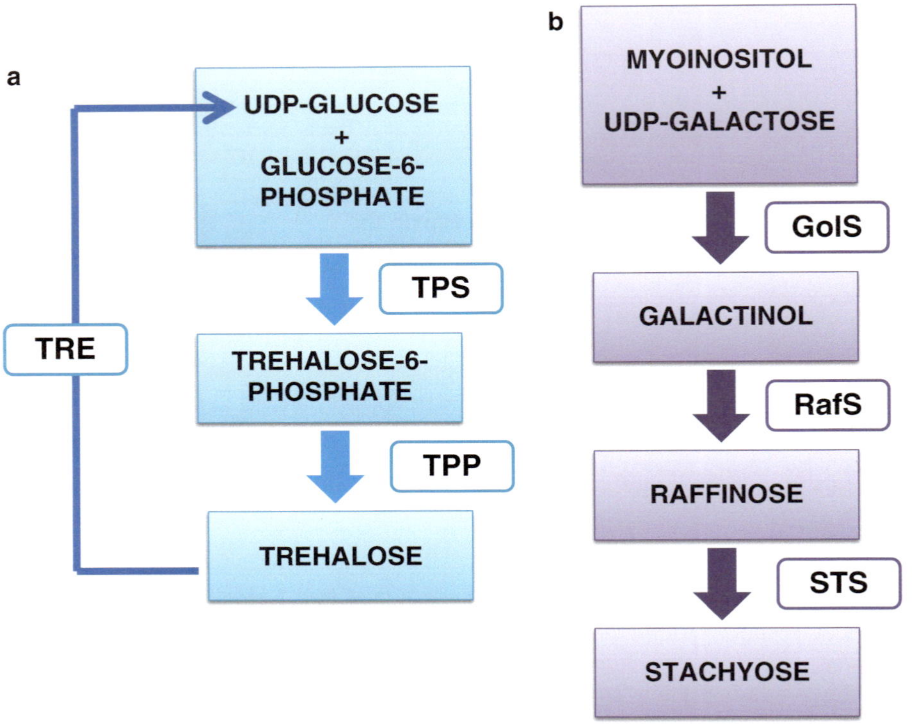

Fig. 7.4 Metabolic pathway of (**a**) trehalose and biosynthesis of (**b**) raffinose family of oligosaccharides. [Treahalose-6-phosphate synthase (TPS), treahalose-6-phosphate phosphatase (TPP), trehalase (TRE), galactinol synthase (GolS), raffinose synthase (RafS) and stachyose synthase (STS)]

accumulate under drought stress and function as an osmolytes, thereby maintaining cell turgor thus stabilising cell. It also depicts antioxidative property against the ROS which is generated within the cell in response to different abiotic stresses (Nishizawa et al. 2008; Van den Ende and Valluru 2009; Bolouri-Moghaddam et al. 2010; Peshev et al. 2013). The enzyme that is involved in the first step of RFO biosynthesis is galactinol synthase, which is located in the cytosol, and it catalyses the conversion of myo-inositol and UDP-galactose into galactinol. The galactinol thus produced is further converted into raffinose and stachyose by the action of enzyme raffinose and stachyose synthase, respectively (Fig. 7.4b). In spite of the presence of other enzymes, galactinol synthase (GolS) is still considered as the key regulator of RFO biosynthetic pathway because the activity of this particular enzyme will determine the accumulation level of certain RFOs (Keller and Pharr 1996; Schneider and Keller 2009). The expression of GolS genes can be regulated by three different transcription factors. In a study with *Arabidopsis*, it was observed that during high temperature, salinity and water-deficit stress, the expression of *AtGolS1* and *AtGolS2* genes was regulated by heat shock factors (HSF), whereas when the plant is exposed to low temperature stress, the expression of the gene *AtGolS3* is regulated by dehydration-responsive element-binding factor 1A/cold-responsive element-binding factor (DREB1A/CBF) transcription factor. And the third transcription factor that regulates the expression of the *GolS* and *RafS* genes is WRKY (Panikulangara et al. 2004; Wang et al. 2009).

7.3.4 Polyols

In halophytic plants, algae and some insects, polyols such as glycerol, mannitol, sorbitol, ononitol and pinitols have been found to accumulate in response to freezing stress functioning as osmoprotectants (Yancey et al. 1982). Glucose-6-phosphate serves as the precursor for synthesis of a number of metabolites that provides stress protection. Bacterial NADPH-dependant mannitol 1-phosphate dehydrogenase, enzyme that plays a key role in mannitol synthesis when overexpressed in tobacco and *Arabidopsis*, shows tolerance against salinity stress (Tarczynski et al. 1992; Thomas et al. 1995). It was observed that mannitol contributes to 30–40 % of the change in the osmotic potential during the period of stress (Karakas et al. 1997). Dual functioning of polyols in stress protection as well as in maintenance of redox control has been reported (Shen et al. 1999). Plants tolerant to abiotic stress are found to accumulate several different polyols such as cyclitols other than mannitol and sorbitol. Genetic engineering of the genes involved in the synthesis of these cyclitols and modulation of their regulation can be an important aspect of developing abiotic stress-tolerant crops. Myo-inositol is important cellular metabolites with diverse cellular functions including stress responses and the regulation of cell death. It has been observed that myo-inositol is converted into osmoprotectants D-ononitol and D-pinitol by a two-step pathway which is primarily regulated by abiotic stresses (Vernon and Bohnert 1992; Adams et al. 1992; Rammesmayer et al. 1995; Nelson et al. 1998).

7.3.5 Polyamines

Polyamines are nitrogenous compound that are ubiquitous in all living cells and have been reported to participate in different cellular processes (Evans and Malmberg 1989; Galston et al. 1997; Walden et al. 1997; Igarashi and Kashiwagi 2000; Thomas and Thomas 2001; Bais and Ravishankar 2002; Paschalidis and Roubelakis-Angelakis 2005a, b; Liu et al. 2006; Kusano et al. 2007a, b; Zhao and Yang 2008). Among the various classes of compatible solutes, polyamines play an important role in mitigating extreme environmental stresses. Diamine putrescine, triamine spermidine and tetramine spermine are the three major types of polyamines abundant in different plant species. Apart from these three major polyamines, some other types of polyamine such as homospermine and cadaverine are also found in some plants. Accumulation of polyamines in different plant tissues during normal developmental phases as well as during abiotic stress has been well documented. Their biological activities have been attributed to their cationic nature which enables them to interact with all the negatively charged cellular components such as DNA, RNA, proteins and phospholipids (Kaur-Sawhney et al. 2003; Liu et al. 2006; Pang et al. 2007; Groppa and Benavides 2008; Kusano et al. 2008). Changes in the endogenous level of polyamines have been widely studied when plant is exposed to a single stress or combination of multiple stresses. Increase in the concentration of polyamines within the cell can be due to de novo synthesis of polyamines or due to reduction in polyamine degradation pathways although the exact mechanisms still remain un-elucidated.

7.3.6 Osmolytes and Their Role in Plant Abiotic Stress Response

Adjustment of the metabolic profile in response to adverse environmental condition involves dynamic and multifaceted responses (Gong et al. 2005; Hannah et al. 2006; Zuther et al. 2007; Janz et al. 2010). Synthesis of these metabolites, commonly known as osmolytes, is generally triggered by primary osmotic stresses; however, in certain cases it may also result due to secondary stress signals such as reactive oxygen species, phytohormones and intracellular second messengers. Accumulation of osmolytes causes a rise in cellular osmolarity that increases water influx and decreases the efflux, which in turn provides the turgor required for cell expansion. Another striking feature of these compatible solutes is its ability to provide structural stability to proteins, thus preventing protein misfolding or aggregation during the period of abiotic stresses (Lamitina et al. 2006). The metabolic alteration that occurs as a result of prolonged exposure to abiotic stresses varies from species to species. Some plants were found to accumulate amino acids, amines, sugars and sugar alcohols in high concentration when exposed to either salinity, drought or low temperature for several days. Proline, a well-known compatible solute, has been found to accumulate when plants are subjected to drought or salinity or temperature stresses (Kaplan et al. 2004; Gong et al. 2005; Cramer et al. 2007; Gagneul et al. 2007; Kempa et al. 2008; Sanchez et al. 2008; Usadel et al. 2008; Urano et al. 2009; Lugan et al. 2010). It has been reported that proline plays an integral role in protecting and preserving functional and structural integrity of the enzyme M4 lactate dehydrogenase. The activity of photosynthetic enzyme RuBisCO is severely hampered in plants subjected to salt stress, as salt inhibits the carboxylase activity of RuBisCO enhancing the oxygenase activity. The presence of high concentration of proline has been observed to suppress the oxygenase activity, thus playing a crucial role in protecting photosynthetic activity during stress (Bohnert et al. 1992; Chadalavada et al. 1994; Sivakumar et al. 2000; Hamilton and Heckathorn 2001). Apart from this, proline biosynthesis is also involved in regulation of cytosolic acidity by maintaining NAD^+/NADH ratio. It serves as a source for carbon, nitrogen and energy during the period of stress recovery. The NADP which is regenerated during proline biosynthetic pathway is later utilised in the pentose phosphate pathway during the period of stress (Alia et al. 1991; Fahrendorf et al. 1995). Therefore, proline accumulated during the period of stress preserves the photosynthetic and the respiratory activity of the plant, thereby enabling the plant to survive even the most stressful conditions. Along with proline accumulation, another compound that has been widely studied in different plant species is glycine betaine. The functional property of glycine betaine resembles that of proline. It plays a pivotal role in providing effective protection against salt, drought and extreme temperature stresses. Accumulation of glycine betaine is restricted amongst specific taxonomical classes which generally contain low amount of glycine betaine, but the level drastically rises when the plant is subjected to abiotic stress. Plants that can accumulate glycine betaine naturally show an adept quality of stress tolerance. However, there are several economically important plants such as rice, tomato, arabidopsis and potato which cannot accumulate glycine betaine both in normal and stress condition (Yang et al. 2005). These species are the potential target for bioengineering of betaine synthesising genes. Exogenous application of glycine betaine is another potent way to increase the stress withstanding capacity in plants. Soya bean which is a low accumulator of glycine betaine can accumulate up to 5 μmol/g dry weight of glycine betaine during normal condition. During drought stress, foliar application of glycine betaine increases its content up to 60 μmol/g dry weights, thus improving its photosynthetic activity, rate of nitrogen fixation, leaf area development and seed yield in soya bean plant (Makela et al. 1996a, b; Agboma et al. 1997a, b). There are several other reports that indicate the role of exogenous glycine betaine in diminishing the harmful effect of drought stress in plants such as tobacco, barley, sorghum and wheat. Rice, which is normally designated as a non-accumulator of glycine betaine, was found to accumulate very low amount of glycine betaine in some of its cultivar including KMDL105, Homjan, Annapurna and Dongjin when exposed to salt stress (Oh et al. 2003; Cha-um et al. 2004, 2007). These evidences gathered by different workers successfully establish the role of glycine betaine in abiotic stress tolerance in plants. A considerable percentage of assimilated CO_2 comprises of polyols and sugars that functions either as compatible solutes or low molecular weight chaperones or reactive oxygen scavenger in plants when it is exposed to stress. Like the other major compatible solutes, their functions also involve osmotic adjustment and osmoprotection. Freezing and desiccation cause destabilisation of cellular membranes. The main aim of the cells is to keep all its membrane and protein in proper functional state. Soluble sugar here plays a crucial part in keeping the membranes in their proper state which is essential for survival under adverse condition.

During drought stress, sugar can replace water, thereby keeping the membrane surface hydrated, and also protect the membrane from fusion by maintaining space between phospholipid molecule (Uemura and Steponkus 1997; Mundree et al. 2002; Van den Ende and Valluru 2009; Ma et al. 2009; Djilianov et al. 2011). This process is known as 'sugar vitrification' which includes formation of amorphous glass that prevents membrane fusion. Some plants accumulate sucrose under drought stress. Sucrose, being an easily metabolisable sugar, generally serves as a source of immediate energy source. Apart from sucrose many other sucrose-derived sugar arise during cold acclimatisation in different plant species (Mundree et al. 2002). Accumulation of trehalose, another disaccharide, was first observed in resurrection plants; however, their presence is not restricted only in the resurrection plants. In *Arabidopsis*, trehalose content increases doubly when exposed to high temperature for 4 h, and there is an eightfold increase when exposed to low temperature for 4 days (Kaplan et al. 2004). Rice plants when exposed to low temperature show an increase in the trehalose content (Pramanik and Imai 2005). Under salt stress trehalose was found to accumulate in the nodule of *Medicago truncatula* (Lopez et al. 2008). The above findings clearly indicate that trehalose plays an active role in abiotic stress responses in plants. The common feature of all these compatible solutes discussed above is that these compounds accumulate in high concentrations without hampering the metabolism (Bohnert and Jensen 1996). They have negligible effect on the cytosolic pH. Syntheses of these compounds are mostly achieved by stress-triggered diversion in the regular metabolite biosynthetic pathways (Rhodes and Hanson 1993). Overexpression of these osmolyte biosynthetic genes in different target crops successfully yields stress-resistant variety. Some of them are cited in Table 7.2.

7.4 Polyamines: A Mega-Modulator in Plant Abiotic Stress Signalling

As discussed earlier, polyamines are low molecular weight ubiquitous aliphatic amines involved in the regulation of plant growth and development. They are involved in various fundamental processes in plants such as root growth, flower and fruit development, central dogma, cell wall and membrane stabilisation, ribosome biogenesis, chromatin reorganisation, enzyme activity modulation and programmed cell death (Evans and Malmberg 1989; Galston et al. 1997; Walden et al. 1997; Igarashi and Kashiwagi 2000; Thomas and Thomas 2001; Bais and Ravishankar 2002; Paschalidis and Roubelakis-Angelakis 2005a, b; Liu et al. 2006; Kusano et al. 2007a, b; Zhao and Yang 2008; Gupta et al. 2013). Polyamines have also been implicated in a wide range of environmental stress tolerance in plants which includes drought (Yamaguchi et al. 2007; Yang et al. 2007), salinity (Duan et al. 2008; Kuznetsov and Shevyakova 2010), oxidative stress (Rider et al. 2007), heavy metal toxicity (Groppa et al. 2003; Groppa and Benavides 2008), low temperature stress (Imai et al. 2004; Nayyar 2005; Cuevas et al. 2008) and ROS homeostasis (Saha et al. 2015). Detailed studies have revealed that abiotic stress plays a key role in modulating polyamine metabolic pathways. To understand their specific role, study on effect of exogenous application of polyamines to plants before and during abiotic stress has been conducted. It has been observed that exogenous polyamines increase their endogenous level, thus conferring tolerance in plants exposed to abiotic stresses establishing the role of polyamines as a potent stress alleviator in plants (Velikova et al. 2000; Navakouidis et al. 2003; Wang et al. 2007).

7.4.1 Polyamine Metabolic Pathways in Plant

Polyamine biosynthetic pathway (Fig. 7.5) begins with decarboxylation of ornithine and arginine by

Table 7.2 Overexpression of osmolytes biosynthetic genes in transgenic plants showing stress tolerance

Transgenic plant	Gene and gene source	Gene product	Osmolytes	References
Tobacco	*CDH* and *BADH* from *E. coli*	Choline dehydrogenase and betaine aldehyde dehydrogenase	Glycine betaine	Lilius et al. (1996), Holmstrom et al. (2000)
	BvCMO from *Beta vulgaris*	Choline monooxygenase	Glycine betaine	Zhang et al. (2008)
	BADH from *Spinacia oleracea* and *Oryza sativa*	Betaine aldehyde dehydrogenase	Glycine betaine	Yang et al. (2008), Hasthanasombut et al. (2010)
	*P5CS*from *Vigna aconitifolia*	Pyrroline-5-carboxylate synthetase	Proline	Kishor et al. (1995)
	MtlD from *E. coli*	Mannitol-1-phosphate dehydrogenase	Mannitol	Thomas et al. (1995)
	TPP and *TPS1* from *Saccharomyces cerevisiae*	Trehalose-6-phosphate phosphatise and treahalose-6-phosphate synthase	Trehalose	Holmstrom et al. (1996), Karim et al. (2007)
Rice	*Cod A* from spinach and *Arthrobacter*	Choline oxidase	Glycine betaine	Su et al. (2006), Kathuria et al. (2009)
Arabidopsis	*P5CS1* from *Arabidopsis thaliana*	Pyrroline-5-carboxylate synthetase	Proline	Mattioli et al. (2008)
	TPS1-TPS2 from *Saccharomyces cerevisiae*	Treahalose-6-phosphate synthase	Trehalose	Yeo et al. (2000), Miranda et al. (2007)
Wheat	*MtlD* from *E. coli*	Mannitol-1-phosphate dehydrogenase	Mannitol	Abebe et al. (2003)
Tomato	*TPS1* from *Saccharomyces cerevisiae*	Treahalose-6-phosphate synthase	Trehalose	Cortina and Culiáñez-Macià (2005)

ornithine or arginine decarboxylase (ODC and ADC) to yield diamine putrescine. Conversion of arginine to putrescine involves three enzymatic steps catalysed by ADC, agmatine iminohydrolase (AIH) and N-carbamoylputrescine amidohydrolase (CPA) sequentially. While in the case of ODC pathway, conversion of ornithine into putrescine involves only one enzymatic reaction and is catalysed by ornithine decarboxylase or ODC (ODC – EC 4.1.1.17). Putrescine thus formed is then sequentially converted into higher polyamine spermidine and spermidine into spermine by addition of aminopropyl group. These two reactions are carried out by the action of enzyme spermidine synthase and spermine synthase. The aminopropyl group that is added to the polyamines is generated from S-adenosylmethionine (SAM), and the reaction is catalysed by SAM decarboxylase. Genes of the enzymes involved in these biosynthetic pathways have been cloned and characterised in different plant species and arabidopsis. In arabidopsis, gene that codes for ornithine decarboxylase is absent; thus, in the case of arabidopsis, only ADC pathway is active. There are six enzymes responsible for polyamine biosynthesis and are encoded by ten different genes: two genes that code for arginine decarboxylase (ADC – EC 4.1.1.19) (*ADC1* and *ADC2*), a single gene for agmatine iminohydrolase (AIH – EC 3.5.3.12) (*AIH*) and N-carbamoylputrescine amidohydrolase (CPA – EC 3.5.1.53) (*CPA*), two genes that code for spermidine synthase (SPDS – EC 2.5.1.16) (*SPDS1* and *SPDS2*) and spermine synthase (SPMS – EC 2.1.5.22) (*SPMS* and *ACL5*) each and four copies of SAM decarboxylase-coding genes (SAMDC –

EC 4.1.1.50) (*SAMDC* 1–4) (Watson and Malmberg 1996; Watson et al. 1997; Hashimoto et al. 1998; Hanzawa et al. 2000; Franceschetti et al. 2001; Hanzawa et al. 2002; Panicot et al. 2002; Piotrowski et al. 2003; Janowitz et al. 2003; Urano et al. 2003, 2004; Knott et al. 2007). Apart from these above-mentioned names, gene that codes for thermospermine synthase is also present in a single copy. Studies have shown that *ADC1* is constitutively expressed in all plant tissue while *ADC2* functions when plant is exposed to abiotic stress such as drought and wounding (Soyka and Heyer 1999; Peréz-Amador et al. 2002). The levels of polyamines within the cells are regulated by its sequential degradation. The cellular polyamine catabolic pathways (Fig. 7.5) involve activity of two major amine oxidases, diamine oxidase (DAO) and polyamine oxidase (PAO) (Bagni and Tassoni 2001; Cona et al. 2006). In numerous study, putrescine concentrations have been a target for successful genetic manipulation, and in several cases, the level of putrescine increases several folds (Bhatnagar et al. 2002). The rise in putrescine level increases the rate of polyamine catabolism (Bhatnagar et al. 2002). Putrescine is converted by DAO (DAO – EC 1.4.3.6) into pyrroline, ammonia and H_2O_2, while spermidine and spermine are oxidised by PAO (PAO – EC 1.5.3.14) which converts spermidine into pyrroline, 1,3-diamine propane (DAP) and H_2O_2 and spermine into aminopropylpyrroline, DAP and H_2O_2 (Martin-Tanguy 2001; Sebela et al. 2001;Cona et al. 2003, 2006). It was found that rice produces high amount of spermidine and spermine when exposed to drought stress for 3 days; however, the content decreases sharply after 6 days exposure which may be due to the action of PAO (Capell et al. 2004).

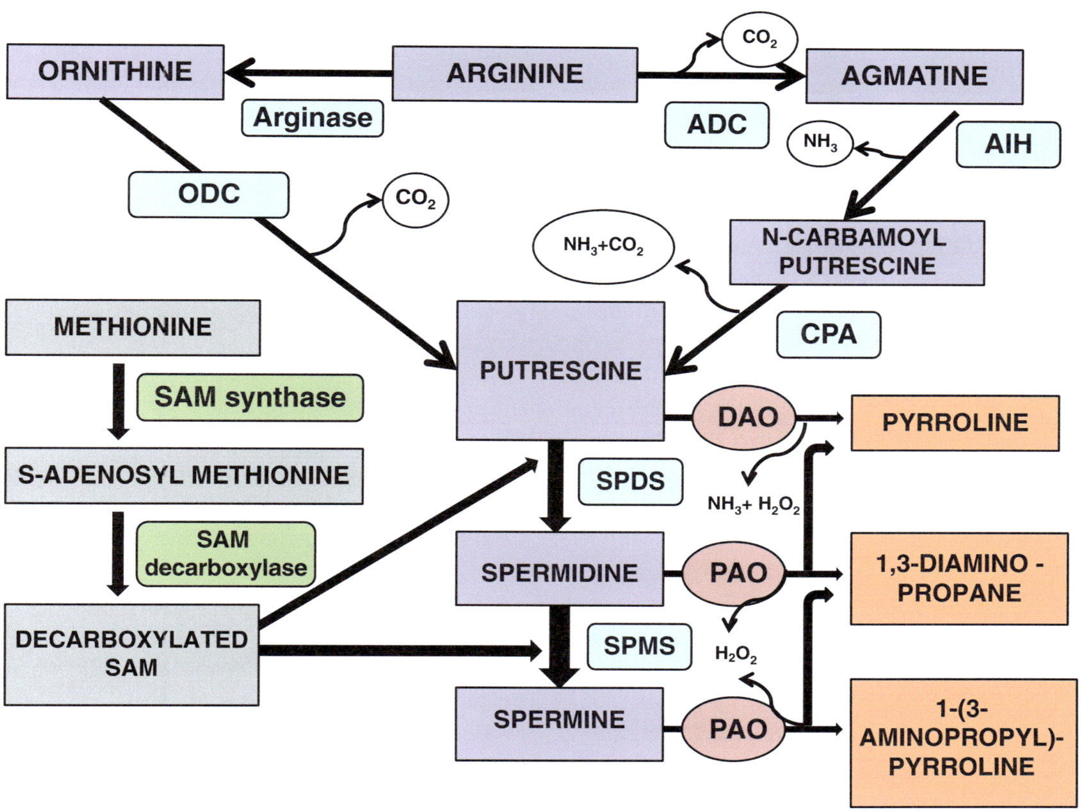

Fig. 7.5 Polyamine anabolic and catabolic pathways. [Ornithine or arginine decarboxylase (ODC and ADC), agmatine iminohydrolase (AIH), N-carbamoylputrescine amidohydrolase (CPA), spermidine synthase (SPDS), spermine synthase (SPMS), S-adenosylmethionine (SAM), diamine oxidase (DAO), polyamine oxidase (PAO)]

7.4.2 Polyamine and Plant Response to Abiotic Stress

The role of polyamine in abiotic stress tolerance was first observed when putrescine level increases due to K^+ ions deficiency in plants exposed to single or multiple stresses (Watson and Malmberg 1996). Several studies were carried out subsequently which established the role of polyamines in conferring stress tolerance. Cuevas et al. (2008) reported an increase in putrescine level during chilling stress in *Arabidopsis*. A similar kind of result was observed by Nadeau et al. (1987) in wheat during cold hardening. Roy et al. (2005) showed that exogenous spermidine treatment helps in the recovery of salinity stress-induced plasma membrane damage in salt-sensitive and salt-tolerant rice cultivar. In a study with *Vitis vinifera*, an increase in putrescine level was observed within 1 day of salinity and osmotic stress exposure; however, an increase in spermidine and spermine level occurs much later, almost 3–4 days (approx.) after stress exposure (Liu et al. 2011). Substantial increase in free and conjugated polyamines was recorded in heat-tolerant rice and cotton plants along with the increase in polyamine oxidising and biosynthetic enzymes during heat stress (Evans and Malmberg 1989; Cona et al. 2006). Abiotic stress causes variations in thylakoid-associated polyamines in many plants which establish its role in protecting photosynthetic apparatus during external adverse conditions (Kaumar et al. 1997; Jantaro et al. 2003; Sfichi et al. 2004; Alcázar et al. 2006).

Exogenous application of putrescine prevents chlorophyll loss and also preserves the membrane structure of thylakoid. Moreover, application of putrescine was found to enhance photochemical efficiency of PSII (Popovic et al. 1979; Cohen et al. 1979; Zhang et al. 2009a). Differential role of putrescine, spermidine and spermine was demonstrated by Sung et al. (2011) in mitigating oxidative stress induced by salinity in marine green macroalgae, *Ulva fasciata*, by modulating antioxidative enzyme activity. There are enough evidences that establish the role of polyamine as an antioxidant, free radical scavenger and membrane stabilisers (Velikova et al. 2000; Roy et al. 2005; Groppa and Benavides 2008). In salt-stressed leaf tissue of brassica, putrescine increases the activity of antioxidative enzymes and carotenoid, thus protecting it from oxidative damage by reducing H_2O_2 content and decreasing the level of lipid peroxidation (Verma and Mishra 2005). A similar kind of result was observed when exogenous polyamine was applied on a 14-day-old chickpea plant exposed to cold and drought stress (Nayyar and Chander 2004). In *Mesembryanthemum crystallinum*, spermine inhibits DNA oxidative degradation carried out by hydroxyl radical, thus proving the role of polyamine as a potent ROS scavenger (Kuznetsov and Shevyakova 2007). Literature survey reveals that polyamines function as 'antioxidant' in both free and conjugated forms. Genes involved in polyamine biosynthetic pathways are under strict regulation and show differential expression in different plant species when exposed to abiotic stress. Adverse environmental conditions such as drought, high salinity, mechanical injury and potassium deficiency strongly induce expression of *ADC2*in *Arabidopsis* (Peréz-Amador et al. 2002; Urano et al. 2003; Armengaud et al. 2004; Hummel et al. 2004; Alcázar et al. 2006). Similarly, expression of *SPMS* also increases during drought or salinity stress, thus causing accumulation of polyamines in plant tissues (Urano et al. 2003; Alcázar et al. 2006). No significant change in *CPA* and *AIH* level was observed during any stress. Constitutive expression of *SPDS2* was observed during all the stresses; however, in contrast *SPDS1* exhibited an increased expression when the plants were exposed to dehydration. Cold stress positively induces the expression pattern of *SAMDC1*, whereas expression of *SAMDC2* increases both in cold and slight exposure to salinity. *ACL5* do not show any corresponding change in their expression profile when treated with a wide range of stress elements (Vergnolle et al. 2005; Alcázar et al. 2006). Increase in the expression and activity of these genes results in the accumulation of polyamine within the plant cell, and many a time their ability to withstand stress is correlated with their ability to synthesise and accumulate polyamines

(Kasinathan and Wingler 2004; Liu et al. 2007). Moschou et al. (2008b) suggested that the ratio of polyamine anabolism and catabolism is a crucial factor in polyamine-mediated stress tolerance. Transcriptomic analysis of transgenic plants overexpressing polyamine biosynthetic enzymes reveals changes in expression of several stress-related genes suggesting that polyamines might play a role as signalling molecules in stress responses. However, the exact role of polyamines and their mode of action still remain to be elucidated (Hussain et al. 2011; Saha et al. 2015).

7.4.3 Polyamines: Potential Compatible Solutes in Plants

Changes in the endogenous polyamine level in response to stress provide us with clue regarding their possible implications. However, the exact mechanism involved in stress tolerance remains to be elucidated. Exogenous applications of polyamines during the period of stress represent an excellent model for answering all the basic fundamental questions regarding their biological activity (Bhatnagar et al. 2002; Alcázar et al. 2006; Liu et al. 2007). It has been widely established that exogenous application of polyamines can in varying degree preserve the integrity of cell membrane, reverse growth or minimise growth inhibition by stress, modulate expression of osmotically induced genes, function as an ROS scavenger, reduce accumulation Na^+ and Cl^- ions in different organs and also increase the activity of antioxidative enzymes (Ali 2000; Iqbal and Ashraf 2005; Tang and Newton 2005; Ndayiragiji and Lutts 2006; Afzal et al. 2009; Yiu et al. 2009; Zhang et al. 2009b). Polyamine plays a pivotal role in modulating the endogenous osmolyte level in plants when subjected to abiotic stress. The concept of polyamine as a 'compatible solute' is still controversial in plants and needs detailed investigation. However, in numerous cases, stress-induced polyamines are considered as better stabilisers than the other protecting biomolecules, thus preventing membrane degradation and denaturation under stress conditions (Liu et al. 2007). Polyamines share several properties such as hydrophobicity, protection of macromolecules, active oxygen scavengers and maintenance of cellular pH which bears similarity with that of a standard compatible solutes such as proline, glycine betaine, etc. (Wi et al. 2006; Liu et al. 2007). However, the concentration of stress-induced polyamines within the cells is much lower than the standard compatible solutes supporting the idea that polyamines are not 'compatible solutes'. But then β-alanine betaine, produced from β-alanine which is a product of polyamine catabolism, is required for osmoregulation in some halophytes (Duhazé et al. 2002). Moreover, proline is one of the most common compatible solutes, and its accumulation during adverse condition is closely associated with polyamine catabolism (Delauney and Verma 1993; Aziz et al. 1998; Bouchereau et al. 1999; Sharma and Dietz 2006; Bartels and Hussain 2008; Urano et al. 2009). Exogenous application of putrescine in low concentration shows alleviated level of proline accumulation in a study carried out by Bouchereau et al. (1999). They found that DFMA (difluoromethylarginine), which blocks putrescine synthesis from arginine in tomato explants under both control and stress conditions, also inhibits proline accumulation. A similar kind of observation was reported by Jimenez-Bremont et al. (2006). As proline and polyamine biosynthetic pathways use the same precursor glutamate, ornithine and arginine, their pathways might be finely coordinated (Simon-Sarkadi et al. 2005, 2006; Sharma and Dietz 2006; Seki et al. 2007). However, there are no such evidences regarding this coordinated accumulation (Radukina et al. 2007). It can be predicted that there are some signalling molecules that regulate all these pathways in a synchronised manner. Ornithine which is the key regulator of both polyamine and proline biosynthetic pathways may play the role of coordinator (Mohapatra et al. 2010). There is no strong evidence to validate this concept. Nevertheless, the probability of polyamine functioning as a compatible solutes or osmolytes can never be ruled out because of the functional similarity they share and requires much detailed study to establish this concept.

7.5 Role of Polyamines in Maintaining Homeostasis

Plants, when exposed to abiotic stress, shows a wide range of responses at cellular, molecular and whole plant level. The responses are mainly of three kinds, that is, maintenance of homeostasis, detoxification of harmful elements and recovery of growth. In this part we will discuss about both the direct and indirect role of polyamine in maintaining homeostasis in plants during the period of abiotic stress.

7.5.1 Macromolecule Homeostasis

The hydration spheres of macromolecules are often affected due to the rise in the concentration of charged element in the cytoplasm, thus altering their conformations or charge interactions. Compatible solutes play a significant role in maintaining the conformation of these macromolecules. The major functions of osmolytes are to maintain osmotic balance within the cell (Yancey et al. 1982). However, common osmolytes such as sugar, polyols, amino acids and their derivatives that are most prevalent in the cell do not show any significant involvement in the functioning of macromolecules even at much higher concentration. One osmolyte that has received most attention is proline. Accumulation of proline has been reported in various plants subjected to abiotic stress, and their protective role was confirmed with the help of transgenic study, which demonstrates that proline increases the tolerance of plants when exposed to adverse condition (Xin and Browse 2000; Nanjo et al. 1999; Hong et al. 2000). In transgenic studies it appears that the degree of proline or any other osmolyte accumulation in response to stress is not sufficient for overall osmotic adjustments; however, the possibilities of high level of accumulation in specific organelles or subcellular compartment cannot be ruled out, thus establishing its role in maintaining macromolecule conformation (Zhu 2001). Although polyamines have yet not been shown to play any direct role in maintaining macromolecule conformation, however, there are reports which indicate its role in increasing proline level within the cell, which in turn helps in macromolecule homeostasis (Kumar 2009; Khan et al. 2010). However, this is a hypothesis which requires further validation.

7.5.2 Ion Homeostasis

The process by which cell maintains their ionic balance in response to the external environment is referred as ion homeostasis, and it plays a major role in adaptation when exposed to stress. Cellular uptake, sequestration, export and long-distance transports are the processes involved in ion homeostasis. Studies have shown differential regulation of a number of ATPases, water channel proteins and ion transporters during salinity and osmotic stress. The role of various ion transporters in combating salinity stress has been a focus for a long time. Salinity stress, most commonly mediated by high NaCl concentration in the environment, hindered the steady ionic state not only for the Na^+ and Cl^- but also for K^+ and Ca^{2+} (Niu et al. 1995). The ion levels act as key regulators during stress alleviations. External Na^+ acts antagonistically preventing intracellular K^+ influx, reducing acquisition of this essential nutrient by cells. High NaCl causes cytosolic accretion of Ca^{2+} and adaptive responses during both biotic and abiotic stress. These intracellular Na^+ ions are eliminated from the cell and sequestered within the vacuole predominantly by Na^+/H^+ antiporters. There are a number of genes that code for Na^+/H^+ antiporter in arabidopsis genome. Overexpression of AtNHX1 in transgenic arabidopsis confers salt tolerance. Barragán et al. (2012) demonstrated that tonoplast-localised NHX proteins (NHX1 and NHX2, the two major tonoplast-localised NHX isoforms) are essential for active K^+ uptake in the tonoplast, for turgor regulation and for stomatal function.

Numerous genes and proteins, such as *HKT* and *HAK*, encoding K^+ transporters and channels have been identified and cloned in various plant species. They play an essential role in salt tolerance by regulating Na^+ and K^+ transport. In *Arabidopsis* leaf, class 1 type of HKT transporters has been identified that protect the plant from salinity stress by preventing excess accumulation of salt in the cytoplasm. Similar kind of result was obtained in a study with rice where class 1 type of HKT transporters removes Na^+ ion from the xylem, thus protecting the photosynthetic tissue from saline environment. The aim of our work is to identify the role of polyamines in regulating ion homeostasis. Polyamines have been reported to modulate ion transport by altering the activity of the plasma membrane and tonoplast-localised cation channels (Demidchik and Maathuis 2007; Zepeda-Jazo et al. 2008; Bose et al. 2011; Zepeda-Jazo et al. 2011). It was proposed that polyamines interact with membrane phospholipids, thereby increasing membrane stability which in turns helps to alter ion transports across the membrane. This phenomenon increases the activity of the ROS-induced conductance (ROSIC) and Ca^{2+}/ATPase and H^+/ATPase transporters during salinity stress (Roy et al. 2005; Pottosin et al. 2014). Ion channel regulations are essential for kinase signalling and maintaining K^+/Na^+ ratio within the cell (Essah et al. 2003; Pottosin et al. 2014). Polyamines function in limiting Na^+ influx and salt-induced K^+ efflux by blocking KIRC (inward rectifying potassium channel) and NSCCs (non-selective cation channels) in leaves and roots (Liu et al. 2000; Shabala et al. 2007; Zhao et al. 2007). Inward flow of Ca^{2+} and Na^+ was directed by voltage-independent non-selective cation channel (VI-NSCC). Apoplastic polyamines and Ca^+ have been reported to inhibit Na^+ influx. Moreover, transgenic plant overexpressing cation/H^+ exchanger (CAX), deficient in spermine synthesis, hinders vacuolar sequestration of Na^+. This view has been strengthened by Yamaguchi et al. (2006) in an experiment with arabidopsis. He demonstrated that a double-knockout arabidopsis which fails to synthesise spermine shows reduced growth in calcium-depleted MS media suggesting possible role of polyamines in Ca^+ homeostasis. Additional information was provided by Roy et al. (2005) and Roychoudhury et al. (2011) regarding protective role of exogenously applied spermidine and spermine in maintaining K^+/Na^+ balance in rice seedling treated with salt. Thus, all the evidences establish the role of polyamines in ion homeostasis.

7.6 Osmolytes and Polyamines as ROS Scavengers

Production of ROS is an unavoidable consequence of aerobic metabolism. Major sites for ROS generations in plants are mitochondria, chloroplast, peroxisomes and plasma membranes. Reduction of molecular oxygen by high energy exposure or electron transfer reactions leads to the production of ROS. They include free radicals such as superoxide anion (O^{-2}), hydroxyl radical (OH^-) as well as non-radical molecules like hydrogen peroxide (H_2O_2) and singlet oxygen ($^1O_2$). Amongst these ROS species, hydroxyl radical is considered as one dangerous of them all, attacking almost all the molecules in its vicinity (Foyer and Harbinson 1994; Foyer 1997; Del Río et al. 2006; Blokhina and Fagerstedt 2010; Heyno et al. 2011). Unfavourable environmental conditions such as drought, salinity, chilling, heat, metal toxicity and UV radiation which cause disruption in cellular homeostasis ultimately lead to the enhanced generation of ROS (Mittler 2002; Sharma and Dubey 2005; Sharma and Dubey 2007; Hu et al. 2008; Han et al. 2009; Maheshwari and Dubey 2009; Tanou et al. 2009; Mishra et al. 2011; Srivastava and Dubey 2011; Saha et al. 2015). Increase in the cellular ROS level also simultaneously increases the level of lipid peroxidation and protein oxidation, causes damage to nucleic acid, inhibits enzyme functionality and also activates programmed cell death in plants. But in spite of their destructive activity, they are also considered as one of the major signalling molecules within the living cell. So it is important to maintain equilibrium between ROS production and scavenging in order to protect the cell from oxidative damage cause due to excess level of ROS generation when exposed to abiotic stress. In plants, two vital pathways for ROS scavenging exist, enzymatic and non-enzymatic. Enzymatic antioxidants include superoxide dismutase (SOD), catalase (CAT), ascorbate peroxidase (APX), guaiacol peroxidase (GPX), monodehydroascorbate reductase (MDHAR), dehydroascorbate (DHAR) and glutathione reductase (GR), while non-enzymatic compounds include ascorbate (AsA), glutathione (GSH), carotenoids, tocopherols and phenolics (Noctor and Foyer 1998; Neill et al. 2002; Yan et al. 2007). All these above-mentioned compounds have well established themselves as a potent 'antioxidant'. However, our aim is to study the role of osmolytes, which lacks the tag of 'antioxidant' in maintaining cellular redox homeostasis. Proline a well-known osmolyte also functions as a singlet oxygen quencher, thus performing its role in ROS scavenging.

Several studies have revealed that exogenous proline treatment can diminish ROS level in fungi and yeast, thus preventing programmed cell death; it can also reduce the level of lipid peroxidation in algal cell exposed to heavy metal stress (Smirnoff and Cumbes 1989; Chen and Dickman 2005). Rice plants pretreated with proline show tolerance towards mercuric toxicity through ROS scavenging. Alia and Mohanty (1997) observed that the damaging effect induced by the singlet oxygen and hydroxyl radical on the photosystem II in an isolated thylakoid membrane can be reduced by exogenous proline application. Apart from its direct ROS scavenging features, proline can stabilise and protect the enzymes involved in ROS scavenging and also can turn on alternative ROS-detoxifying pathways. In tobacco plants exposed to salinity stress, activities of ROS scavenging enzymes such as peroxidase, superoxide dismutase and catalase are found to be increased by proline (Hoque et al. 2008; Islam et al. 2009). Proline plays a distinctive protective function in mitochondria. Oxidation of proline serves as a source of electron for the respiratory chain, thus contributing energy required for growth resumption. Thus, it can be said that proline metabolism is therefore an important regulator of cellular ROS balance and is also crucial in regulating numerous additional biochemical pathways (Hamilton and Heckathorn 2001). Recent studies have proposed the role of plant oligosaccharides as 'antioxidant' (Nishizawa et al. 2008; Sinkevich et al. 2010; Foyer and Shigeoka 2011; Stoyanova et al. 2011). Nishizawa et al. (2008) compared hydroxyl radical scavenging capacity of galactinol and raffinose with typical antioxidants such as ascorbate and glutathione in an in vitro test in arabidopsis, and it was found to be more superior. He also demonstrated the role of intracellular level of galactinol and stachyose in protecting the cellular component from oxidative damage. Furthermore, concentration of raffinose in chloroplast of stressed plants was found to be more or less similar with the level of glutathione and ascorbate suggesting the role of raffinose as a direct scavenger of hydroxyl radical in chloroplasts (Stoyanova et al. 2011; Foyer and Shigeoka 2011).

ROS scavenging properties have also been reported in sugar alcohol such as mannitol, sorbitol, inositol, etc. Genetically engineered tobacco with increased mannitol concentration showed stress tolerance by increasing the scavenging capacity, thus protecting cell from oxidative damage. Moreover, mannitol accumulations do not show any harmful effect on plants, thus proving that sugar does not show any negative feedback (Bolouri-Moghaddam et al. 2010). Trehalose another potent osmolytes also can act as a ROS scavenger. In a study with yeast, it was observed that the exogenous hydrogen peroxide application increases the intracellular trehalose concentration, thus protecting the cell from oxidative damage and thereby reducing the level of lipid peroxidation (Stoyanova et al. 2011).

Transgenic rice plant that accumulates high amount of trehalose shows tolerance towards salinity, drought and low temperature stresses (Garg et al. 2002). There are several studies that demonstrate the role of polyamine as an antioxidant (Larher et al. 2003; Velikova et al. 2000; Saha et al. 2015). Work on *Brassica juncea* treated with putrescine revealed a reduction in H_2O_2 and lipid peroxidation level during salinity stress (Verma and Mishra 2005). In another experiment it was observed that putrescine increases the activity of antioxidant enzymes and carotenoid in leaf of *Brassica juncea* exposed to saline environment. On the basis of this result, they concluded that polyamines might function as antioxidant under certain conditions. Saleethong et al. (2011) demonstrated the effect of 1 mM exogenous spermidine on rice plant during salinity stress in both salt-sensitive (KMDL 105) and salt-tolerant (pokkali) cultivar and observed that the exogenous spermidine successfully enhances the tolerance towards stress by scavenging free radical and membrane stabilisation. Application of exogenous spermidine increases the antioxidant enzyme activities in the root of salt-sensitive cucumber cultivar when exposed to salinity (Duan et al. 2008). So it can be concluded that polyamines either function as antioxidant itself or can efficiently modulate or alter the activity of antioxidative enzymes as well as the non-enzymatic antioxidants, thus playing a substantial role in ROS scavenging.

7.7 Bioengineering for Abiotic Stress Tolerance in Plant: An Attempt to Develop Stress-Resistant Transgenic Plants

Genetic engineering of polyamine biosynthetic and metabolic genes has paved a pathway for detailed investigation regarding the role of polyamines during stress responses (Table 7.3) (Groppa and Benavides 2008). It has already been well established that environmental stresses such as salinity, drought, temperature, heavy metal and other abiotic stresses differentially increase the polyamine content within the cell (Song et al. 2002; Capell et al. 2004; Kasukabe et al. 2004; Hummel et al. 2004; Imai et al. 2004; Roy et al. 2005; Liu et al. 2006). In order to gain proper insight about the role of polyamines, genes involved in polyamine biosynthetic pathways have been either upregulated or downregulated in different plant species (Minocha and Sun 1997; Kakkar and Sawhney 2002; Thu-Hang et al. 2002). Overexpression of genes involved in polyamine biosynthetic pathways shows an elevated level of tolerance towards abiotic stresses in plants like *Oryza sativa*, *Nicotiana tabacum* and *Arabidopsis thaliana*.

In a study with transgenic rice which harbours *Datura stramonium*, *ADC* gene accumulates higher amount of putrescine during drought stress than its wild counterpart. The putrescine thus produced is then converted to spermidine and spermine which further enhances its ability to withstand stress (Capell et al. 2004; Peremarti et al. 2009). A similar kind of work was carried out by Peremarti et al. (2009) where transgenic rice plants were generated which constitutively expresses *SAMDC* gene from *Datura stramonium* to dissect the role of lower polyamine putrescine from higher polyamine spermidine and spermine. Both transgenic and wild type show similar kind of features during the period of drought stress. However, transgenic plants recover more quickly on rewatering than its wild-type counterpart. Roy and Wu (2002) in their experiment observed a rise in the level of spermidine and spermine almost by three- to fourfold when *SAMDC* gene of tritordeum was introduced in rice plants, thus conferring salt tolerance (Roy and Wu 2002). In another instance, Franceschetti et al. (2004) overexpressed *SAMDC* gene of arabidopsis in tobacco plants which shows an increase in the level of decarboxylated SAM which in turn regulate polyamine accumulation exhibiting tolerance towards multiple stress. *ADC* gene when overexpressed in transgenic rice exhibits tolerance towards drought stress. Further in-depth analyses depicted the role of putrescine in growth recovery, while spermidine and spermine function as a ROS detoxifier (Larher et al. 2003). This result confirmed the role of individual polyamines in stress alleviation. Ectopic expression of *SPDS1* gene from *Malus sylvestris* in *Lycopersicon esculentum* had resulted up to 1.5–1.6-fold increase in spermidine level and showed an elevated expression of *SlApx* gene, whereas in wild-type plants, the expression declined after one-month treatment of 150 mM NaCl (Neily et al. 2011). Overexpression of *SPDS* (spermidine synthase) from *Cucurbita ficifolia* in sweet potato, tobacco and arabidopsis manifested a wide spectrum of tolerance towards drought, chilling, freezing, salinity and oxidative stress (Kasukabe et al. 2004; Wi et al. 2006). Detailed study with transgenic arabidopsis harbouring *SPDS* gene cloned from *Cucurbita ficifolia* plant demonstrated higher accumulation of spermidine that was correlated with the abundantly expressed transcripts of DREB and

Table 7.3 Bioengineering of polyamine biosynthetic genes conferring abiotic stress tolerance

Transgenic plant	Gene overexpressed	Molecular functions	Observation	References
Rice	*ADC* from oat	Arginine decarboxylase	1. Accumulation of high amount of putrescine 2. Increase in ADC activity 3. Confers salt tolerance	Roy and Wu (2001)
	ADC and *SAMDC* from *Datura*	Arginine decarboxylase and S-adenosylmethionine decarboxylase	1. Drought tolerance 2. Transgenic plant shows rapid recovery from drought stress on rewatering	Capell et al. (2004), Peremarti et al. (2009)
	SAMDC from tritordeum	S- adenosylmethionine decarboxylase	1. Salt tolerant 2. Transgenic plant shows normal growth with increased biomass	Roy and Wu (2002)
Tobacco	*ODC* from mouse	Ornithine decarboxylase	1. Enhanced polyamine metabolism with higher putrescine level 2. Salt tolerant	Kumria and Rajam (2002)
	SAMDC from human	S- adenosylmethionine decarboxylase	1. Increased putrescineand spermidine level 2. Drought, salt and biotic stress tolerant	Waie and Rajam (2003)
	SAMDC from carnation	S- adenosylmethionine decarboxylase	1. Increase in photosynthetic rate and seed rate 2. Tolerance to multiple abiotic stresses	Wi et al. (2006)
Arabidopsis	*ADC* and *SAMDC1* from arabidopsis	Arginine decarboxylase and S-adenosylmethionine decarboxylase	1. Overproduction of spermine 2. Reduce stomatal conductance and transpiration rate 3. Tolerance towards wide variety of abiotic stress specially drought and freezing	Alcázar et al. (2006, 2010a, b)
European pear	*SPDS1* and *SAMDC2* from apple	Spermidine synthase and S- adenosylmethionine decarboxylase	1. Elevated accumulation of polyamines 2. Multiple abiotic stress tolerance	He et al. (2008), Wen et al. (2008)
Tomato	*SAMDC* from yeast	S- adenosylmethionine decarboxylase	1. Enhanced antioxidant activity and protection from lipid peroxidation 2. Tolerance towards high temperature	Cheng et al. (2009)

stress-protective proteins like RD29A, thus showing long-time viability of seedling during salt stress in transgenic plants over the wild-type plant (Kasukabe et al. 2004).

A rapid increase in free polyamines (by 4–7 %) and higher photosynthetic rate and biomass by 10–26 % and 4–21 %, respectively, were observed when full-length S-adenosylmethionine synthetase (*SsSAMS2*) gene from *Suaeda salsa* was introduced in tobacco plant (Qi et al. 2010). Wen et al. (2008) demonstrated that overexpression of apple *MdSPDS1* gene in European pear

(*Pyrus communis L.*) increases tolerance to multiple stresses by altering polyamine level. Similarly, transgenic eggplants exhibited an increase in tolerance towards drought, salinity, low and high temperature and heavy metal stress, when oat *ADC* gene is over expressed in it (Prabhavathi and Rajam 2007). Thus, to meet the ever-increasing demand of food crop across the world, transgenic approach involving polyamine biosynthetic genes is considered as a proficient strategy to improve crop tolerance.

7.8 Conclusion and Future Perspective

The menace of abiotic stress is essentially a matter of great concern. In order to survive such hostile condition, sessile plants have implemented several defence mechanisms, many of which have not yet been identified at molecular level. Synthesis of osmolytes is considered as one of the various major strategies, employed by plants to avoid stress. The role of almost all the osmolytes has been thoroughly studied till date. However, our main focus is to analyse the role of polyamine as an 'osmolyte' and to draw an interrelationship between the major compatible solutes and polyamines. There are enough evidences available which demonstrate the role of polyamine as a potent stress alleviator. But those evidences are not sufficient for establishing the role of polyamine as a compatible solute or modulator of endogenous compatible solute level. Genetic engineering of polyamine biosynthetic genes in different crop plants has been a success in generating stress-tolerant variety. But still there exist lacunae in deciphering the exact mechanism by which polyamine helps to withstand stress. Understanding the function of this unique molecule will improve our understanding regarding the complexities of plant defence mechanisms. With advancement of biotechnological approaches, our future focus should include in-depth analysis of the functional mechanisms of polyamines and their exact role in plant protection against abiotic stresses.

Acknowledgments The authors acknowledge the support of technical facilities available at Presidency University, Kolkata. Financial assistance from DBT (RGYI) (Govt. of India) and W.B. State DST (Govt. of West Bengal) to BG and KG and DST-SERB (Govt. of India) to KG is also gratefully acknowledged.

References

Abebe T, Arron C, Martin B, Cushman JC (2003) Tolerance of mannitol-accumulating transgenic wheat to water stress and salinity. Plant Physiol 131:1748–1755

Abraham E, Rigo G, Szekely G, Nagy R, Konez C, Szabados L (2003) Light dependent induction of proline biosynthesis by abscisic acid and salt stress is inhibited by brassinosteriods in *Arabidopsis*. Plant Mol Biol 51:363–372

Adams P, Thomas JC, Vernon DM, Bohnert HJ, Jensen RG (1992) Distinct cellular and organismic responses to salt stress. Plant Cell Physiol 33:1215–1223

Afzal I, Munir F, Ayub CM, Basra SMA, Hameed A, Nawaz A (2009) Changes in antioxidant enzymes, germination capacity and vigour of tomato seeds in response of priming with polyamines. Seed Sci Technol 37:765–770

Agboma PC, Peltonen-Sainio P, Hinkkanen R, Pehu E (1997a) Effect of foliar application of glycine betaine on yield components of drought-stressed tobacco plant. Explementary Agric 33:345–352

Agboma PC, Sinclair TR, Jokinen K, Peltonen-Sainio P, Pehu E (1997b) An evaluation of the effect of exogenous glycine betaine on the growth and yield of soybean: timing of application, watering regimes and cultivars. Field Crops Res 54:51–64

Ahmad P, Jaleel CA, Salem MA, Nabi G, Sharma S (2010) Roles of Enzymatic and non-enzymatic antioxidants in plants during abiotic stress. Crit Rev Biotechnol 30:161–175

Alcázar R, Altabella T, Marco F, Bortolotti C, Reymond M, Koncz C, Carrasco P, Tiburcio A (2010a) Polyamines: molecules with regulatory functions in plant abiotic stress tolerance. Planta 231:1237–1249

Alcázar R, Planas J, Saxena T, Zarza X, Bortolotti C, Cuevas J, Bitrian M, Tiburcio AF, Altabella T (2010b) Putrescine accumulation confers drought tolerance in transgenic Arabidopsis plants overexpressing the homologous arginine decarboxylase 2 gene. Plant Physiol Biochem 48:547–552

Alcázar R, Marco F, Cuevas JC, Patron M, Ferrando A, Carrasco P, Tiburcio AF, Altabella T (2006) Involvement of polyamines in plant response to abiotic stress. Biotechnol Lett 28:1867–1876

Alet AI, Sanchez DH, Cuevas JC, Marina M, Carrasco P, Altabella T, Tiburcio AF, Ruiz OA (2012) New insights into the role of spermine in *Arabidopsis thaliana* under long-term salt stress. Plant Sci 182:94–100

Ali RM (2000) Role of putrescine in salt tolerance of *Atropa belladonna* plant. Plant Sci 152:173–179

Alia PS, Mohanty P (1991) Proline enhances primary photochemical activities in isolated thylakoid membranes of *Brassica juncea* by arresting photoinhibitory damage. Biochem Biophys Res Commun 181:1238–1244

Alia PS, Mohanty P (1997) Involvement of proline in protecting thylakoid membranes against free radical-induced photodamage. J Photochem Photobiol 38:253–257

Armengaud P (2004) Transcriptional regulation of proline biosynthesis in *Medicago truncatula* reveals developmental and environmental specific features. Physiol Plant 120:442–450

Armengaud P, Breitling R, Amtmann A (2004) The potassium-dependent transcriptome of *Arabidopsis* reveals a prominent role of jasmonic acid in nutrient signaling. Plant Physiol 136:2556–2576

Asada K (1994) Production and action of active oxygen species in photosynthetic tissues. In: Foyer CH, Mullineaux PM (eds) Causes of photooxidative stress and amelioration of defence systems in plants. CRC Press, Boca Raton, pp 77–104

Ashraf M, Foolad MR (2007) Roles of glycine betaine and proline in improving plant abiotic stress resistance. Environ Exp Bot 59:206–216

Avonce N, Mendoza-Vargas A, Morett E, Iturriaga G (2006) Insights on the evolution of trehalose biosynthesis. BMC Evol Biol 6:109

Aziz A, Martin-Tanguy J, Larher F (1998) Stress-induced changes in polyamine and tyramine levels can regulate proline accumulation in tomato leaf discs treated with sodium chloride. Plant Physiol 104:195–202

Bagni N, Tassoni A (2001) Biosynthesis, oxidation and conjugation of aliphatic polyamines in higher plants. Amino Acids 20:301–317

Bais HP, Ravishankar GA (2002) Role of polyamines in the ontogeny of plants and their biotechnological applications. Plant Cell Tissue Organ Cult 69:1–34

Barragán V, Leidi EO, Andrés Z, Rubio L, Luca AD, Fernández JA, Cubero B, Pardo JM (2012) Ion exchangers NHX1 and NHX2 mediate active potassium uptake into vacuoles to regulate cell turgor and stomatal function in Arabidopsis. Plant Cell 24(3):1127–1142

Bartels D, Hussain SS (2008) Current status and implications of engineering drought tolerance in plants using transgenic approaches. CAB Rev Perspect Agri Vet Sci Nutri Natur Sci 3:020

Bhatnagar P, Minocha R, Minocha S (2002) Genetic manipulation of the metabolism of polyamines in poplar cells: the regulation of putrescine catabolism. Plant Physiol 128:1455–1469

Blázquez MA, Santos E, Flores CL, Martínez-Zapater JM, Salinas J, Gancedo C (1998) Isolation and molecular characterization of the *Arabidopsis TPS1* gene, encoding trehalose-6-phosphate synthase. Plant J 13:685–689

Blokhina O, Fagerstedt KV (2010) Reactive oxygen species and nitric oxide in plant mitochondria: origin and redundant regulatory systems. Physiol Plant 138:447–446

Blunden G, Gordon SM (1986) Betaines and their sulphonio analogues in marine algae. Prog Phycol Res 4:39–80

Bohnert HJ, Jensen RG (1996) Strategies for engineering water stress tolerance in plants. Trends Biotechnol 14:89–97

Bohnert HJ, Vernon DM, DeRocher EJ, Michalowski CB, Cushman JC (1992) Biochemistry and molecular biology of CAM. In: Wray Z (ed) Inducible plant proteins. Cambridge University Press, Cambridge, UK, pp 113–137

Bolouri-Moghaddam MR, Le Roy K, Xiang L, Rolland F, Van den Ende W (2010) Sugar signalling and antioxidant network connections in plant cells. FEBS J 277:2022–2037

Bose J, Pottosin I, Shabala SS, Palmgren MG, Shabala S (2011) Calcium efflux systems in stress signaling and adaptation in plants. Front Plant Sci 2:85

Bouchereau A, Aziz A, Larher F, Martin-Tanguy J (1999) Polyamines and environmental challenges: recent development. Plant Sci 140:103–125

Boyer JS (1982) Plant productivity and environment. Science 218:443–448

Bray EA, Bailey-Serres J, Weretilnyk E (2000) Responses to abiotic stresses. In: Gruissem W, Buchannan B, Jones R (eds) Biochemistry and molecular biology of plants. American Society of Plant Physiologists, Rockville, pp 1158–1249

Brown AD (1976) Microbial water stress. Bacteriol Rev 40:803–846

Capell T, Bassie L, Christou P (2004) Modulation of the polyamine biosynthetic pathway in transgenic rice confers tolerance to drought stress. Proc Natl Acad Sci U S A 101:9909–9914

Chadalavada SV, Rajendrakumar RBVB, Reddy AR (1994) Proline–protein interactions: protection of structural and functional integrity of M4 lactate dehydrogenase. Biochem Biophys Res Commun 201:957–963

Cha-um S, Supaibulwatana K, Kirdmanee C (2004) Biochemical and physiological responses of Thai jasmine rice (*Oryza sativa* L. ssp. indica cv. KDML105) to salt-stress. Sci Asia 30:247–253

Cha-um S, Supaibulwatana K, Kirdmanee C (2007) Glycinebetaine accumulation, physiological characterizations, and growth efficiency in salt tolerant and salt sensitive lines of indica rice (*Oryza sativa* L. ssp. indica) response to salt stress. J Agron Crop Sci 193:157–166

Chen C, Dickman MB (2005) Proline suppresses apoptosis in the fungal pathogen *Colletotrichum trifolii*. Proc Natl Acad Sci U S A 102:3459–3464

Cheng L, Zou Y, Ding S, Zhang Y, Yu X, Cao J, Lu G (2009) Polyamine accumulation in transgenic tomato enhances the tolerance to high temperature stress. J Integr Plant Biol 51:489–499

Cohen AS, Popovic RB, Zalik S (1979) Effects of polyamines on chlorophyll and protein content, photochemical activity, and chloroplast ultrastructure of barley leaf discs during senescence. Plant Physiol 64:717–720

Cona A, Cenci F, Cervelli M, Federico R, Mariottini P, Moreno S, Angelini R (2003) Polyamine oxidase, a hydrogen peroxide-producing enzyme, is up-regulated by light and down-regulated by auxin in the outer tissues of the maize mesocotyle. Plant Physiol 131:803–813

Cona A, Rea G, Angelini R, Federico R, Tavladoraki P (2006) Functions of amine oxidases in plant development and defence. Trends Plant Sci 11:80–88

Cortina C, Culiáñez-Macià A (2005) Tomato abiotic stress enhanced tolerance by trehalose biosynthesis. Plant Sci 169:75–82

Cramer GR, Ergul A, Grimplet J, Tillett RL, Tattersall EA, Bohlman MC, Vincent D, Sonderegger J, Evans J, Osborne C, Quilici D, Schlauch KA, Schooley DA, Cushman JC (2007) Water and salinity stress in grapevines: early and late changes in transcript and metabolite profiles. Funct Integr Genomics 7:111–134

Cuevas JC, Lopez-Cobollo R, Alcázar R, Zarza X, Koncz C, Altabella T, Salinas J, Tiburcio AF, Ferrando A (2008) Putrescine is involved in *Arabidopsis* freezing tolerance and cold acclimation by regulating abscisic acid levels in response to low temperature. Plant Physiol 148:1094–1105

Cushman J, Bohnert HJ (2000) Genomic approaches to plant stress tolerance. Curr Opin Plant Biol 3:117–124

Del Río LA, Sandalio LM, Corpas FJ, Palma JM, Barroso JB (2006) Reactive oxygen species and reactive nitrogen species in peroxisomes. Production, scavenging, and role in cell signalling. Plant Physiol 141:330–335

Delauney AJ, Verma DPS (1993) Proline biosynthesis and osmoregulation in plants. Plant J 4:215–223

Demidchik V, Maathuis FJM (2007) Physiological roles of nonselective cation channels in plants: from salt stress to signalling and development. New Phytol 175:387–404

Dey A, Gupta K, Gupta B (2014) Role of polyamines in plant-pathogen interactions. In: Anjum NA, Gill SS, Gill R (eds) Plant adaptation to environmental change: significance of amino acids and their derivatives. CABI, Oxfordshire, p 222, Chapter 12. Doi: 10.1079/9781780642734.0222

Djilianov D, Ivanov S, Moyankova D, Miteva L, Kirova E, Alexieva V, Joudi M, Peshev D, Van den Ende W (2011) Sugar ratios, glutathione redox status and phenols in the resurrection species Haberlea rhodopensis and the closely related non-resurrection species Chirita eberhardtii. Plant Biol 13:767–776

Dos Santos VC, Bragança CR, Passos FJ, Passos FM (2013) Kinetics of growth and ethanol formation from a mix of glucose/xylose substrate by *Kluyveromyces marxianus* UFV-3. Antonie Van Leeuwenhoek 103:153–161

Drennan PM, Smith MT, Goldsworthy D, van Staden J (1993) The occurrence of trehalose in the leaves of the desiccation-tolerant angiosperm *Myrothamnus flabellifolius* Welw. J Plant Physiol 142:493–496

Duan J, Li J, Guo S, Kang Y (2008) Exogenous spermidine affects polyamine metabolism in salinity-stressed *Cucumis sativus* roots and enhances short-term salinity tolerance. J Plant Physiol 165:1620–1635

Duhazé C, Gouzerh G, Gagneul D, Larher F, Bouchereau A (2002) The conversion of spermidine to putrescine and 1,3-deaminopropane in the roots of *Limonium tataricum*. Plant Sci 163:639–646

Elbein AD (1974) The metabolism of alpha, alpha-trehalose. Adv Carbohydr Chem Biochem 30:227–256

Essah PA, Davenport R, Tester M (2003) Sodium influx and accumulation in *Arabidopsis*. Plant Physiol 133:307–318

Evans PT, Malmberg RL (1989) Do polyamines have roles in plant development? Annu Rev Plant Physiol Plant Mol Biol 40:235–269

Fahrendorf T, Li W, Shrrosh BS, Dixon RA (1995) Stress response in alfalfa (*Medicago sativa* L.) XIX. Transcriptional activation of oxidative pentose phosphate pathway genes at the onset of the isoflavonoid phytoalexin response. Plant Mol Biol 28:885–900

Ford CW (1984) Accumulation of low molecular solutes in water stress tropical legumes. Phytochemistry 23:1007–1015

Foyer CH (1997) Oxygen metabolism and electron transport in photosynthesis. In: Scandalios J (ed) Molecular biology of free radical scavenging systems. Cold Spring Harbor Laboratory Press, New York, pp 587–621

Foyer CH, Harbinson J (1994) Oxygen metabolism and the regulation of photosynthetic electron transport. In: Foyer CH, Mullineaux P (eds) Causes of photooxidative stresses and amelioration of defence systems in plants. CRC Press, Boca Raton, pp 1–42

Foyer CH, Shigeoka S (2011) Understanding oxidative stress and antioxidant functions to enhance photosynthesis. Plant Physiol 155:93–100

Franceschetti M, Hanfrey C, Scaramagli S, Torrigiani P, Bagni N, Burtin D, Michael AJ (2001) Characterization of monocot and dicot plant S-adenosyl-L-methionine decarboxylase gene families including identification in the mRNA of a highly conserved pair of upstream overlapping open reading frames. Biochem J 53:403–409

Franceschetti M, Fornale S, Tassoni A, Zuccherelli K, Mayer MJ, Bagni N (2004) Effects of spermidine synthase overexpression on polyamine biosynthetic pathway in tobacco plants. J Plant Physiol 161:989–1001

Funck D, Stadelhofer B, Koch W (2008) Ornithine-delta-aminotransferase is essential for arginine catabolism but not for proline biosynthesis. BMC Plant Biol 8:40

Gage DA, Rathinasabapathi B (1999) Role of glycine betaine and dimethylsulfoniopropionate in water stress tolerance. In: Shinozaki K, Yamaguchi-shinozaki K (eds) Molecular response to cold, drought, heat and salt stress in higher plants. RG Landes Company, Georgetown, pp 125–152

Gagneul D, Ainouche A, Duhaze C, Lugan R, Larher FR, Bouchereau A (2007) A reassessment of the function of the so-called compatible solutes in the halophytic *Plumbaginaceae Limonium* latifolium. Plant Physiol 144:1598–1611

Galston AW, Kaur-Sawhney R, Altabella T, Tiburcio AF (1997) Plant polyamines in reproductive activity and response to abiotic stress. Bot Acta 110: 197–207

Garg AK, Kim JK, Owens TG (2002) Trehalose accumulation in rice plants confers high tolerance levels to different abiotic stresses. Proc Natl Acad Sci U S A 99:15898–15903

Geeta (2014) Genetic engineering for osmolyte overproduction for abiotic stress tolerance in plants. Int J Curr Res 6:6130–6135

Gill SS, Tuteja N (2010) Polyamines and abiotic stress tolerance in plants. Plant Signal Behav 51:26–33

Ginzberg I, Stein H, Kapulnik Y, Szabados L, Strizhov N, Schell J, Koncz C, Zilberstein A (1998) Isolation and characterization of two different cDNAs of delta1-pyrroline-5-carboxylate synthase in alfalfa, transcriptionally induced upon salt stress. Plant Mol Biol 38:755–764

Goddijn OJM, van Dun K (1999) Trehalose metabolism in plants. Trends Plant Sci 4:315–319

Gong Q, Li P, Ma S, Rupassara SI, Bohnert HJ (2005) Salinity stress adaptation competence in the extremophile *Thellungiella halophila* in comparison with its relative Arabidopsis thaliana. Plant J 44:826–839

Greenway H, Munns R (1980) Mechanisms of salt tolerance in nonhalophytes. Annu Rev Plant Physiol 31:149–190

Groppa MD, Benavides MP (2008) Polyamines and abiotic stress: recent advances. Amino Acids 34:35–45

Groppa MD, Benavides MP, Tomaro ML (2003) Polyamine metabolism in sunflower and wheat leaf discs under cadmium or copper stress. Plant Sci 16:481–488

Gupta K, Dey A, Gupta B (2013) Plant polyamines in abiotic stress responses. Acta Physiol Plant 35:2015–2036

Gupta B, Huang B (2014) Mechanism of salinity tolerance in plants: physiological, biochemical, and molecular characterization. Int J Genom 2014, Article ID 701596, 18 pp. Doi: 10.1155/2014/701596

Gupta B, Gupta K, Huang B (2014) Role of polyamines in plant abiotic stress responses. In: Pessarakli M (ed) Handbook of crop stress physiology (3rd edn). CRC Press (Taylor & Francis), Boca Raton, pp 369–388, Chapter 19. Doi: 10.1201/b16675-24

Hamilton EW, Heckathorn SA (2001) Mitochondrial adaptations to NaCl complex I is protected by antioxidants and small heat shock proteins, whereas complex II is protected by proline and betaine. Plant Physiol 126:1266–1274

Han C, Liu Q, Yang Y (2009) Short-term effects of experimental warming and enhanced ultraviolet-B radiation on photosynthesis and antioxidant defence of *Picea asperata* seedlings. Plant Growth Regul 58:153–162

Hannah MA, Wiese D, Freund S, Fiehn O, Heyer AG, Hincha DK (2006) Natural genetic variation of freezing tolerance in *Arabidopsis*. Plant Physiol 142:98–112

Hanzawa Y, Takahashi T, Michael AJ, Burtin D, Long D, Pineiro M, Coupland G, Komeda Y (2000) *ACAULIS5*, an *Arabidopsis* gene required for stem elongation, encodes a spermine synthase. EMBO J 19:4248–4256

Hanzawa Y, Imai A, Michael AJ, Komeda Y, Takahashi T (2002) Characterization of the spermidine synthase-related gene family in *Arabidopsis thaliana*. FEBS Lett 527:176–180

Hasegawa PM, Bressan RA, Zhu JK, Bohnert HJ (2000) Plant cellular and molecular responses to high salinity. Annu Rev Plant Physiol Plant Mol Biol 51:463–499

Hashimoto T, Tamaki K, Suzuki K, Yamada Y (1998) Molecular cloning of plant spermidine synthases. Plant Cell Physiol 39:73–79

Hasthanasombut S, Ntui V, Supaibulwatana K, Mii M, Nakamura I (2010) Expression of Indica rice *OsBADH1* gene under salinity stress in transgenic tobacco. Plant Biotechnol Rep 4:75–83

He L, Ban Y, Inoue H, Matsuda N, Liu J, Moriguchi T (2008) Enhancement of spermidine content and antioxidant capacity in transgenic pear shoots overexpressing apple spermidine synthase in response to salinity and hyperosmosis. Phytochemistry 69:2133–2141

Heyno E, Mary V, Schopfer P, Krieger-Liszkay A (2011) Oxygen activation at the plasma membrane: relation between superoxide and hydroxyl radical production by isolated membranes. Planta 234:35–45

Holmstrom KO, Mantyla E, Welin B, Mandal A, Palva ET (1996) Drought tolerance in tobacco. Nature 379:683–684

Holmstrom KO, Somersalo S, Mandal A, Palva ET, Welin B (2000) Improved tolerance to salinity and low temperature in transgenic tobacco producing glycine betaine. J Exp Bot 51:177–185

Hong ZL, Lakkineni K, Zhang ZM, Verma DPS (2000) Removal of feedback inhibition of Delta(1)-pyrroline-5-carboxylate synthetase results in increased proline accumulation and protection of plants from osmotic stress. Plant Physiol 122:1129–1136

Hoque MA, Banu MNA, Okuma E, Amako K, Nakamura K, Shimoishi Y, Murata Y (2007) Exogenous proline and glycinebetaine increase NaCl-induced ascorbate-glutathione cycle enzyme activities, and proline improves salt tolerance more than glycinebetaine in tobacco Bright Yellow-2 suspension-cultured cells. J Plant Physiol 164:1457–1468

Hoque MA, Okuma E, Nakamara Y, Shimoishi Y, Murata Y (2008) Proline and glycinebetaine enhance antioxidant defence and methylglyoxal detoxification systems and reduce NaCl induced damage in cultured tobacco cells. J Plant Physiol 165:813–824

Hu WH, Song SX, Shi K, Xia XJ, Zhou YH, Yu JQ (2008) Changes in electron transport, superoxide dismutase and ascorbate peroxidase isoenzymes in chloroplasts and mitochondria of cucumber leaves as influenced by chilling. Photosynthetica 46:581–588

Hu Y, Chen L, Wang H, Zhang L, Wang F, Yu D (2013) *Arabidopsis* transcription factor WRKY8 functions antagonistically with its interacting partner VQ9 to modulate salinity stress tolerance. Plant J 74:730–745

Hummel I, Gouesbet G, El Amrani A, Ainouche A, Couee I (2004) Characterization of the two arginine decarboxylase (polyamine biosynthesis) paralogues of the endemic subantarctic cruciferous species *Pringlea antiscorbutica* and analysis of their differential expression during development and response to environmental stress. Gene 342:199–209

Hussain SS, Ali M, Ahmad M, Siddique KH (2011) Polyamines: natural and engineered abiotic and biotic stress tolerance in plants. Biotechnol Adv 29:300–311

Igarashi K, Kashiwagi K (2000) Polyamines: mysterious modulators of cellular functions. Biochem Biophys Res Commun 271:559–564

Imai R, Ali A, Pramanik HR, Nakaminami K, Sentoku N, Kato H (2004) A distinctive class of spermidine synthase is involved in chilling response in rice. J Plant Physiol 161:883–886

Islam MM, Hoque MA, Okuma E, Banu MNA, Shimoishi Y, Nakamura Y, Murata Y (2009) Exogenous proline and glycinebetaine increase antioxidant enzyme activities and confer tolerance to cadmium stress in cultured tobacco cells. J Plant Physiol 166(15):1587–1597

Iqbal M, Ashraf M (2005) Changes in growth, photosynthetic capacity, and ionic relations in spring wheat (*Triticum aestivum* L.) due to pre-sowing seed treatment with polyamines. Plant Growth Regul 46:19–30

Janowitz T, Kneifel H, Piotrowski M (2003) Identification and characterization of plant agmatine iminohydrolase, the last missing link in polyamine biosynthesis of plants. FEBS Lett 544:258–261

Jantaro S, Mäenpää P, Mulo P, Incharoensakdi A (2003) Content and biosynthesis of polyamines in salt and osmotically stressed cells of Synechocystis sp. PCC 680. FEMS Microbiol Lett 228:129–135

Janz D, Behnke K, Schnitzler JP, Kanawati B, Schmitt-Kopplin P, Polle A (2010) Pathway analysis of the transcriptome and metabolome of salt sensitive and tolerant poplar species reveals evolutionary adaption of stress tolerance mechanisms. BMC Plant Biol 10:150

Jimenez-Bremont JF, Becerra-Flora A, Hernandez-Lucero E, Rodriguez-Kessler M, Acosta-Gallegos JA, Ramirez-Pimentel JG (2006) Proline accumulation in two bean cultivars under salt stress and the effect of polyamines and ornithine. Biol Plant 50:763–766

Kakkar RK, Sawhney VK (2002) Polyamine research in plants a changing perspective. Physiol Plant 116:281–292

Kamal-Eldin A, Appelqvist LA (1996) The chemistry and antioxidant properties of tocopherols and tocotrienols. Lipids 31:671–701

Kaplan F, Kopka J, Dale WH, Zhao W, Schiller KC, Gatzke N, Sung DY, Guy CL (2004) Exploring the temperature-stress metabolome of *Arabidopsis*. Plant Physiol 136:4159–4168

Karakas B, Ozias-Akins P, Stushnoff C, Suefferheld M, Rieger M (1997) Salinity and drought tolerance in mannitol-accumulating transgenic tobacco. Plant Cell Environ 20:609–616

Karim S, Aronsson H, Ericson H, Pirhonen M, Leyman B, Welin B, Mantyla E, Palva ET, Van Dijck P, Holmstrom KO (2007) Improved drought tolerance without undesired side effects in transgenic plants producing trehalose. Plant Mol Biol 64:371–386

Kasinathan V, Wingler A (2004) Effect of reduced arginine decarboxylase activity on salt tolerance and on polyamine formation during salt stress in *Arabidopsis thaliana*. Plant Physiol 121:101–107

Kasukabe Y, He L, Nada K, Misawa S, Ihara I, Tachibana S (2004) Overexpression of spermidine synthase enhances tolerance to multiple environmental stresses and up-regulates the expression of various stress-regulated genes in transgenic *Arabidopsis thaliana*. Plant Cell Physiol 45:712–722

Kathuria H, Giri J, Nataraja KN, Murata N, Udayakumar M, Tyagi AK (2009) Glycinebetaine-induced water-stress tolerance in codA-expressing transgenic indica rice is associated with upregulation of several stress responsive genes. Plant Biotechnol J 7:512–526

Kaumar A, Altabella T, Taylor MA, Tiburcio AF (1997) Recent advances in polyamine research. Trends Plant Sci 2:124–130

Kaur-Sawhney R, Tiburcio AF, Altabella T, Galston A (2003) Polyamines in plants: an overview. J Cell Mol Biol 2:1–12

Keller F, Pharr DM (1996) Metabolism of carbohydrates in sinks and sources: galactosyl-sucrose oligosaccharides. In: Zamski E, Schaffer AA (eds) Photoassimilate distribution in plants and crops. Marcel Dekker, New York, pp 115–184

Kempa S, Krasensky J, Dal Santo S, Kopka J, Jonak C (2008) A central role of abscisic acid in stress-regulated carbohydrate metabolism. PLoS One 3, e3935

Kerepesi I, Galiba G (2000) Osmotic and salt stress-induced alteration in soluble carbohydrate content in wheat seedlings. Crop Sci 40:482–487

Kesari R, Lasky JR, Villamor JG, Des Marais DL, Chen Y-JC, Liu T-W, Lin W, Juenger TE, Verslues PE (2012) Intron-mediated alternative splicing of *Arabidopsis P5CS1* and its association with natural variation in proline and climate adaptation. Proc Natl Acad Sci U S A 109:9197–9202

Khan S, Ahmed N, Ahmed F, Kumar R (2010) Naturally occurring organic osmolytes: From cell physiology to disease prevention. IUBMB Life 62:891–895

Kishor PBK, Hong Z, Miao GH, Hu CAA, Verma DPS (1995) Overexpression of 1-pyrroline-5-carboxylate synthetase increases proline production and confers osmotolerance in transgenic plants. Plant Physiol 108:1387–1394

Klähn S, Hagemann M (2011) Compatible solute biosynthesis in cyanobacteria. Environ Microbiol 13:551–562

Knott JM, Römer P, Sumper M (2007) Putative spermine synthases from *Thalassiosira pseudonana* and *Arabidopsis thaliana* synthesize thermospermine rather than spermine. FEBS Lett 581:3081–3086

Kumar R (2009) Role of naturally occurring osmolytes on the protein and stability. Arch Biochem Biophys 491:1–6

Kumria R, Rajam MV (2002) Ornithine decarboxylase transgene in tobacco affects polyamine metabolism, in vitro morphogenesis and response to salt stress. J Plant Physiol 159:983–990

Kusano T, Yamaguchi K, Berberich T, Takahashi Y (2007a) Advances in polyamine research in 2007. J Plant Res 120:345–350

Kusano T, Yamaguchi K, Berberich T, Takahashi Y (2007b) The polyamine spermine rescues Arabidopsis from salinity and drought stresses. Plant Signal Behav 2:250–251

Kusano T, Berberich T, Tateda C, Takahashi Y (2008) Polyamines: essential factors for growth and survival. Planta 228:367–381

Kuznetsov VV, Shevyakova NI (2007) Polyamines and stress tolerance of plants. Plant Stress 1:50–71

Kuznetsov V, Shevyakova NI (2010) Polyamines and plant adaptation to saline environments. In: Remawat KG (ed) Desert plants. Springer, Berlin-Heidelberg, pp 261–298

Lamitina T, Huang CG, Strange K (2006) Genome-wide RNAi screening identifies protein damage as a regulator of osmoprotective gene expression. Proc Natl Acad Sci U S A 103:12173–12178

Larher FR, Aziz A, Gibon Y, Trotel-Aziz P, Sulpice R, Bouchereau A (2003) An assessment of the physiological properties of the so-called compatible solutes using in vitro experiments with leaf discs. Plant Physiol Biochem 41:657–t66

Le Rudulier D, Strom AR, Dandekar AM, Smith LT, Valentine RC (1984) Molecular biology of osmoregulation. Science 224:1064–1068

Lehmann S, Funck D, Szabados L, Rentsch D (2010) Proline metabolism and transport in plant development. Amino Acids 39:949–962

Leyman B, Van Dijck P, Thevelein JM (2001) An unexpected plethora of trehalose biosynthesis genes in *Arabidopsis thaliana*. Trends Plant Sci 6:510–513

Leyman B, Avonce N, Ramon M, Van Dijck P, Iturriaga G, Thevelein JM (2006) Trehalose-6-phosphate synthase as an intrinsic selection marker for plant transformation. J Biotechnol 121:309–317

Lilius G, Holmberg N, Bulow L (1996) Enhanced NaCl stress tolerance in transgenic tobacco expressing bacterial choline dehydrogenase. Nat Biotechnol 14:177–180

Liu K, Fu HH, Bei QX, Luan S (2000) Inward potassium channel in guard cells as a target for polyamine regulation of stomatal movements. Plant Physiol 124:1315–1325

Liu JH, Nada K, Honda C, Kitashiba H, Wen XP, Pang XM, Moriguchi T (2006) Polyamine biosynthesis of apple callus under salt stress: importance of Arginine decarboxylase pathway in stress response. J Exp Bot 57:2589–2599

Liu JH, Kitashiba H, Wang J, Ban Y, Moriguchi T (2007) Polyamines and their ability to provide environmental stress tolerance to plants. Plant Biotechnol 24:117–126

Liu JH, Nakajima I, Moriguchi T (2011) Effects of salt and osmotic stresses on free polyamine content and expression of polyamine biosynthetic genes in *Vitis vinifera*. Biol Plant 55:340–344

López M, Tejera NA, Iribarne C, Lluch C, Herrera-Cervera JA (2008) Trehalose and trehalase in root nodules of Medicago truncatula and Phaseolus vulgaris in response to salt stress. Physiol Plant 134(4):575–582

Lugan R, Niogret MF, Leport L, Guegan JP, Larher FR, Savoure A, Kopka J, Bouchereau A (2010) Metabolome and water homeostasis analysis of *Thellungiella salsuginea* suggests that dehydration tolerance is a key response to osmotic stress in this halophyte. Plant J 64:215–229

Ma H, Liu WW, Chen X, Wu YJ, Yu ZL (2009) Enhanced enzymatic saccharification of rice straw by microwave pretreatment. Bioresour Technol 100:1279–1284

Maheshwari R, Dubey RS (2009) Nickel-induced oxidative stress and the role of antioxidant defence in rice seedlings. Plant Growth Regul 59:37–49

Makela P, Mantila J, Hinkkanen R, Pehu E, Peltonen-Sainio P (1996a) Effects of foliar applications of glycine betaine on stress tolerance, growth, and yield of spring cereals and summer turnip rape in Finland. J Agr Crop Sci 176:223–234

Makela P, Peltonen-Sainio P, Jokinen K, Pehu E, Setala H, Hinkkanen R, Somersalo S (1996b) Uptake and translocation of foliar applied glycine betaine in crop plants. Plant Sci 121:221–230

Martin-Tanguy J (2001) Metabolism and function of polyamines in plants: recent development (new approaches). Plant Growth Regul 34:135–148

Mattioli R, Marchese D, D'Angeli S, Altamura MM, Costantino P, Trovato M (2008) Modulation of intracellular proline levels affects flowering time and inflorescence architecture in *Arabidopsis*. Plant Mol Biol 66(3):277–288

McCue KF, Hanson AD (1990) Drought and salt tolerance: towards understanding and application. Trends Biotechnol 8:358–362

Minocha SC, Sun DY (1997) Stress tolerance in plants through transgenic manipulation of polyamine biosynthesis. Plant Physiol 114:S297

Miranda JA, Avonce N, Suarez R, Thevelein JM, Van Dijck P, Iturriaga G (2007) A bifunctional TPS-TPP enzyme from yeast confers tolerance to multiple and extreme abiotic-stress conditions in transgenic Arabidopsis. Planta 226:1411–1421

Mishra S, Jha AB, Dubey RS (2011) Arsenite treatment induces oxidative stress, upregulates antioxidant system, and causes phytochelatin synthesis in rice seedlings. Protoplasma 248:565–577

Mittler R (2002) Oxidative stress, antioxidants and stress tolerance. Trends Plant Sci 7:405–410

Mizoi J, Shinozaki K, Yamaguchi-Shinozaki K (2011) AP2/ERF family transcription factors in plant abiotic stress responses. Biochim Biophys Acta 1819:86–96

Mohapatra S, Minocha R, Long S, Minocha SC (2010) Transgenic manipulation of a single polyamine in poplar cells affects the accumulation of all amino acids. Amino Acids 38:1117–1129

Moschou PN, Paschalidis KA, Roubelakis-Angelakis KA (2008) Plant polyamine catabolism: the state of the art. Plant Signal Behav 3:1061–1066

Mundree SG, Baker B, Mowla S, Peters S, Marais S, Willigen CV, Govender K, Maredza A, Muyanga S, Farrant JM, Thomson JA (2002) Physiological and molecular insights into drought tolerance. Afr J Biotechnol 1:28–38

Nadeau P, Delaney S, Chouinard L (1987) Effects of cold hardening on the regulation of polyamine levels in wheat (*Triticum aestivum* L.) and alfalfa (*Medicago sativa* L.). Plant Physiol 84:73–77

Nakayama N, Saneoka H, Moghaieb REA, Premachandra GS, Fujita K (2007) Response of growth, photosynthetic gas exchange, translocation of 13C-labelled photosynthate and N accumulation in two soybean (*Glycine max* L. Merrill) cultivars to drought stress. Int J Agric Biol 9:669–674

Nanjo T, Kobayashi M, Yoshiba Y, Kakubari Y, Yamaguchi-Shinozaki K, Shinozaki K (1999) Antisense suppression of proline degradation improves tolerance to freezing and salinity in *Arabidopsis thaliana*. FEBS Lett 461:205–210

Navakouidis E, Lütz C, Langebartels C, Lütz-Meindl U, Kotzabasis K (2003) Ozone impact on the photosynthetic apparatus and the protective role of polyamines. Biochim Biophys Acta 1621:160–169

Nayyar H (2005) Putrescine increases floral retention, pod set and seed yield in cold stressed chickpea. J Agron Crop Sci 191:340–345

Nayyar H, Chander S (2004) Protective effects of polyamines against oxidative stress induced by water and cold stress in chickpea. J Agron Crop Sci 190:355–365

Nayyar H, Kaur S, Singh S, Kumar S, Singh KJ, Dhir KK (2005) Involvement of polyamines in the contrasting sensitivity of chickpea (*Cicer arietinum* L.) and soybean (*Glycine max* (L.) Merrill.) to water deficit stress. Bot Bull Acad Sin 46:333–338

Ndayiragiji A, Lutts S (2006) Exogenous putrescine reduces sodium and chloride accumulation in NaCl-treated calli of the salt-sensitive rice cultivar I Kong Pao. Plant Growth Regul 48:51–63

Neill S, Desikan R, Hancock J (2002) Hydrogen peroxide signalling. Curr Opin Plant Biol 5:388–395

Neily MH, Baldet P, Arfaoui I, Saito T, Li QL, Asamizu E, Matsukura C, Moriguchi T, Ezura H (2011) Overexpression of apple spermidine synthase 1 (MdSPDS1) leads to significant salt tolerance in tomato plants. Plant Biotechnol 28(1):33–42

Nelson DE, Rammesmayer G, Bohnert HJ (1998) Regulation of cell-specific inositol metabolism and transport in plant salinity tolerance. Plant Cell 10:753–764

Nishizawa A, Yabuta Y, Shigeoka S (2008) Galactinol and raffinose constitute a novel function to protect plants from oxidative damage. Plant Physiol 147:1251–1263

Niu X, Bressan R, Hasegawa P, Pardo J (1995) Ion homeostasis in NaCl stress environments. Plant Physiol 109:735–742

Noctor G, Foyer CH (1998) Ascorbate and glutathione: keeping active oxygen under control. Annu Rev Plant Physiol Plant Mol Biol 49:249–279

Nounjan N, Nghia PT, Theerakulpisut P (2012) Exogenous proline and trehalose promote recovery of rice seedlings from salt-stress and differentially modulate antioxidant enzymes and expression of related genes. J Plant Physiol 169:596–604

Nyyssola A, Kerovuo J, Kaukinen P, von Weymarn N, Reinikainen T (2000) Extreme halophiles synthesize betaine from glycine by methylation. J Biol Chem 275:22196–22201

Oh MJ, Chun HS, Lee CB (2003) Differences in photosynthetic characterization of salt tolerance for two rice (*Oryza sativa*) cultivars. J Plant Biol 46:17–22

Pang XM, Zhang ZY, Wen XP, Ban Y, Moriguchi T (2007) Polyamine, all-purpose players in response to environment stresses in plants. Plant Stress 1:173–188

Panicot M, Minguet EG, Ferrando A, Alcázar R, Blázquez MA, Carbonell J, Altabella T, Koncz C, Tiburcio AF (2002) A polyamine metabolon involving aminopropyl transferase complexes in *Arabidopsis*. Plant Cell 14:2539–2551

Panikulangara TJ, Eggers-Schumacher G, Wunderlich M, Stransky H, Sch€offl F (2004) Galactinol synthase 1: a novel heat shock factor target gene responsible for heat-induced synthesis of raffinose family oligosaccharides in *Arabidopsis*. Plant Physiol 136:3148–3158

Paschalidis KA, Roubelakis-Angelakis KA (2005a) Sites and regulation of polyamine catabolism in the tobacco plant. Correlations with cell division/expansion, cell cycle progression, and vascular development. Plant Physiol 138:2174–2184

Paschalidis KA, Roubelakis-Angelakis KA (2005b) Spatial and temporal distribution of polyamine levels and polyamine anabolism in different organs/tissues of the tobacco plant. Correlations with age, cell division/expansion, and differentiation. Plant Physiol 138:142–152

Peremarti A, Bassie L, Christou P, Capell T (2009) Spermine facilitates recovery from drought but does not confer drought tolerance in transgenic rice plants expressing Datura stramonium S-adenosylmethionine decarboxylase. Plant Mol Biol 70:253–264

Peréz-Amador MA, Leon J, Green PJ, Carbonell J (2002) Induction of the Arginine decarboxylase *ADC2* gene provides evidence for the involvement of polyamines in the wound response in *Arabidopsis*. Plant Physiol 130:1454–1463

Peshev D, Vergauwen R, Moglia A, Hideg E, Van den Ende W (2013) Towards understanding vacuolar antioxidant mechanisms: a role for fructans? J Exp Bot 64:1025–1038

Piotrowski M, Janowitz T, Kneifel H (2003) Plant C–N hydrolases and the identification of a plant N-carbamoylputrescine amidohydrolase involved in polyamine biosynthesis. J Biol Chem 278:1708–1712

Popovic RB, Kyle DJ, Cohen AS, Zalik S (1979) Stabilization of thylakoid membranes by spermine during stress induced senescence of barley leaf discs. Plant Physiol 64:721–726

Pottosin I, Velarde-Buendía AM, Bose J, Zepeda-Jazo I, Shabala S, Dobrovinskaya O (2014) Cross-talk between reactive oxygen species and polyamines in regulation of ion transport across the plasma membrane: implications for plant adaptive responses. J Exp Bot 65:1271–1283

Prabhavathi VR, Rajam MV (2007) Polyamine accumulation in transgenic eggplant enhances tolerance to multiple abiotic stresses and fungal resistance. Plant Biotechnol 24:273–282

Pramanik MHR, Imai R (2005) Functional identification of a trehalose 6-phosphate phosphatase gene that is involved in transient induction of trehalose biosynthesis during chilling stress in rice. Plant Mol Biol 58:751–762

Qi YC, Wang FF, Zhang H, Liu WQ (2010) Overexpression of suadea salsa S-adenosylmethionine synthetase gene promotes salt tolerance in transgenic tobacco. Acta Physiol Plant 32:263–269

Radukina NL, Ivanov YV, Kartashov AV, Shevyakova NI, Rakitin VY, Khryanin VN, Kuznetsov VV (2007) Inducible and constitutive mechanisms of salt stress resistance in *Geum urbanum*. Russ J Plant Physiol 54:612–618

Rammesmayer G, Pichorner H, Adams P, Jensen RG, Bohnert HJ (1995) Characterization of IMT1, myo-inositol O-methyltransferase, from *Mesembryanthemum crystallinum*. Arch Biochem Biophys 322:183–188

Rawia Eid A, Taha LS, Ibrahiem SMM (2011) Alleviation of adverse effects of salinity on growth, and chemical constituents of marigold plants by using glutathione and ascorbate. J Appl Sci Res 7:714–721

Rhodes D, Hanson AD (1993) Quaternary ammonium and tertiary sulfonium compounds in higher plants. Annu Rev Plant Physiol Plant Mol Biol 44: 357–384

Ribarits A, Abdullaev A, Tashpulatov A, Richter A, Heberle-Bors E, Touraev A (2007) Two tobacco proline dehydrogenases are differentially regulated and play a role in early plant development. Planta 225:1313–1324

Richards AB, Krakowka S, Dexter LB (2002) Trehalose: a review of properties, history of use and human tolerance, and results of multiple safety studies. Food Chem Toxicol 40:871–898

Rider JE, Hacker A, Mackintosh CA, Pegg AE, Woster PM, Casero RA Jr (2007) Spermine and spermidine mediate protection against oxidative damage caused by hydrogen peroxide. Amino Acids 33:231–240

Roy M, Wu R (2001) Arginine decarboxylase transgene expression and analysis of environmental stress tolerance in transgenic rice. Plant Sci 160:869–875

Roy M, Wu R (2002) Overexpression of S-adenosylmethionine decarboxylase gene in rice increases polyamine level and enhances sodium chloride-stress tolerance. Plant Sci 163:987–992

Roy P, Niyogi K, Sengupta DN, Ghosh B (2005) Spermidine treatment to rice seedlings recovers salinity stress induced damage of Plasma membrane and PM-bound H + ATPase in salt-tolerant and salt sensitive rice cultivars. Plant Sci 168:583–591

Roychoudhury A, Basu S, Sengupta DN (2011) Amelioration of salinity stress by exogenously applied spermidine or spermine in three varieties of indica rice differing in their level of salt tolerance. J Plant Physiol 168:317–328

Saha J, Brauer EK, Sengupta A, Popescu SC, Gupta K, Gupta B (2015) Polyamines as redox homeostasis regulators during salt stress in plants. Front Environ Sci 3:21

Sairam RK, Tyagi A (2004) Physiology and molecular biology of salinity stress tolerance in plants. Curr Sci 84:407–421

Saleethong P, Sanitchon J, Kong-ngern K, Theerakulpisut P (2011) Pretreatment with spermidine reverses inhibitory effects of salt stress in two rice (*Oryza sativa* L.) cultivars differing in salinity tolerance. Asian J Plant Sci 10:245–254

Sanchez DH, Siahpoosh MR, Roessner U, Udvardi M, Kopka J (2008) Plant metabolomics reveals conserved and divergent metabolic responses to salinity. Physiol Plant 132:209–219

Schneider T, Keller F (2009) Raffinose in chloroplasts is synthesized in the cytosol and transported across the chloroplast envelope. Plant Cell Physiol 50:2174–2182

Sebela M, Radova A, Angelini R, Tavladoraki P, Frebort I, Pec P (2001) FAD-containing polyamine oxidases: a timely challenge for researchers in biochemistry and physiology of plants. Plant Sci 160:197–207

Seki M, Umezawa T, Urano K, Shinozaki K (2007) Regulatory metabolic networks in drought stress responses. Curr Opin Plant Biol 10:296–302

Sfichi L, Ioannidis N, Kotzabasis K (2004) Thylakoid-associated polyamines adjust the UV-B sensitivity of the photosynthetic apparatus by means of light-harvesting complex II changes. Photochem Photobiol 80:499–506

Shabala S, Cuin TA, Pottosin I (2007) Polyamines prevent NaCl- induced K+ efflux from pea mesophyll by blocking non-selective cation channels. FEBS Lett 581:1993–1999

Sharma S, Dietz K (2006) The significance of amino acids and amino acid-derived molecules in plant responses and adaptation to heavy metal stress. J Exp Bot 57:711–726

Sharma P, Dubey RS (2005) Drought induces oxidative stress and enhances the activities of antioxidant enzymes in growing rice seedlings. Plant Growth Regul 46:209–221

Sharma P, Dubey RS (2007) Involvement of oxidative stress and role of antioxidative defence system in growing rice seedlings exposed to toxic concentrations of aluminium. Plant Cell Rep 26:2027–2038

Shen B, Hohmann S, Jensen RG, Bohnert H (1999) Roles of sugar alcohols in osmotic stress adaptation. Replacement of glycerol by mannitol and sorbitol in yeast. Plant Physiol 121:45–52

Shevyakova NI, Cheremisina AI, Kuznetsov VV (2011) Phytoremediation potential of amaranthus hybrids: antagonism between nickel and iron and chelating role of polyamines. Russ J Plant Physiol 58:634–642

Shinozaki K, Yamaguchi-Shinozaki K (2000) Molecular responses to dehydration and low temperature: difference and cross-talk between two stress signaling pathways. Curr Opin Plant Biol 3:217–223

Shu S, Guo SR, Yuan LY (2012) A review: polyamines and photosynthesis. In: Najafpour MM (ed) Advances in photosynthesis – fundamental aspects. InTech, Rijeka, pp 439–464

Simon-Sarkadi L, Kocsy G, Várhegyi Á, Galiba G, De Ronde J (2005) Genetic manipulation of proline accumulation influences the concentrations of other amino acids in soybean subjected to simultaneous drought and heat stress. J Agric Food Chem 53:7512–7517

Simon-Sarkadi L, Kocsy G, Várhegyi Á, Galiba G, De Ronde JA (2006) Stress-induced changes in the free amino acid composition in transgenic soybean plants having increased proline content. Plant Biol 50:793–796

Sinkevich MS, Naraykina NV, Trunova TI (2010) Sugars participate in the antioxidant protection from oxidative stress induced by paraquat in the case of potato transformed with yeast invertase gene. Dokl Akad Nauk 434:570–573

Sivakumar P, Sharmila P, Saradhi PP (2000) Proline alleviates salt-stress induced enhancement in ribulose-1,5-bisphosphate oxygenase activity. Biochem Biophys Res Commun 279:512–515

Smirnoff N, Cumbes QJ (1989) Hydroxyl radical scavenging activity of compatible solutes. Phytochemistry 28:1057–1060

Song J, Nada K, Tachibana S (2002) Suppression of s-adenosylmethionine decarboxylase activities is a major cause for high temperature inhibition of pollen germination and tube growth in tomato (*Lycopersicon esculentum* Mill.). Plant Cell Physiol 43:619–627

Song SY, Chen Y, Chen J, Dai XY, Zhang WH (2011) Physiological mechanisms underlying OsNAC5-dependent tolerance of rice plants to abiotic stress. Planta 234:331–345

Soyka S, Heyer AG (1999) *Arabidopsis* knockout mutation of *ADC2* gene reveals inducibility by osmotic stress. FEBS Lett 458:219–223

Srivastava S, Dubey RS (2011) Manganese-excess induces oxidative stress, lowers the pool of antioxidants and elevates activities of key antioxidative enzymes in rice seedlings. Plant Growth Regul 64:1–16

Storey R, Ahmad N, Wyn Jones RG (1977) Taxonomic and ecological aspects of the distribution of glycinebetaine and related compounds in plants. Oecologia 27:319–332

Stoyanova S, Geuns J, Hideg E, Van den Ende W (2011) The food additives inulin and stevioside counteract oxidative stress. Int J Food Sci Nutr 62:207–214

Strizhov N, Abraham E, Okresz L, Blickling S, Zilberstein A, Schell J, Koncz C, Szabados L (1997) Differential expression of two *P5CS* genes controlling proline accumulation during salt-stress requires ABA and is regulated by *ABA1*, *ABI1* and AXR2 in *Arabidopsis*. Plant J 12:557–569

Su J, Hirji R, Zhang L, He C, Selvaraj G, Wu R (2006) Evaluation of the stress-inducible production of choline oxidase in transgenic rice as a strategy for producing the stress-protectant glycine betaine. J Exp Bot 57:1129–1135

Sun SJ, Guo SQ, Yang X, Bao YM, Tang HJ, Sun H (2010) Functional analysis of a novel Cys2/His2-type zinc finger protein involved in salt tolerance in rice. J Exp Bot 61:2807–2818

Sung MS, Chow TJ, Lee TM (2011) Polyamine acclimation alleviates hyper salinity-induced oxidative stress in a marine green macroalga, *Ulva fasciata*, by modulation of antioxidative enzyme gene expression. J Phycol 47:538–547

Szabados L, Savoure A (2010) Proline: a multifunctional amino acid. Trends Plant Sci 15:89–97

Szekely G, Abraham E, Cseplo A, Rigo G, Zsigmond L, Csiszar J, Ayaydin F, Strizhov N, Jasik J, Schmelzer E, Koncz C, Szabadoz L (2008) Duplicated P5CS genes of Arabidopsis play distinct roles in stress regulation and developmental control of proline biosynthesis. Plant J 53:11–28

Tang W, Newton RJ (2005) Polyamines reduce salt-induced oxidative damage by increasing the activities of antioxidant enzymes and decreasing lipid peroxidation in Virginia pine. Plant Growth Regul 46:31–43

Tanou G, Molassiotis A, Diamantidis G (2009) Induction of reactive oxygen species and necrotic death-like destruction in strawberry leaves by salinity. Environ Exp Bot 65:270–281

Tarczynski MC, Jensen RG, Bohnert HJ (1992) Expression of a bacterial *mtlD* gene in transgenic tobacco leads to production and accumulation of mannitol. Proc Natl Acad Sci U S A 89:2600–2604

Thomas T, Thomas TJ (2001) Polyamines in cell growth and cell death: molecular mechanisms and therapeutic applications. Cell Mol Life Sci 58:244–258

Thomas JC, Sepahi M, Arendall B, Bohnert HJ (1995) Enhancement of seed germination in high salinity by engineering mannitol expression in *Arabidopsis thaliana*. Plant Cell Environ 18:801–806

Thu-Hang P, Bassie L, Safwat G, Trung-Nghia P, Christou P, Capell T (2002) Expression of a heterologous S-adenosylmethionine decarboxylase cDNA in plants demonstrates that changes in S-adenosyl-L-methionine decarboxylase activity determine levels of the higher polyamines spermidine and spermine. Plant Physiol 129:1744–1754

Uemura M, Steponkus PL (1997) Artificial manipulation of the intracellular sucrose content alters the incidence of freeze-induced membrane lesions of isolated protoplasts of Arabidopsis thaliana. Cryobiology 35:336

Urano K, Yoshiba Y, Nanjo T, Igarashi Y, Seki M, Sekiguchi F, Yamaguchi-Shinozaki K, Shinozaki K (2003) Characterization of *Arabidopsis* genes involved in biosynthesis of polyamines in abiotic stress responses and developmental stages. Plant Cell Environ 26:1917–1926

Urano K, Yoshiba Y, Nanjo T, Ito T, Yamaguchi-Shinozaki K, Shinozaki K (2004) Arabidopsis stress-inducible gene for arginine decarboxylase AtADC2 is required for accumulation of putrescine in salt tolerance. Biochem Biophys Res Commun 313:369–375

Urano K, Maruyama K, Ogata Y, Morishita Y, Takeda M, Sakurai N, Suzuki H, Saito K, Shibata D, Kobayashi M, Yamaguchi-Shinozaki K, Shinozaki K (2009) Characterization of the ABA-regulated global responses to dehydration in *Arabidopsis* by metabolomics. Plant J 57:1065–1078

Usadel B, Blasing OE, Gibon Y, Poree F, Hohne M, Gunter M, Trethewey R, Kamlage B, Poorter H, Stitt M (2008) Multilevel genomic analysis of the response of transcripts, enzyme activities and metabolites in *Arabidopsis* rosettes to a progressive decrease of temperature in the non-freezing range. Plant Cell Environ 31:518–547

Van den Ende W, Valluru R (2009) Sucrose, sucrosyl oligosaccharides, and oxidative stress: scavenging and salvaging. J Exp Bot 60:9–18

Van Leeuwenhoek A (1678) Observationes D. Anthonii Leeuwenhoek, de natis e semine genitali animalculis. Philos Trans R Soc Lond 12:1040–1043

Van-Houtte H, Vandesteene L, López-Galvis L, Lemmens L, Kissel E, Carpentier S, Feil R, Avonce N, Beeckman T, Lunn JE, Van Dijck P (2013) Overexpression of the trehalase gene *AtTRE1* leads to increased drought stress tolerance in *Arabidopsis* and is involved in abscisic acid-induced stomatal closure. Plant Physiol 161:1158–1171

Velikova V, Yordanov I, Edreva A (2000) Oxidative stress and some antioxidant systems in acid rain-treated bean plants: protective role of exogenous polyamines. Plant Sci 151:5966

Vergnolle C, Vaultier MN, Taconnat L, Renou JP, Kader JC, Zachowski A, Ruelland E (2005) The cold-induced early activation of phospholipase C and D pathways determines the response of two distinct clusters of genes in *Arabidopsis* cell suspensions. Plant Physiol 139:1217–1233

Verma S, Mishra SN (2005) Putrescine alleviation of growth in salt stressed *Brassica juncea* by inducing antioxidative defence system. J Plant Physiol 162:669–677

Vernon DM, Bohnert HJ (1992) A novel methyl transferase induced by osmotic stress in the facultative halophyte Mesembryanthemum crystallinum. EMBO J 11:2077–2085

Waie B, Rajam MV (2003) Effect of increased polyamine biosynthesis on stress responses in transgenic tobacco by introduction of human S-adenosylmethionine gene. Plant Sci 164:727–734

Walden R, Cordeiro A, Tiburcio AF (1997) Polyamines: small molecules triggering pathways in plant growth and development. Plant Physiol 113:1009–1013

Wang X, Shi G, Xu Q, Hu J (2007) Exogenous polyamines enhance copper tolerance of *Nymphoides peltatum*. J Plant Physiol 164:1062–1070

Wang Z, Zhu Y, Wang L, Liu X, Liu Y, Phillips J, Deng X (2009) A WRKY transcription factor participates in dehydration tolerance in *Boea hygrometrica* by binding to the W-box elements of the galactinol synthase (BhGolS1) promoter. Planta 230:1155–1166

Wani SH, Gosal SS (2011) Introduction of *OsglyII* gene into Indica rice through particle bombardment for increased salinity tolerance. Biol Plant 55: 536–540

Wani SH, Lone AA, da Silva T, Gosal SS (2010) Effects of NaCl stress on callus induction and plant regeneration from mature seeds of rice (*Oryza sativa* L.). Asian Aust J Plant Sci Biotech 4:56–71

Watson MB, Malmberg RL (1996) Regulation of Arabidopsis thaliana (L.) heynh arginine decarboxylase by potassium deficiency stress. Plant Physiol 111:1077–1083

Watson MB, Yu W, Galloway G, Malmberg RL (1997) Isolation and characterization of a second arginine decarboxylase cDNA from *Arabidopsis* (Accession No. AF009647). Plant Physiol 114:1569

Wen XP, Pang XM, Matsuda N, Kita M, Inoue M, Hao YJ, Honda C, Moriguchi T (2008) Overexpression of the apple spermidine synthase gene in pear confers multiple abiotic stress tolerance by altering polyamine titers. Transgenic Res 17:251–263

Wi SJ, Kim WT, Park KY (2006) Overexpression of carnation S-adenosylmethionine decarboxylase gene generates a broad spectrum tolerance to abiotic stresses in transgenic tobacco plants. Plant Cell Rep 25:1111–1121

Xiang Y, Huang Y, Xiong L (2007) Characterization of stress-responsive CIPK genes in rice for stress tolerance improvement. Plant Physiol 144:1416–1428

Xin Z, Browse J (2000) Cold comfort farm: the acclimation of plants to freezing temperatures. Plant Cell Environ 23:893–902

Yamaguchi K, Takahashi Y, Berberich T, Imai A, Miyazaki A, Takahashi T (2006) The polyamine spermine protects against high salt stress in *Arabidopsis thaliana*. FEBS Lett 580:6783–6788

Yamaguchi K, Takahashi Y, Berberich T, Imai A, Takahashi T, Michael AJ, Kusano T (2007) A protective role for the polyamine spermine against drought stress in *Arabidopsis*. Biochem Biophys Res Commun 352:486–490

Yan J, Tsuichihara N, Etoh T, Iwai S (2007) Reactive oxygen species and nitric oxide are involved in ABA inhibition of stomatal opening. Plant Cell Environ 30:1320–1325

Yancey P, Clark ME, Had SC, Bowlus RD, Somero GN (1982) Living with water stress: evolution of osmolyte system. Science 217:1214–1222

Yang X, Lu C (2005) Photosynthesis is improved by exogenous glycine betaine in salt-stressed maize plants. Physiol Plant 124:343–352

Yang JC, Zhang JH, Liu K, Wang ZQ, Liu LJ (2007) Involvement of polyamines in the drought resistance of rice. J Exp Bot 58:1545–1555

Yang X, Liang Z, Wen X, Lu C (2008) Genetic engineering of the biosynthesis of glycine betaine leads to increased tolerance of photosynthesis to salt stress in transgenic tobacco plants. Plant Mol Biol 66:73–86

Yeo ET, Kwon HB, Han SE, Lee JT, Ryu JC, Byu MO (2000) Genetic engineering of drought resistant potato plants by introduction of the trehalose-6-phosphate synthase (*TPS1*) gene from *Saccharomyces cerevisiae*. Mol Cells 10:263–268

Yiu JC, Juang LD, Fang DYT, Liu CW, Wu SJ (2009) Exogenous putrescine reduces flooding induced oxidative damage by increasing the antioxidant properties of Welsh onion. Sci Hortic 120:306–314

Zepeda-Jazo I, Shabala S, Chen Z, Pottosin II (2008) Na^+-K^+ transport in roots under salt stress. Plant Signal Behav 3:401–403

Zepeda-Jazo I, Velarde-Buendía AM, Enríquez-Figueroa R, Bose J, Shabala S, Muñiz-Murguía J, Pottosin II (2011) Polyamines interact with hydroxyl radicals in activating Ca^{2+} and K^+ transport across the root epidermal plasma membranes. Plant Physiol 157:2167–2180

Zhang J, Tan W, Yang XH, Zhang HX (2008) Plastid expressed choline monooxygenase gene improves salt and drought tolerance through accumulation of glycine betaine in tobacco. Plant Cell Rep 27:1113–1124

Zhang RH, Li J, Guo SR, Tezuka T (2009a) Effects of exogenous putrescine on gas exchange characteristics and chlorophyll fluorescence of NaCl-stressed cucumber seedlings. Photosynth Res 100:155–162

Zhang W, Jiang B, Li W, Song H, Yu Y, Chen J (2009b) Polyamines enhance chilling tolerance of cucumber (*Cucumis sativus* L.) through modulating antioxidative system. Sci Hort 122:200–208

Zhang CY, Wang NN, Zhang YH, Feng QZ, Yang CW, Liu B (2013) DNA methylation involved in proline accumulation in response to osmotic stress in rice (*Oryza sativa*). Genet Mol Res 12:1269–1277

Zhao H, Yang H (2008) Exogenous polyamines alleviate the lipid peroxidation induced by cadmium chloride stress in *Malus hupehensis*. Rehd Sci Hortic 116:442–447

Zhao F, Song CP, He J, Zhu H (2007) Polyamines improve K^+/Na^+ homeostasis in barley seedlings by regulating root ion channel activities. Plant Physiol 145:1061–1072

Zhu JK (2001) Plant salt tolerance. Trends Plant Sci 61:66–71

Zuther E, Koehl K, Kopka J (2007) Comparative metabolome analysis of the salt response in breeding cultivars of rice. In: Jenks MA, Hasegawa PM, Jain SM (eds) Advances in molecular breeding toward drought and salt tolerant crops. Springer, Berlin-Heidelberg, New York, pp 285–315

Evaluating the Importance of Proline in Cadmium Tolerance and Its Interaction with Phytohormones

8

Noushina Iqbal, Rahat Nazar, and Shahid Umar

Abstract

Cadmium (Cd) stress is one of the most challenging environmental problems that adversely affects growth and development of plants. Plants adopt various strategies to overcome the adverse effects of Cd stress. Among these, recently phytohormones and osmolytes have been involved in overcoming the adverse effects of Cd stress. Proline is an important osmolyte that maintains cellular homeostasis through osmotic regulation and regulates physiological processes under Cd stress. The role of phytohormones under Cd stress is critical in modulating physiological responses that eventually leads to adaptation of plants to an unfavorable environment. The individual role of both proline and phytohormones has been extensively studied, but a comprehensive study on the interaction of phytohormones with proline under Cd is lacking. The present study focuses on enhancing our understanding on the mechanism of Cd tolerance via proline and phytohormones with emphasis on phytohormones' interaction with proline under Cd stress.

8.1 Introduction

Metal accumulation is among the most harmful environmental hazards arising in the late nineteenth and early twentieth centuries. Cadmium (Cd) is a persistent and bioaccumulative element and is potentially toxic to living organisms. It may be introduced into the environment by many anthropogenic activities, such as mining, fertilizer use, metal-based pesticides, and a wide range of industrial activities, which release metals into the environment (Zawoznik et al. 2007). In the case of plants, metals in the soil can enter the roots through symplastic or apoplastic pathways before entering the xylem and being translocated to the shoot (Lux et al. 2011), although transport through the phloem may also play a key role in delivering metals (Mendoza-Cózatl et al. 2011).

N. Iqbal (✉) • R. Nazar
Department of Botany, Jamia Hamdard University, New Delhi-110062, Delhi, India
e-mail: naushina.iqbal@gmail.com

S. Umar
Department of Botany, Faculty of Science, Jamia Hamdard University, New Delhi-110062, India

N. Iqbal et al. (eds.), *Osmolytes and Plants Acclimation to Changing Environment: Emerging Omics Technologies*, DOI 10.1007/978-81-322-2616-1_8

Cadmium is one of the most toxic heavy metal for plants, animals and humans and enters the environment through phosphate fertilizers and waste disposal (Yilmaz et al. 2006; Mahmood et al. 2009).

The adverse effects of Cd cause various biochemical, physiological, or molecular changes in plants. Cadmium toxicity leads to leaf chlorosis, stunted growth, and unspecific necrosis which causes death of plants (Kapoor and Bhardwaj 2014). It has been reported to inhibit photosynthesis and impair growth in plants (Nazar et al. 2012; Asgher et al. 2013). At the molecular level, Cd toxicity is associated with the formation and disruption of sulfhydryl and metal thiolate bonds, alterations in protein secondary structure, changes in the redox status of the cell, and interference with essential metal uptake, transport, and metabolism (Ouariti et al. 1997; Nies 1999; Sandalio et al. 2001). Cadmium may cause Chl destruction as a consequence of magnesium (Mg) substitution in both Chl a and b pigments (Parmar et al. 2013). It leads to the production of reactive oxygen species (ROS) that causes oxidative stress in plants (Anjum et al. 2014; Asgher et al. 2014). These ROS cause lipid peroxidation, enzyme inactivation, and DNA damage, resulting in dramatic reduction of growth and productivity, finally causing plant death (Sun et al. 2007). Cadmium is reported to disturb the uptake, transport, and use of several essential mineral elements, such as calcium (Ca), Mg, phosphorus (P), potassium (K), and iron (Fe) by inducing deficiency in plants (Metwally et al. 2005; Nedjimi and Daoud 2009; Nazar et al. 2012). Plant nutrients and Cd compete for the same transporters, and, therefore, the presence of Cd results in mineral nutrient deficiency (Nazar et al. 2012). Sun and Shen (2007) reported that in cabbage the key reason for decrease in leaf photosynthesis under Cd stress was the decrease in Mn, Fe, Mg, S, and P concentrations in leaves. The study of Nazar et al. (2012) extensively focuses on the role of Cd in mineral deficiency and the role of nutrients in its alleviation.

The role of phytohormones in the reduction of Cd-induced oxidative stress has also been reported. Cross talk of oxidative stress signaling cascades and endogenous factors, like ethylene, jasmonate, auxin, or abscisic acid (ABA), is pivotal for plant acclimation to stress and development (Potters et al. 2007), where antioxidants modulate ROS production (Considine and Foyer 2014). NahG plants with lower salicylic acid (SA) accumulation level showed more sensitivity to Cd toxicity. Exogenously applied SA prevented the Cd-induced photochemical efficiency decrease and mitigated Cd toxicity. SA pretreatment could alleviate Cd-induced ROS overproduction (Zhang and Chen 2011). Ethylene and gibberellic acid are involved in reducing Cd toxicity (Masood and Khan 2013). Basalah et al. (2013) reported that the application of both nitric oxide (NO) and SA together improved the plant growth and development by reducing the formation of ROS by improving antioxidant enzymes and CA activity and balancing supply of nutrients

In this study, we will focus on the importance of proline in Cd tolerance and its regulation directly or indirectly by phytohormones. Indirectly phytohormones may influence nutrient availability which affects proline content and Cd tolerance. Calcium has been reported to influence proline accumulation (Iqbal et al. 2014). Cadmium entry through the Ca channel in the leaves disturbs the plant–water relationship (Perfus-Barbeoch et al. 2002), causing stomatal closure in many plants, leading to lower transpiration rate and inhibition of photosynthesis through an adverse effect on chlorophyll metabolism. Pankovic et al. (2000) have shown that optimal N (7.5 mM) supply decreased the inhibitory effects of Cd on photosynthesis of sunflower plants by increasing ribulose 1,5-bisphosphate carboxylase (Rubisco) activity or by the increase in soluble protein content. Upon exposure to Cd, plants often synthesize a set of N-containing metabolites through N metabolism, such as proline, GSH, and phytochelatins, which play a significant role in Cd tolerance of plants (Sharma and Dietz 2006). Supplementation of S has also been reported to reduce Cd stress (Khan et al. 2014) through increase in GSH content or proline accumulation. Anjum et al. (2014) recently reported that proline and GSH are involved in metal tolerance and both are related to each other. Both GSH and proline, with molecular formula $C_{10}H_{17}N_3O_6S$ and $C_5H_9NO_2$, respectively, belong to the "glutamate or α-ketoglutarate" family and originate from a common precursor L-glutamate (Moat et al. 2003).

8.2 Proline in Cadmium Tolerance

Cadmium disturbs various biochemical and physiological processes, leading to cell death and inhibition of growth (Sandalio et al. 2001; Popova et al. 2009). One of the main adaptive mechanisms to Cd stress in plants is the accumulation of compatible solutes (Ashraf and Foolad 2007; Mehta and Gaur 1999; Shah and Dubey 1998; Sharma and Dietz 2006). The deleterious effect of Cd on plant growth can be mitigated through proline. Proline accumulates in response to heavy metal exposure among plants (Chen et al. 2001; Radic et al. 2010; Zengin and Munzuroglu 2005). Islam et al. (2009a) reported that exogenous application of proline and betaine restored the membrane integrity and increased the enzymes of ASH–GSH cycle under Cd stress in tobacco cells. They also reported that proline was more effective in protecting Cd-stressed plants as compared to betaine. Proline synthesis in plants occurs mainly from glutamate, which is reduced to L-glutamate γ-semialdehyde (GSA) by the P5CS enzyme and spontaneously converted to pyrroline-5-carboxylate (P5C). P5C reductase (P5CR) reduces the P5C intermediate to proline. The catabolism of proline occurs in mitochondria via the sequential action of proline dehydrogenase or proline oxidase (PDH or POX) producing P5C from proline and P5C dehydrogenase (P5CDH), which regenerates glutamate from 5C. Glutamate plays a central role in overall nitrogen (N) homeostasis. Alternatively, proline can also be synthesized from ornithine, which is transaminated first by ornithine-delta-aminotransferase (OAT) producing GSA and P5C and then converted to proline (Iqbal et al. 2014).

Proline may also protect Cd-stressed plants through increase in antioxidant enzyme activity. An efficient antioxidant defense system provided by proline and betaine might play an important role in tolerance of plants to Cd stress. Proline and betaine enhance antioxidant defense systems in plant responses to various oxidative stresses (Demiral and Türkan 2004; Okuma et al. 2004; Molinari et al. 2007). The antioxidant protection role of proline has also been described in fungal pathogenesis during various oxidative stresses (Chen and Dickman 2005). Proline alleviates Cd toxicity by detoxifying ROS and increasing the activity of superoxide dismutase (SOD) and catalase (CAT) and glutathione content (Xu et al. 2009).

Besides, osmolytes like proline and glycine betaine play key role in cellular compatibility and osmotic adjustment to maintain normal metabolism of plants during stress. Khan et al. (2014) reported that Cd-induced oxidative stress was alleviated by ethylene through increase in proline and GSH content. Wen et al. (2011) reported that proline accumulation is positively correlated with putrescine (Put), spermidine (Spd), and titers of total polyamines in Cd-treated pear shoots. Proline accumulation was also closely linked with polyamines catabolism in plants exposed to abiotic stresses (Bouchereau et al. 1999), owing to the reason that proline and Put share a common substrate, glutamate (Seki et al. 2007; Sharma and Dietz 2006). The application of Zn alleviated Put and proline accumulation and thus helped in the adaptation of *L. minor* to Cd-induced stress (Qiao et al. 2015). The free proline has been found to chelate Cd ion in plants and form a nontoxic Cd–proline complex (Sharma et al. 1998). Proline increased significantly under all applied Cd concentrations (Aly 2012). The accumulation of proline may be due to increased synthesis from glutamate, lower rate of protein oxidation, and slowed incorporation of proline into protein (Bohnert et al. 1995). Hayat et al. (2013) suggested that the application of proline increases endogenous proline content under heavy metal stress conditions. This helps to protect enzymes (Islam et al. 2009), 3-D structure of proteins (Paleg et al. 1981), and organelle and cell membranes by reducing the lipid peroxidation (Okuma et al. 2004). Besides this, it also supplies energy for growth and survival, thereby helping the plant to tolerate stress (Hoque et al. 2007).

Supplementation of proline scavenges ROS and enhances activity of antioxidative enzymes (Islam et al. 2009; Hayat et al. 2013), i.e., CAT, POX, and SOD, by acting at the level of transcription and/or translation and the endogenous level of proline (Hayat et al. 2012). Proline protects plants from Cd-induced membrane degradation and electrolyte leakage (Hasan et al. 2009) and enhances leaf water potential by protecting the membranes from metal-induced oxidative damage (Okuma et al. 2004).

The involvement of proline in Cd tolerance also influences the GSH-mediated Cd tolerance. Siripornadulsil et al. (2002) studied the expression of a moth bean pyrroline-5-carboxylate synthetase (*P5CS*) gene in the green microalga *Chlamydomonas reinhardtii* and its influence on Cd tolerance. They reported that transgenic algae with *P5CS* gene produced more proline that subsequently increased GSH content and phytochelatin to sequester Cd and provide tolerance compared to wild-type algae. The increase in GSH/0.5 GSSG ratios (fourfold) in P5CS cells relative to wild-type cells was identical to the relative increase in bound Cd per cell (fourfold). Thus, proline is involved in Cd tolerance either by maintaining the cell water balance, increasing antioxidants that scavenge ROS, or by influencing the polyamines and GSH content.

8.3 Nutrients in Proline Accumulation and Cadmium Tolerance

Among the various nutrients, N and Ca have been reported to be involved in the regulation of proline metabolism (Iqbal et al. 2014). In the present study, we will discuss the role of N and Ca together with S in regulating proline and Cd tolerance. Nitrogen-induced alleviation of Cd toxicity has been suggested as the increased biomass production by increasing the photosynthetic rate with N application. The increased biomass production sequesters more Cd in vegetative parts and very little moves into the grain. Recently, Zhu et al. (2011) have shown that N fertilizer in the form of 16 mM $(NH_4)_2SO_4$ was effective in Cd alleviation in *Sedum*. The effect of N in alleviating Cd stress is dose dependent and occurs through its promoting effect on chlorophyll synthesis as well as on the activities of antioxidant enzymes, such as SOD, CAT, and POX, which partially alleviate the accumulation of ROS (Lin et al. 2011).

Tarighaleslami et al. (2012) reported that leaf proline increased with the application of N fertilizer. Rais et al. (2013) have shown that both individual and combined applications of N and sulfur (S) resulted in increased nitrate reductase activity (NRA), N content, and proline accumulation and alleviated salt stress effects on photosynthetic efficiency and growth of *Brassica juncea*. In contrast, Lee et al. (2009) reported that the increase in proline concentrations in phloem exudates was closely related to reduction in NRA in the roots, N uptake, and the assimilation of newly absorbed N, suggesting a link of N assimilation with proline. Zakery-Asl et al. (2014) reported that the application of high N in *Phaseolus vulgaris* increased proline in root and foliar organs, and proline biosynthesizing enzymes (P5CS and ornithine-delta-aminotransferase) varied with the N status. Sánchez et al. (2007) found that N deficiency resulted in decreased proline accumulation both in the pods and seeds of *Phaseolus vulgaris* due to the stimulation of the enzyme proline dehydrogenase (PDH). In their study on *Triticum aestivum*, Khan et al. (2013) have shown that increased N assimilation by SA induced enzymes of proline biosynthesis and proline accumulation for increased tolerance to heat stress. In another study, Khan et al. (2014) reported that increased proline content was responsible for Cd tolerance in *Triticum aestivum*. Upon exposure to Cd, plants often synthesize a set of N-containing metabolites through N metabolism, such as proline, GSH, and phytochelatins, which play a significant role in Cd tolerance of plants (Sharma and Dietz 2006). Balestrasse et al. (2005) reported that in the nodules of soybean plants under 200 µM Cd, where the GS/GOGAT cycle is severely inhibited, ammonia increased greatly and the same occurred with proline. Sakakibara (2003) reported that Cd supply caused an increased partitioning of nitrate in roots of *Arabidopsis*, as a result of the induction of NRT1.8 expression and repression of NRT1.5, specific transporters which are responsible for loading/unloading of nitrate into/from the xylem sap, respectively. Retention of nitrate in the roots serves as a signal to enhance nitrate assimilation in these organs. Since glutathione and phytochelatins are N-containing molecules, this could lead to increased synthesis of these chelators to detoxify Cd ions (Gojon and Gaymard 2010). The same can also be true for proline as it is also a N-containing compound.

Combined N and S application resulted in greater proline accumulation and alleviated salt stress effects on photosynthetic efficiency and growth. The effects of one element are positively influenced by the other. It is, therefore, assumed that the effects of N in inducing proline synthesis will be greater in the presence of S, and the adequate supply of S and N may result in reducing the negative effects of salinity stress (Rais et al. 2013). When there is limitation in S-dependent stress tolerance mechanism(s), plant opted to strengthen alternate mechanism of defense by accumulating proline, for the biosynthesis of which glutamate serves as precursor (Nikiforova et al. 2005). Khan et al. (2014) reported that Se and S both reversed Cd-induced oxidative stress by regulating ethylene formation and proline and GSH metabolism. Abdelhamid et al. (2013) studied that S improved proline content by about 20.4 % and 11.2 % relative to the control treatments under 100 % and 120 % irrigation treatments, respectively. Zafar et al. (2014) reported that foliar Zn and S application increased total chlorophyll, proline content, and SOD enzyme activity but decreased membrane stability. Sulfur dioxide has also been shown to affect glutamic acid and glutamine content of pea seedlings (Jager and Pahlich 1972) and asparagine, glutamine, glutamic acid, serine, glycine, alanine, arginine, ornithine, and proline content of spruce needles (Jager and Grill 1975). Sulfur may also regulate the formation of osmolytes by its influence on NRA and N assimilation (Marschner 1995). Khan et al. (2005) have reported that S supply improved photosynthesis and growth through regulating N assimilation. The availability of S regulates the NRA and the accumulation of N (Pal et al. 1976). Anjum et al. (2012) and Fismes et al. (2000) have shown using field-grown oilseed rape that S deficiency can reduce N use efficiency and that N deficiency can also reduce S use efficiency. Thus, it may be said that S supplementation could increase N assimilation and this in turn could increase proline accumulation to combat Cd stress.

Calcium is an important nutrient required for Cd tolerance. Ahmad et al. (2015) reported that the application of Ca enables mustard plant to withstand the deleterious effect of Cd, resulting in improved growth and seed quality of mustard plants. Proline was significantly increased in mustard plants under Cd stress, and exogenously sprayed Ca was found to have a positive impact on proline content in Cd-stressed plants. Supplemental Ca also enhances the accumulation of proline in saline conditions (Shah et al. 1990). Separate NaCl and $CaCl_2$ treatments increased the free amino acid and proline content in cotton plants compared to the control (Amuthavalli et al. 2012). The involvement of Ca in the alleviation of Cd toxicity by reducing the uptake, accumulation, and toxic effects of Cd was reported in *Arabidopsis*and common bean (Suzuki 2005; Ismail 2008). Since plasma membrane surfaces are usually negatively charged, high level of Ca^{2+} would reduce cell-surface negativity and alleviate the harmfulness of Cd toxicity. There is a report that the uptake of Cd is inhibited by the Ca^{2+} channel blockers (Ismail 2008). Cadmium decreases Ca content because of the competition between Cd and Ca at both Ca channels (Nelson 1986) and intracellular Ca-binding proteins (Rivetta et al. 1997). The alleviation of Cd toxicity by Ca supplementation occurs either by decreasing Cd uptake or by reducing Cd toxicity. In *Arabidopsis* seedlings, the content of Cd per gram fresh weight was reduced from 46.7 to 17.4 μg in the presence of 30 mM Ca (Suzuki 2005) by reducing Cd uptake. Wang and Song (2009) reported that 5 mM $CaCl_2$ application reversed the Cd effects on the activity of SOD, CAT, glutathione peroxidase, and APX and reversed the Cd-induced decrease in fresh mass as well as the changes in lipid peroxidation in *Trifolium repens*. The enhanced Cd tolerance in *Lactuca sativa* by 4 mM $CaCl_2$ was the result of increased expression of phytochelatin synthase gene under 0.05 mM Cd treated in *Lactuca sativa* plants (Zhenyan et al. 2005). Xu et al. (2013a) found that the availability of foliar $CaCl_2$ could modulate endogenous proline accumulation under water stress, because the leaves of stressed plants that were pretreated with $CaCl_2$ accumulated less proline than did the nontreated plants. Increased proline accumulation in $CaCl_2$-treated seedling is due to decreased level of proline oxidase (degrading enzyme) and increased levels of the activities of P5C reductase and γ-glutamyl kinase (synthesizing enzyme) (Misra and Gupta 2006). Figure 8.1 shows that Cd stress leads to the production of ROS, which signals for phytohormones' production. These phytohormones regulate the level of N, S, and Ca to increase proline content and Cd tolerance.

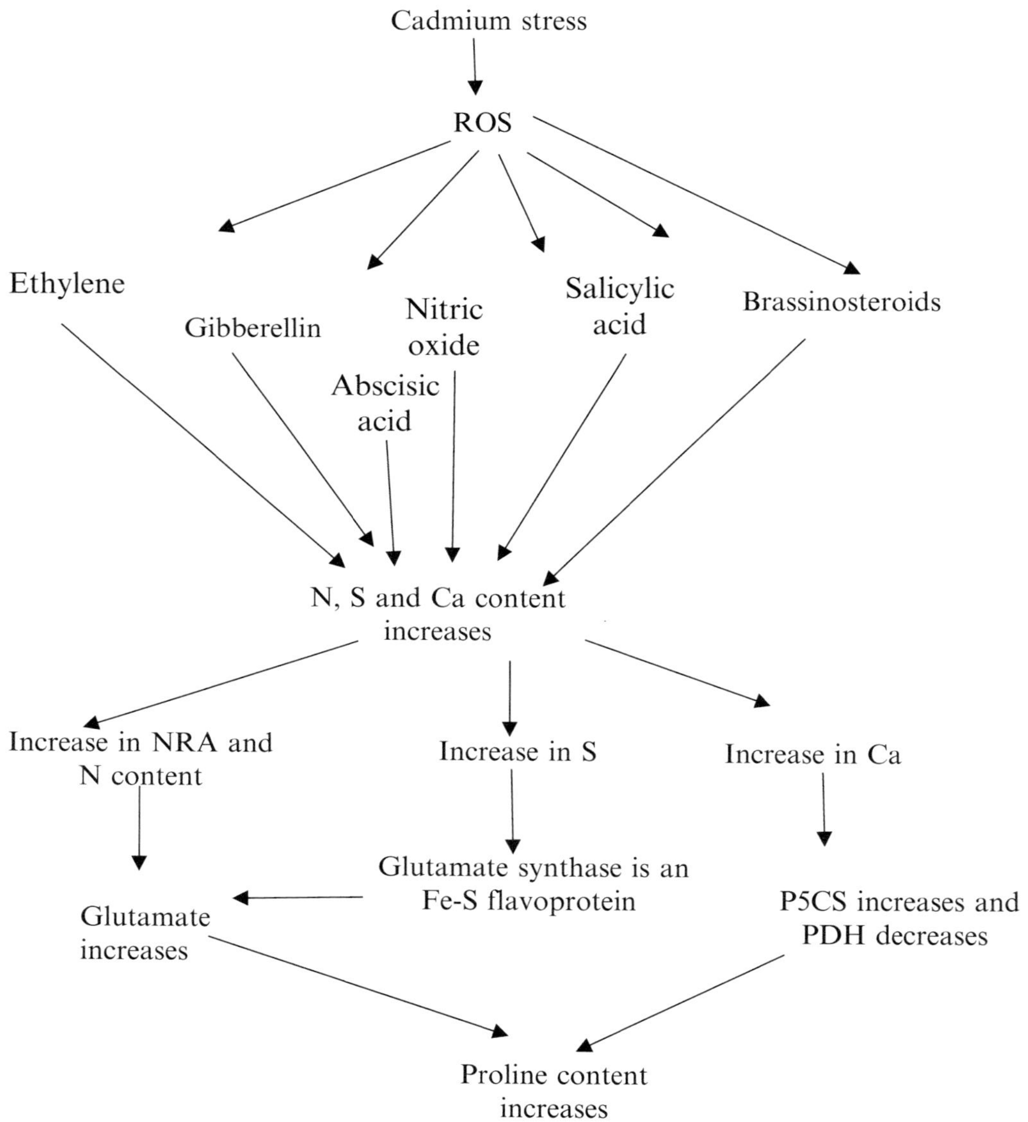

Fig. 8.1 Diagrammatic representation of indirect influence of phytohormones on proline content via regulating the nutrients N, S, and Ca

8.4 Phytohormones in Regulating Proline Content and Cadmium Tolerance

Reports about the regulation of Cd tolerance by phytohormones are sufficient; however, the literature lacks data regarding the regulation of proline by phytohormones for Cd tolerance. In the present study, we have discussed the role of some major phytohormones in regulating proline content either directly or via increasing nutrient to subsequently increase proline content and Cd tolerance. The phytohormone-mediated changes on proline accumulation under Cd stress are shown in Table 8.1. Cadmium stress leads to oxidative burst which activates the MAPK cascade and leads to phytohormones production. Phytohormones may directly increase proline content or indirectly via increase in N, S, and Ca accumulation. Besides, osmolytes like proline are directly activated under Cd stress as plants' adaptive strategy to maintain osmoregulation and Cd tolerance (Fig. 8.2). Major phytohormones and their interactions are discussed below.

8.5 Ethylene and Cadmium Tolerance

Ethylene regulates multiple stress responses (Cao et al. 2009) such as the signal transduction events during Cd-induced programmed cell death (PCD) (Iakimova et al. 2008) and induction of plant-specific transcription factors which have been linked to Cd tolerance (Youm et al. 2008). Experiments with young soybean seedlings revealed that an increase in ethylene concentration was accompanied not only with the induction of the genes encoding the enzymes of the ethylene biosynthesis pathways but also the genes related to the proteins involved in the polyamine metabolism, NO generation, and MAPK cascades (Chmielowska-Bąak et al. 2013). It has been reported that ethylene perception and not production regulates developmental responses under Cd stress. Inhibition of ethylene perception can minimize the effect of Cd toxicity through the increased membrane stability and delay leaf senescence (Iakimova et al. 2008). Masood et al. (2012) reported that the S-mediated alleviation of Cd stress was via ethylene which regulated S

Table 8.1 Phytohormone-mediated changes in proline accumulation in plants under Cd stress

Phytohormones	Proline content	Plant	References
Ethylene	Increase	*Triticum aestivum*	Khan et al. (2014)
Ethylene-insensitive plant	Increase	*Lycopersicon esculentum Nr mutant*	Gratão et al. (2012)
Salicylic acid	Decrease	*Perennial ryegrass*	Wang et al. (2013)
	Increase	*Cucumus melo*	Zhang et al. (2015)
	Decrease	*Cicer arietinum*	Çanakci and Dursun (2013)
Abscisic acid	Decrease	*Oryza sativa*	Hsu and Kao (2003)
	Increase	*Phaseolus vulgaris*	Bahmani et al. (2012)
Nitric oxide	Decrease	*Perennial ryegrass*	Wang et al. (2013)
	Decrease	*Perennial ryegrass*	Bai et al. (2015)
	Increase	*Triticum aestivum*	Basalah et al. (2013)
Brassinosteroid	Increase	*Raphanus sativus*	Anuradha and Rao (2007)
	Increase	*Phaseolus vulgaris*	Rady (2011)

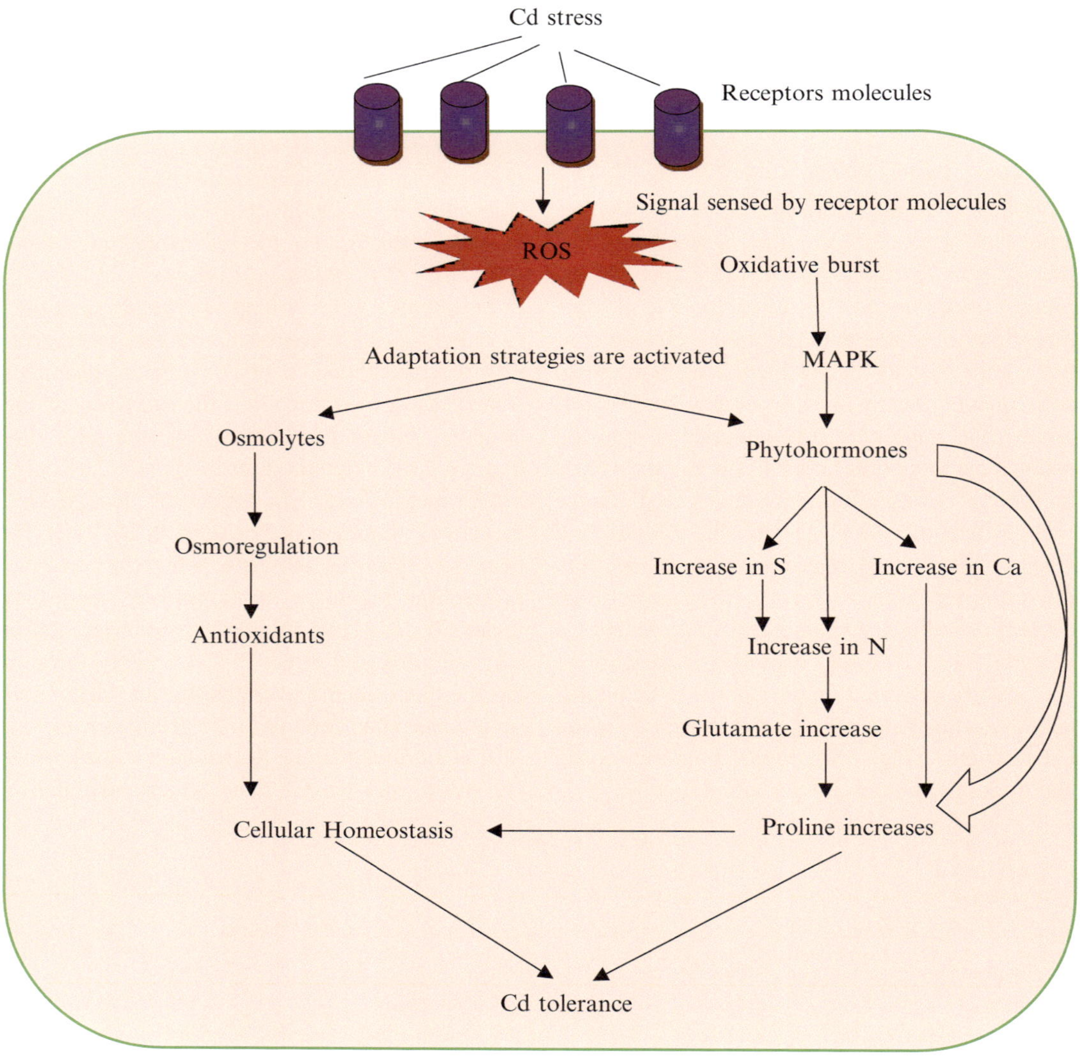

Fig. 8.2 Diagrammatic representation of the role of phytohormones in regulating proline content and Cd tolerance

assimilation and GSH content and tolerance. Inhibition of ethylene was found to inhibit Cd tolerance.

Besides ethylene (Zhang et al. 2010) and auxin (Pustovoitova et al. 2000) can regulate proline accumulation in stress conditions. Gavelienė et al. (2014) reported that ethylene production increased in cold-acclimated conditions and under the effect of exogenous proline and glutamine. Gratão et al. (2012) reported an exaggerated accumulation of proline in leaves and fruits of ethylene-insensitive (Nr) and auxin-insensitive diageotropica (dgt) tomato mutants, being dose-dependent increasing with Cd concentrations. This response, probably, can be associated to the less pronounced decline in chlorophyll and lower MDA contents in these genotypes confirming the protective role of proline against Cd-induced stress. Khan et al. (2013) showed interaction between proline metabolism and ethylene formation for heat stress tolerance in *Triticum aestivum*. Khan et al. (2014) reported that the inhibition of Cd-induced stress ethylene and oxidative stress by Se and S may result in the increase in sensitivity of plants to ethylene and promote proline and antioxidant metabolism. Inhibiting ethylene biosynthesis using AVG reversed the effects of Se and S on Cd-induced proline metabolism, GSH production, and net photosynthesis.

Besides the direct influence of ethylene on proline under Cd stress, it may indirectly regulate proline metabolism through increase in N assimilation. Increased NR activity and N content on the application of ethephon have been reported in mustard (Iqbal et al. 2012). Iqbal et al. (2012) also reported increase photosynthetic nitrogen use efficiency (PNUE) on ethephon application. Iqbal et al. (2011) reported increased NUE in mustard plants grown under varying N levels and supplemented with ethephon. The production of compatible solutes such as proline has been shown to be related to N assimilation and salt tolerance (Iqbal et al. 2014, 2015). Under Cd stress, N content decreases while ethylene increases N assimilation and hence proline content. Cadmium decreases N metabolism leading to reduced nitrate assimilation by plants (Chaffei et al. 2004). Cd sensitivity of plants is affected by both the N forms supplied (Hassan et al. 2005; Xie et al. 2009) and N availability (Pankovic et al. 2000; Finkemeier et al. 2003). Asgher et al. (2013) reported that the tolerant mustard cultivar had greater N assimilation and proline content than sensitive mustard cultivar. Thus, it could be assumed that ethephon supplementation to Cd-stressed plants could increase plants' ability to tolerate Cd through increase in N assimilation and proline accumulation.

Cadmium causes an increase in cytosolic calcium levels in various animal cell types (Yamagami et al. 1998; Ye et al. 2007; Yang et al. 2008). The observed Cd-induced intracellular Ca release depends on phospholipase C (PLC) activation and an increase in cellular inositol triphosphate concentration. The involvement of Ca ions in plant response to Cd can also be presumed on the basis of the fact that $CdCl_2$ caused an increase in the expression of genes coding calmodulin-like protein in black night shade and of the calcium-binding protein HvC2d1 in barley plants (Oulhajd et al. 2006). Iqbal et al. (2013) reported an interaction between ethylene and Ca. Ca is required for a variety of ethylene-dependent processes (Raz and Fluhr 1992). Apoplastic Ca^{2+} is required for the induction of ACC oxidase (ACO) by ethylene in etiolated internodes of *Pisum sativum* seedlings (Kwak and Lee 1997). Intracellular Ca^{2+} enhances the gene induction by ethylene of Ps-ACO1 in the *Pisum sativum* radicle and is, therefore, a positive regulator of the response. CaM gene expression in the embryonic axis of chickpea seeds is correlated with ACO gene expression, ethylene biosynthesis, and germination (Gómez-Jiménez et al. 1998). Calcium acts as a second signal and is involved in the ethylene-mediated response. Iqbal et al. (2014) reported interaction between Ca and proline. Yoo et al. (2005) reported that the overexpression of a *Glycine max*calmodulin isoform in *Arabidopsis* markedly triggered the expression of *P5CS1* and increased proline content by threefold in transgenic plants. Similarly, Shah et al. (2001) reported that Ca participated in the regulation of proline accumulation at the level of mRNA translation under salinity stress in the suspension cultures of *Oryza sativa*. Thus, ethylene under Cd stress may regulate Ca to increase proline accumulation and tolerance, or the increased Ca under Cd stress may act as signal for ethylene to increase proline content.

Ethylene influences S assimilation and increases photosynthetic sulfur use efficiency (PSUE) in mustard plants (Iqbal et al. 2012). Masood et al. (2012) reported Cd tolerance via ethylene by the regulation of S assimilation. The role of S in proline accumulation has been discussed above. Thus, Cd tolerance may be mediated by ethylene via increase in proline content through the regulation of N, S, and Ca.

8.6 Gibberellin and Cadmium Tolerance

Gibberellin is involved in Cd tolerance. GA_3 has promoted effects on cell division and cell enlargement and plays a prominent role in the Cd detoxification via improvement of Ca^{2+} and other mineral nutrients' uptake and antioxidant enzyme activities and decreasing lipid peroxidation (Maggio et al. 2010; Siddiqui et al. 2011). Zhu et al. (2012) reported that GA-alleviated Cd toxicity is mediated through the reduction of the Cd-dependent NO accumulation and expression of Cd^{2+} uptake-related gene IRT1 in *Arabidopsis*. GA could be a potential modulator of Cd stress alleviation through the increase in SUE of crops (Masood and Khan 2013). Reports are available on the mitigating effects of GA in Cd-stressed plants (Rubio et al. 1994; Abd El-Hamid et al. 2003; Picazo et al. 2007; Meng et al. 2009). The application of 10 M GA together with Cr alleviated inhibited levels of growth, N assimilation, and antioxidant system compared to Cr treatments alone, whereas in pea seedling, it has been shown that exogenous application of GA in excess (100 μM GA) induces increased ethylene production, generation of ROS, and alteration in antioxidant system (Fuchs and Lieberman 1968; Mori and Schroeder 2004; Celik et al. 2007). Celik et al. (2007) have shown that 75 ppm of GA caused tissue damage by enhanced lipid peroxidation and decreased levels of glutathione-S-transferase and glutathione. Ghorbanli et al. (1999) showed that the addition of 10 mg m^{-3} gibberellin reduced the negative effects of Cd^{2+} in shoot and root growth of soybean plants. Although direct effect of GA on proline under Cd stress is lacking, it may regulate proline content by interacting with ethylene. Iqbal et al. (2012) reported interaction between ethylene and GA for salt tolerance. Masood and Khan (2013) discussed about the interaction between ethylene and GA for Cd tolerance through increase in SUE in mustard. Studies have also shown that GA and ethylene metabolism genes are expressed in the majority of plant organs and both GA and ethylene precursor ACC are synthesized ubiquitously (Dugardeyn et al. 2008). De Grauwe et al. (2008) reported that a functional GA response pathway is required for the increased ethylene biosynthesis of eto2-1 (ethylene overproducing mutant) since gai eto2-1 (gibberellin insensitive; ethylene overproducing double mutant) does not overproduce ethylene, showing dependence of ethylene on GA.

Besides, GA may influence N assimilation and Cd tolerance. It has been reported that elevated N levels increase endogenous proline content (Jang et al. 2008; Iqbal et al. 2014) and GA_3 has been found to increase N content. GA_3 application significantly increased N content and NUE in *Brassica juncea* grown with different levels of S (Khan et al. 2005). GA_3 in combination with glucose had antagonistic effect on NRA in the presence and absence of exogenous nitrate (Garg 2013).

GA may regulate Ca level for proline accumulation and Cd tolerance. Ca^{2+} plays an important role in heavy metal detoxification (Antosiewicz and Hennig 2004; Jáuregui-Zùñiga et al. 2005). GA_3 induces Ca^{2+} and other nutrients' uptake that may be involved in plant tolerance to stress by regulating antioxidant metabolism and reducing the lipid peroxidation of cell membrane (Gilroy and Jones 1992; Jiang and Huang 2001; Siddiqui et al. 2008; Khan et al. 2010; Badr-uz-Zaman et al. 2010); furthermore, these may also be involved in signal transduction (McAinsh et al. 1997) and gene expression (Braam 1992). Siripornadulsil et al. (2002) reported that the free proline acts as an antioxidant in Cd-stressed cells, and proline levels are correlated with the GSH redox state and malondialdehyde levels in heavy metal-treated algae. Khan et al. (2010) reported that the production of proline was higher in GA_3- and Ca^{2+}-exposed plants. The production of proline is an important mechanism by which plants respond to and apparently detoxify toxic heavy metals (Wu et al. 1995; El-Enany and Issa 2001; Choudhary et al. 2007).

Masood and Khan (2013) reported the role of GA in S assimilation. They reported that the application of 10 μM GA increased SUE of mustard (*Brassica juncea* L. Czern & Coss) plants treated with optimal S (100 mg S kg^{-1} soil) in comparison with the control plants. SUE was not increased at excess S (200 mg S kg^{-1} soil). The increase in SUE by GA was through the increase in growth, CO_2 exchange rate, and use efficiency of N (Khan et al. 2005). Although there is no direct influence of GA on proline, however, we know that GA plays an important role in Cd tolerance, and it may regulate proline content through its influence of N, S, and Ca. Further studies need to be carried out to study proline accumulation in

Cd-stressed plants that are treated with the phytohormones GA.

8.7 Abscisic Acid and Cadmium

Heavy metals such as Cd, Ni, Zn, and Al (Hollenbach et al. 1997; Foy 1998; Fediuc et al. 2005) have been shown to increase ABA contents in plants. Fediuc et al. (2005) demonstrated that Cd-induced ABA accumulation was observed in roots, but not in shoots of *Typha* and *Phragmites* plants. Exogenous application of ABA reduced transpiration rate, decreased Cd content, and enhanced Cd tolerance of TN1 seedlings (Hsu and Kao 2003). Treatment of potato (*Solanum tuberosum*) plants with Cd or ABA enhanced the content of *PCS* 1 transcript and activity of phytochelatin synthase in roots (Stroiński et al. 2013). In the presence of Cd, the application with low doses of (0.1–0.5 μM) ABA inhibited Cd uptake by roots and decreased Cd level in *Arabidopsis* wild-type plants (Fan et al. 2014).

Proline accumulation increases in many species in response to osmotic stress stimulated by drought, salt, cold, and ABA. It appears to be mediated by both ABA-dependent and ABA-independent signaling pathways (Hare et al. 1999). It has been proposed that ABA is responsible for inducing proline production in stressed plants (Makela et al. 2003) and the effects of the exogenous ABA application on proline production have been examined. Aspinall and Paleg (1981) reported that ABA triggers proline accumulation. Hsu and Kao (2003) reported that ABA content increases in Cd-treated TNG67 leaves, which did not result in proline accumulation; conversely in Cd-treated TN1 leaves, proline increased but not ABA. The increased SOD activity and proline and ABA content prevented the occurrence of oxidative damage under Cd stress in *Phaseolus vulgaris* seedlings (Bahmani et al. 2012). Pretreatment with ABA enhanced Cd tolerance and reduced Cd-induced NH_4^+ accumulation in TN1 seedlings, and exogenous application of the ABA biosynthesis inhibitor, fluridone, decreased Cd tolerance and increased NH_4^+ content in leaves of TNG67. ABA pretreatment reduced the decrease in glutamine synthase (GS) activity in the third leaves of TN1 seedlings (Hsu and Kao 2004). Protection of GS activity may increase GSH content and GSH is related to proline accumulation (Nazar et al. 2014). The plant responses to heavy metal treatment are accompanied by the accumulation of ABA, free proline, and soluble proteins in leaf tissues. Cd stress leads to the accumulation of free ABA in leaf tissue which plays an important role in plant tolerance to heavy metal and also in the accumulation of proline (Talanova et al. 2000). The role of proline in Cd tolerance is already discussed. Cadmium and Pb caused a decrease in seedling biomass and an increase in free proline and ABA concentration in cotyledon leaves of cucumber. The accumulation of proline and ABA in leaf tissues is seen as an adaptive response to the Pb and Cd treatment of plant roots (Talanova et al. 2000). However, since ABA content increased under Cd stress and ABA is reported to increase proline accumulation, therefore, the possibility of regulation of proline by ABA cannot be ignored and needs to be exploited.

The relationship between ABA and proline for Cd tolerance can be exploited through the influence of ABA on N metabolism and Ca metabolism. ABA in N signaling is becoming increasingly evident (Kiba et al. 2011). ABA stimulates NRA in cotyledons of *Phaseolus aconitifolius* (Sankla and Huber 1975). Decreased NRA with ABA application has also been reported. NRA levels were suppressed by ABA in *Hordeum vulgare* leaves (Lu et al. 1992) and in *Gossypium hirsutum* (Pandey 2001). Signora et al. (2001) reported that ABA-insensitive mutants (abi4-1, abi4-2, and abi5-1) and ABA-deficient mutants (aba1-1, aba2-3, aba2-4, and aba3-2) were less sensitive to the inhibitory effects of high nitrate, signifying a relationship between ABA and N. A recent study on *Medicago truncatula* latd mutant provided evidence for a link between ABA and N signaling (Yendrek et al. 2010).

The role of Ca in proline accumulation has been reported, and ABA triggers an increase in cytosolic Ca in guard cells ([Ca^{2+}]cyt) (Schroeder and Hagiwara 1990; Grabov and Blatt 1998) that has been proposed to include Ca^{2+} influx across the plasma membranes (Schroeder and Hagiwara 1990; Grabov and Blatt 1998; Schwartz 1985). MacRobbie (1992) studied the relationship between ABA and Ca in stomatal closure and reported that ABA increases cytoplasmic Ca. Calcium-permeable channels activated by ROS have recently been shown to function in the ABA signaling network in *Arabidopsis*guard cells. ABA activation of these Ca^{2+} channels requires the presence of NAD(P)H in the cytosol (Murata et al. 2001). Kurtyka et al. (2008) reported that the positive effects of highest concentration of Ca on the growth in the presence of Cd relay on the maintenance of high concentration of potassium in the roots. ABA has also been shown to trigger changes in the amplitude and the oscillation of the free cytosolic Ca^{2+} level, and these oscillations induced by ABA together with Ca^{2+}/CaM are involved in the stomatal movement of *Zea mays* seedlings, and CaM participated in the ABA signal transduction process (Guo et al. 2008). Kim et al. (2002) reported that Ca^{2+} blocked both Cd^{2+} transport into rice roots and Cd^{2+} toxicity on root growth by competition of Cd^{2+} with Ca^{2+} for transport into root cells via transporters that pass Ca^{2+} (Clemens et al. 1998). Thus, increased Ca by ABA may lead to Cd tolerance either by reducing Cd uptake or increasing proline accumulation.

Cao et al. (2014) reported the transcription of ABA3 and NCED3, encoding another key enzyme of the ABA biosynthesis pathway, to be regulated by S supply in wild-type seedlings. In contrast, ABA upregulated the transcript level of SULTR3;1 and other S metabolism-related genes. They provided evidence for a significant coregulation of S metabolism and ABA biosynthesis that operates to ensure sufficient cysteine for AO maturation and highlights the importance of sulfur for stress tolerance of plants. Deficiency of N, P, K, S, and Fe in the nutrient solution resulted in marked increases in the levels of ABA extracted from root, leaf, stem, and flower at the three developmental stages of *Zea mays*. Excessive concentrations of these macroelements resulted in a decrease in ABA levels in all parts of plants at all three stages as compared with their respective controls (Battal et al. 2003). Kim et al. (1997) reported that the concentrations of O-acetyl-L-serine increased in accordance with the increase of exogenous ABA concentrations. ABA therefore affects S assimilation and may lead indirectly to proline accumulation and Cd tolerance.

8.8 Salicylic Acid and Cadmium Tolerance

Salicylic acid-induced protection of plants from oxidative injury caused by Cd is mainly linked to enhanced accumulation of antioxidant enzymes (Wang et al. 2006). High endogenous levels of SA after treatment with Cd suggest that SA may act directly as an antioxidant to scavenge the ROS or indirectly modulate redox balance through the activation of antioxidant responses. SA plays a protective role on photochemical activity of chloroplast membranes and photosynthetic carboxylation reactions in Cd-stressed pea plants (Popova et al. 2009). SA pretreatment attenuated the decrease in Chl content, photosynthetic capacity, and PSII photochemistry efficiency in Cd-stressed plants. SA alleviated the inhibitory effects of Cd on protein content and could be used as a stabilizer of membrane integrity due to lipid protection of Cd-induced oxidative stress to improve plant resistance to Cd stress (Moradkhani et al. 2013). SA-mediated alleviation of Cd toxicity was attributed to the reduction of Cd uptake, improvement of photosynthetic capacity, and enhancement of antioxidant activities in *Cannabis sativa* (Shi et al. 2009). Exogenous SA at 3 mM enhanced OsWRKY45 expression, which resulted in increased endogenous content of SA and prevented membrane damage by lowering H_2O_2 content in Cd-exposed *Oryza sativa* (Chao et al. 2010). SA application can decrease the Cd-accrued inhibition in CO_2 fixation by enhancing the activity of ribulose 1,5-bisphosphate carboxylase/oxygenase (Rubisco) and phosphoenolpyruvate carboxylase (PEPC) that in turn may protect plants against Cd-induced oxidative stress (Krantev et al. 2008).

Exogenous SA significantly reduced superoxide anion production and lipid peroxidation, followed by increase in the activities of antioxidant enzyme and content of soluble protein and free proline under Cd stress (Zhang et al. 2015). Proline accumulation may be involved in the mechanism of osmoregulation. Reduction in proline concentration by SA pretreatment has been reported in *Medicago sativa* plants under mercury stress (Zhou et al. 2009). Lipid peroxidation and proline content were lowered by SA treatment, while soluble carbohydrates were increased in the leaves of soybean under Cd stress (Marefat et al. 2012). Çanakci and Dursun (2013) reported that the leaves of *C. arietinum* seedlings pretreated with SA alleviated the toxic effects of Cd by increasing the growth parameters, photosynthetic pigments, and GSH content and decreasing proline and MDA content. Nazar et al. (2015) reported that SA application alleviated the drought-induced decrease in growth and photosynthesis through increased proline content. SA increased enzyme activities, levels of nonenzymatic antioxidants, and GSH in roots/shoots, suggesting alleviation of Cd-induced oxidative damage in rice (Singh and Shah 2015). Salicylic acid pretreatment decreased the toxic level of Cd manifested by lower lipid peroxidation, lesser production of H_2O_2, and reduction in free proline content (Moussa and El-Gamal 2010).

SA may influence proline metabolism under Cd stress through increase in N assimilation and/or N content. The relationship of proline with N is already discussed. Koç et al. (2013) reported that all concentrations of $CdCl_2$ with 0.5 mM SA treatments caused an increase in the amounts of malondialdehyde, H_2O_2, and NO_3^- on the fifth day of the treatment in tomato. Hayat et al. (2014) reported in *Cicer arietinum* that SA increases the fixation and assimilation of N regardless of whether the plants are grown in the presence or absence of cadmium. Nazar et al. (2011) reported that SA increases N and S assimilation under salt stress. The activity of NR and GS increased in SA treatments under drought stress in *Cucumis sativa* (Jing-hong et al. 2012). Singh and Chaturvedi (2012) found that optimal physiological concentrations of SA increased N use efficiency in *Cucumis sativa*. Hayat et al. (2012) reported that the activities of nitrogenase (EC 1.18.6.1), NR (EC 1.6.6.1), GS (EC 6.3.1.2), glutamate synthase (GOGAT) (EC 1.4.7.1), and glutamate dehydrogenase (GDH) (EC 1.4.1.3) increased with the foliar application of SA in *Cicer arietinum*.

Ali et al. (2015) reported that both Cd and SA had the same inhibitory effects on Ca and Mg concentrations in *Brassica napus*. SA treatment results in the release of Ca from internal stores, which in turn leads to PAL activity increase, RA accumulation, and a large amount of Ca^{2+} influx from apoplast after 10 h of SA induction (Guo et al. 2015). SA or Ca alone as well as in combination markedly improved proline concentration (Al-Whaibi et al. 2012). Proline content increased with the treatment of SA or Ca to NaCl-stressed plants (Manaa et al. 2014). The synthesis of proline in the presence of salinity and SA is regulated by the activation of pyrroline-5-carboxylate reductase and γ-glutamyl kinase and inhibition of proline oxidase and proline dehydrogenase (Misra and Saxena 2009). Bender et al. (2014) through the immunoblot analysis revealed that the CML43 protein accumulates following treatment with SA. Guo et al. (2009) reported that SA pretreatment significantly decreased Cd accumulations and increased the chlorophyll level, growth, and nutrient element content (K, Ca, Mg, and Fe) in plants, accompanying with the reduction in malondialdehyde and H_2O_2 contents.

SA also affect S assimilation for Cd tolerance. The influence of S on proline content either alone or through its interaction with N has been discussed. Nazar et al. (2011) reported an increased GSH content that helps in maintaining the proper functioning of enzymes of ascorbate–glutathione pathway and results in the increased activity of GR and APX with the application of 0.5 mM SA under salt stress. The addition of 0.1 mM salicylate to the nutrient solution led to an increase of mRNA levels of the APR isoforms in *Arabidopsis* (Koprivova et al. 2008). SA increased glutathione status in tomato crop, imparted resistance against *M. incognita*, and augmented crop yield and functional food quality (Meher et al. 2011).

8.9 Nitric Oxide and Cadmium

Nitric oxide is an important phytohormone involved in abiotic stress tolerance. Exogenous supply of sodium nitroprusside (SNP), a NO donor, alleviated the short- and long-term Cd stress in chickpea (*Cicer arietinum*) by decreasing Cd accumulation, membrane injury, lipid peroxidation, ROS, and H_2O_2 contents and by elevating GSH/GSSG ratio and activity of antioxidant enzymes (Kumari et al. 2010). Cadmium treatment is reported to increase NO content in cells of different plant species (Groppa et al. 2008; Besson-Bard et al. 2008; Bartha et al. 2005). Laspina et al. (2005) demonstrated that NO, when applied before Cd exposure, could significantly reduce Cd-induced oxidative damage in *H. annuus* plants by preventing growth inhibition and chlorophyll degradation and recovery of CAT activity and GSH levels, along with an enhancement in AsA content and APX activity, as components of the antioxidant machinery which allowed plants to better cope with Cd stress. NO pretreatment alleviated the toxic effect of Cd^{2+} by preventing the oxidative stress development (Groppa et al. 2008). Exogenous NO enhances Cd tolerance in *O. sativa* by increasing pectin and hemicellulose content in the root cell walls. Panda et al. (2011) showed that SNP treatment regulated the stress metabolism in seedlings under Cd toxicity by generating NO. This led to reduced H_2O_2 and MDA levels, along with increase in the activities of GR and SOD, indicating that SNP in combination with Cd treatment protects the seedlings under stress conditions. In a recent study, Xiong et al. (2010) reported that exogenous application of NO decreases ROS accumulation both in rice roots and leaves under Cd stress. Ma et al. (2010) investigated the roles and mechanisms of NO-induced protection under Cd^{2+} cytotoxicity in tobacco cell. In this case, NO played a positive role in CdC12-induced PCD by modulating Cd^{2+} uptake, promoting Cd^{2+} accumulation in BY-2 cells. They observed that SNP addition had accelerated PCD, whereas NO synthase inhibitors Nω-nitro-l-arginine methyl ester hydrochloride (l-NAME) and 2-(4-carboxyphenyl)-4,4,5,5-tetra methyl imidazoline-1-oxyl-3-oxide (cPTIO) alleviated this toxicity. Xu et al. (2010) reported that exogenous NO improved Cd tolerance in plants by reducing oxidative damage, maintaining auxin equilibrium, and enhancing ion absorption. De Michele et al. (2009) showed that modulation in NO levels has an effect on PC content in *Arabidopsis* cell suspension cultures. Pretreatment with l-NG-monomethylarginine (l-NMMA) induced an increase in PC levels, along with the prevention of H_2O_2 increase and changes in senescence-associated gene 12 (SAG12) expression, indicating that NO is required for Cd2+-induced cell death. In some plant species, metallothioneins (MTs) play an important role in Cd detoxification (Lux et al. 2011).

The NO toxic effect resulting from the oxidative state may be alleviated by several antioxidative defense systems, including the induction of polyamines and proline (Bouchereau et al. 1999; Kuthanova et al. 2004; Xu et al. 2011). Basalah et al. (2013) studied that under Cd stress, proline accumulation increased which was further increased by either SA or NO. However, combined application of SA and NO proved beneficial in alleviating the adverse effect of Cd by improving proline accumulation. Simaei et al. (2011) reported that proline is able to scavenge hydroxyl radical and stabilize the structure and function of macromolecules such as DNA, protein, and membranes via interactions with these macromolecules. Fan et al. (2012) reported that NO could alleviate salinity damage in cucumber seedlings by regulating proline metabolism through increase in the P5CS activity and decrease in the PDH activity. Filippou et al. (2013) reported that long-term (24 h), higher SNP concentration resulted in decreased photosynthetic rate and stomatal conductance followed by intracellular putrescine and proline accumulation, as a result of an increase in biosynthetic arginine decarboxylase (ADC) and D1-pyrroline-5-carboxylate synthetase (P5CS) enzymatic activity, respectively. He et al. (2014) reported that addition of 30 μM SNP increased accumulation of proline in both root and shoot and significantly reduced Cd accumulation implicating the protective role of SNP in preventing Cd tolerance.

Bai et al. (2015) reported that exogenous NO mitigates the damage associated with Cd and Cu stress by inducing proline degradation, especially at Cd+Cu treatment. Lopez-Carrion et al. (2008) also reported that exogenous NO reduced proline accumulation in cabbage plants under saline stress by increasing the activity of proline dehydrogenase. The application of exogenous NO at sowing, seedling, and flowering stages increased Cd-induced decrease in the content of proline, especially at seedling stage. However, SNP supplementation at pod setting did not improve proline content compared with Cd treatment (Xu et al. 2013b). Verma et al. (2013) studied that the roots pretreated with 5 mM SNP for 6 h when exposed to 25 μM Cd for 24 h reduced the level of proline, nonprotein thiols, SOD, APX, and CAT in comparison with only Cd treatments. NO upregulates the expression of enzymes involved in proline biosynthesis (Δ1-pyrroline-5-carboxylate synthetase, or P5CS) (Zhang et al. 2008). NO application increases the production of proline and glutathione. Both of these act as nonenzymatic antioxidant and can directly detoxify oxygen free radicals, and glutathione is an important antioxidant and redox buffer when plants are under stress (Xu et al. 2010). Wang et al. (2013) reported that exogenous NO mitigated Cd-induced damage by reducing oxidative stress and inducing proline degradation.

Another possible mechanism through which NO regulates proline metabolism is via increase in N content under cadmium stress. Nitrogen management is a promising agronomic strategy to minimize Cd contamination in crops. Du et al. (2008) have found that NRA significantly enhanced by the addition of the NO donors in *Brassica oleracea* and suggested that the effect of NO on NRA might be due to an enhancement of electron transfer from heme to nitrate through the activation of heme and molybdenum centers in the NR. Neill et al. (2003) showed that NR can induce the increment of NO.

The supply with exogenous Ca^{2+} in pea plants exposed to Cd reduced the Cd-dependent O^{2-} accumulation and restored NO production to the level observed in control plants (Rodríguez-Serrano et al. 2009). Exogenous NO enhanced Cd tolerance in *Medicago truncatula* by increasing the uptake of mineral nutrients such as K and Ca and synthesis of proline and GSH (Xu et al. 2010). However, Besson-Bard et al. (2009) suggested that NO contributes to Cd^{2+} toxicity by favoring Cd^{2+} versus Ca^{2+} uptake and by initiating a cellular pathway resembling those activated upon iron deprivation. Concurrent increases in NO concentration and free cytosolic Ca^{2+} levels were found to occur during signal transduction initiated by abiotic and biotic stressors (Khatai et al. 2004). NO treatment stimulated an increase in intracellular Ca^{2+} levels in *V. faba* and *Nicotiana tabacum* guard cells (Garcia-Mata et al. 2003; Lamotte et al. 2004). The accumulation of free cytosolic Ca^{2+} induced by osmotic stress and cryptogein elicitor in tobacco cells was also influenced by NO (Lamotte et al. 2004), suggesting that it can function as a Ca^{2+}-activating intracellular compound in plant cells.

NO is reported to increase GSH content in various plants under Cd stress (Hsu and Kao 2004; Laspina et al. 2005; Groppa et al. 2008). Redox-dependent modulation by eATP and NO signaling may be an adaptive strategy under environmental stress conditions (Terrile et al. 2010). However, the literature lacks data regarding the direct influence of NO on S content or S assimilation.

8.10 Brassinosteroid and Cadmium

Brassinosteroids (BRs) are involved in reducing the adverse effects of Cd stress in plants. They may reduce the toxic effects of Cd on photochemical processes by diminishing the damage on photochemical reaction centers and the activity of oxygen evolving center as well as by maintaining efficient photosynthetic electron transport (Janeczko et al. 2005). The application of BRs enhanced transcript level of the antioxidant system in *L. esculentum* under Cd–phenanthrene co-contamination (Ahammed et al. 2013). Cd exposure-mediated activation in BR signaling pathway may influence plant responses to Cd through upregulation and downregulation of many Cd stress-responsive genes (Villiers et al. 2012). Brassinosteroids somehow activate the uptake of nitrate (Mai et al. 1989) and also favor transcription and/or translation (Kalinich et al. 1985; Bajguz 2000).

Anuradha and Rao (2007) reported that brassinosteroids increased the free proline content and ameliorated the toxic effect of Cd on seed germination and seedling growth of *Raphanus sativus* L. homobrassinolide (HBL) or epibrassinolide (EBL) spray caused a further increase in proline content and antioxidative enzyme activities, which were already enhanced by Cd stress. This effect of brassinosteroids (HBL/EBL) was more pronounced in K-25 than in Sarvodya, representing the tolerance and adoptable behavior of K-25 (Hayat et al. 2010).

Brassinosteroid increased root nodulation, zeatin content, and nitrogenase activity in unstressed *Phaseolus vulgaris* plants and also ameliorated stress-induced decline in the same parameters. Moreover, EBL was relatively more effective than HBL (Upreti and Murti 2004). Improved Cd tolerance in *Phaseolus vulgaris* was possible as a result of 24-EBL (5 μM)-mediated increased activity of antioxidative enzymes and proline content and subsequent improvements in the membrane stability index (MSI) and relative leaf water content (RLWC) (Rady 2011).

EBL also increases the N metabolism of *Vigna radiata* plants under different concentrations of Ni (Yusuf et al. 2014). Dalio et al. (2013) reported that the decline in several components of N metabolism, induced by salt, was attenuated by 24-epibrassinolide application and accentuated by clotrimazole, indicating the importance of brassinosteroid synthesis for plants growing under salinity in pigeon pea. Yadav et al. (2012) reported that brassinosteroid isomers 28-HBL and 24-EBL increased NRA and the antioxidant system with respect to the control. However, the 10^{-8} M concentration of EBL produced higher NRA and better stimulated the antioxidant system in tomato. Exogenously EBL compensated for the damage/losses by $Ca(NO_3)_2$ stress to some extent through the regulation of N metabolism and metabolites (Yuan et al. 2012).

Oh et al. (2012) proposed a possible new link between Ca^{2+} and BR signaling. Zhao et al. (2013) reported that a BR-dependent elevation in cyclic GMP may be involved in the Ca^{2+} signaling cascade initiated by this hormone. Working with *Arabidopsis*, they used genetic and biochemical approaches to evaluate if BR-dependent Ca^{2+} signaling is involved in the regulation of IAA1 (as well as other BR-responsive gene) expression and plant phenotypes impacted by BR. A report on hormone signaling in wheat (*Triticum aestivum*; Singla et al. 2006) indirectly suggested that expression of a Ca^{2+}-regulated auxin-responsive gene in wheat orthologous to *Arabidopsis* INDOLE-3-ACETIC ACID-INDUCIBLE 1 (IAA1) is also induced by BR.

Recently, 24-EBL (10^{-7}, 10^{-9}, 10^{-11} M)-mediated increased activity of antioxidative enzymes such as SOD, CAT, APOX, GPX, GR, MDHAR, and DHAR and also the contents of GSH were argued to help radish plants to counteract the consequences of Hg (Kapoor et al. 2014).

8.11 Conclusion and Future Prospects

Plants adopt different strategies to reduce the severity of Cd stress. The accumulation of proline is one of the adaptive mechanisms that plants operate to minimize the adverse effects of Cd stress. Proline availability is regulated both by nutrients and phytohormones. The phytohormones may influence proline directly or may indirectly affect proline via the regulation of nutrient content. The key regulatory points of proline metabolism need to be identified to target upregulation of proline biosynthesis using phytohormones. This could improve cadmium tolerance in plants via proline. However, a single process may be the influence of a single hormone or the coordination between multiple hormones. Plant hormones interact with each other in such a way that the output of plant hormone action depends on specific hormone combinations rather than on the independent activities of each hormone. Therefore, proline content may be regulated by multiple hormones instead of a single hormone under Cd stress, and a crosstalk of this interaction needs to be studied. The present study deals with the role of individual phytohormone and proline in Cd tolerance, but the studies need to be focused on enhancing our understanding on the interaction between different phytohormones in regulating proline metabolism for Cd tolerance. Such knowledge will enhance our understanding for the development of Cd tolerant plants via manipulating proline metabolism.

Acknowledgment The first and the second author are thankful to the University Grants Commission, New Delhi, for D. S. Kothari Postdoctoral Fellowship (File No. F.4-2/2006 (BSR)/13-848/2013(BSR) and File No. F.4-2/2006(BSR)/13-711/2012(BSR)).

References

Abdelhamid MT, Rady M, Osman A, Abdalla MA (2013) Exogenous application of proline alleviates salt-induced oxidative stress in *Phaseolus vulgaris* L. plants. J Hort Sci Biotechnol 88:439–446

Ahammed GJ, Choudhary SP, Chen S, Xia X et al (2013) Role of brassinosteroids in alleviation of phenanthrene-cadmium co contamination-induced N photosynthetic inhibition and oxidative stress in tomato. J Exp Bot 64:199–213

Ahmad P, Sarwat M, Bhat NA, Wani MR, Kazi AG, Tran LSP (2015) Alleviation of cadmium toxicity in *Brassica juncea* L. (Czern. & Coss.) by calcium application involves various physiological and biochemical strategies. PLoS One 10, e0114571

Ali E, Modzeka A, Hussain N, Shamsi IH, Jiang L (2015) The alleviation of cadmium toxicity in oilseed rape (*Brassica napus*) by the application of salicylic acid. Plant Growth Regul 75:641–655

Al-Whaibi MH, Siddiqui MH, Basalah MO (2012) Salicylic acid and calcium induced protection of wheat against salinity. Protoplasma 249:769–778

Aly AA (2012) Application of DNA [RAPD] and ultrastructure to detect the effect of cadmium stress in Egyptian clover and Sudan grass plantlets. J Stress Physiol Biochem 8:241–257

Amuthavalli P, Anbu D, Sivasankaramoorthy S (2012) Effect of calcium chloride on growth and biochemical constituents of cotton (*Gossypium hirsutum* L.) under salt stress. Int J Res Bot 2:9–12

Anjum NA, Gill SS, Umar S, Ahmad I, Duarte AC, Pereira E (2012) Improving growth and productivity of Oleiferous Brassicas under changing environment: significance of nitrogen and sulphur nutrition, and underlying mechanisms. Sci World J 2012:657808

Anjum NA, Aref IM, Duarte AC, Pereira E, Ahmad I, Iqbal M (2014) Glutathione and proline can coordinately make plants withstand the joint attack of metal(loid) and salinity stresses. Front Plant Sci 5:Article 662 | 1–4

Antosiewicz DM, Hennig J (2004) Overexpression of LCT1 in tobacco enhances the protective action of calcium against cadmium toxicity. Environ Poll 129:237–245

Anuradha A, Rao SSR (2007) The effect of brassinosteroids on radish (*Raphanus sativus* L.) seedlings growing under cadmium stress. plant. Soil Environ 53:465–472

Asgher M, Khan MIR, Iqbal N, Masood A, Khan NA (2013) Cadmium tolerance in mustard cultivars: dependence on proline accumulation and nitrogen assimilation. J Funct Environ Bot 3:30–42

Asgher M, Khan NA, Khan MIR, Fatma M, Masood A (2014) Ethylene production is associated with alleviation of cadmium-induced oxidative stress by sulfur in mustard types differing in ethylene sensitivity. Ecotoxicol Environ Saf 106:54–61

Ashraf M, Foolad MR (2007) Roles of glycine betaine and proline in improving plant abiotic stress resistance. Environ Exp Bot 59:206–216

Aspinall D, Paleg LG (1981) Proline accumulation: biochemical aspects. In: Paleg LG, Aspinall D (eds) The physiology and biochemistry of drought resistance in plants. Academic, Sydney, pp 243–259

Badr-uz-Zaman, Salim M, Asghar R (2010) Role of Ca^{2+} on growth of *Brassica campestris* L. and B. *juncea* (L.) Czern & Coss under Na^{+} stress. J Integ Plant Biol 52:549–555

Bahmani R, Bihamta MR, Habibi D, Forozesh P, Ahmadvand S (2012) The effect of cadmium stress on growth, SOD activity, proline and ABA content in bean seedlings (*Phaseolus vulgaris* L.). The 1st international and the 4th National Congress on Recycling of Organic Waste in Agriculture 26–27 April 2012 in Isfahan, Iran

Bai XY, Dong YJ, Xu LL, Kong J, Liu S (2015) Effects of exogenous nitric oxide on physiological characteristics of perennial ryegrass under cadmium and copper stress. Russ J Plant Physiol 62:237–245

Bajguz A (2000) Effect of brassinosteroids on nucleic acids and protein content in cultured cells of *Chlorella vulgaris*. Plant Physiol Biochem 38:209–215

Balestrasse KB, Gallego SM, Benavides MP, Tomaro ML (2005) Polyamines and proline are affected by cadmium stress in nodules and roots of soybean plants. Plant Soil 270:343–353

Bartha B, Kolbert Z, Erdei L (2005) Nitric oxide production induced by heavy metals in *Brassica juncea* L. Czern. and *Pisum sativum* L. Acta Biol Szeged 49:9–12

Basalah MO, Ali HM, Al-Whaibi MH, Siddiqui MH, Sakran AM, Al Sahl AA (2013) Nitric oxide and salicylic acid mitigate cadmium stress in wheat seedlings. J Pure Appl Microbiol 7:139–148

Battal P, Turker M, Tileklioglu B (2003) Effects of different mineral nutrients on abscisic acid in maize (*Zea mays*). Ann Bot Fenn 40:301–308

Bender KW, Dobney S, Ogunrinde A, Chiasson D, Mullen RT, Teresinski HJ, Singh P, Munro K, Smith SP, Snedden WA (2014) The calmodulin-like protein CML43 functions as a salicylic-acid-inducible root-specific Ca^{2+} sensor in *Arabidopsis*. Biochem J 457:127–136

Besson-Bard A, Pugin A, Wendehenne D (2008) New insights into nitric oxide signaling in plants. Annu Rev Plant Biol 59:21–39

Besson-Bard A, Gravot A, Richaud P, Auroy P, Duc C, Gaymard F, Taconnat L, Renou JP, Pugin A, Wendehenne D (2009) Nitric oxide contributes to cadmium toxicity in *Arabidopsis* by promoting cadmium accumulation in roots and by up-regulating genes related to iron uptake. Plant Physiol 149:1302–1315

Bohnert HJ, Nelson DE, Jensen RG (1995) Adaption to environmental stress. Plant Cell 7:1099–1111

Bouchereau A, Aziz A, Larher F, Martin-Tanguy J (1999) Polyamines and environmental challenges: recent development. Plant Sci 140:103–125

Braam J (1992) Regulated expression of the calmodulin-related TCH genes in cultured *Arabidopsis* cells: induction by calcium and heat shock. Proc Natl Acad Sci 89:3213–3216

Çanakci S, Dursun B (2013) Amelioration of Cd toxicity by pretreatment of salicylic acid in *Cicer arietinum* L. seedlings. J Environ Biol 34:1089–1094

Cao S, Chen Z, Liu G, Jiang L, Yuan H, Ren G, Bian X, Jian H, Ma X (2009) The *Arabidopsis* ethylene-insensitive 2 gene is required for lead resistance. Plant Physiol Biochem 47:308–312

Cao MJ, Wang Z, Zhao Q, Mao JL, Speiser A, Wirtz M, Hell R, Zhu JK, Xiang CB (2014) Sulfate availability affects ABA levels and germination response to ABA and salt stress in *Arabidopsis thaliana*. Plant J 77:604–615

Celik I, Tuluce Y, Isik I (2007) Evaluation of toxicity of abscisic acid and gibberellic acid in rats: 50 days drinking water study. J Enzym Inhib Med Chem 22:219–226

Chaffei C, Pageau K, Suzuki A, Gouia H, Brulfert J, Ghorbal MH, Daubresse CM (2004) Cadmium toxicity induced changes in nitrogen management in *Lycopersicon esculentum* leading to a metabolic safeguard through an amino acid storage strategy. Plant Cell Physiol 45:1681–1693

Chao YY, Chen CY, Huang WD, Ching CH (2010) Salicylic acid mediated hydrogen peroxide accumulation and protection against Cd toxicity in rice leaves. Plant Soil 329:327–337

Chen C, Dickman MB (2005) Proline suppresses apoptosis in the fungal pathogen *Colletotrichum trifolii*. Proc Natl Acad Sci USA 102:3459–3464

Chen CT, Chen LM, Lin CC, Kao HC (2001) Regulation of proline accumulation in detached rice leaves exposed to excess copper. Plant Sci 160:283–290

Chmielowska-Bąak J, Lefèvre I, Lutts S, Deckert J (2013) Short term signaling responses in roots of young soybean seedlings exposed to cadmium stress. J Plant Physiol 170:1585–1594

Choudhary M, Jetley UK, Khan MA, Zutshi S, Fatma T (2007) Effect of heavy metal stress on proline, malondialdehyde, and superoxide dismutase activity in the cyanobacterium *Spirulina platensis S5*. Ecotox Environ Safet 66:204–209

Clemens S, Antosiewicz DM, Ward JM, Schachtman DP, Schroeder JI (1998) The plant cDNA LCT1 mediates the uptake of calcium and cadmium in yeast. Proc Natl Acad Sci U S A 95:12043

Considine MJ, Foyer CH (2014) Redox regulation of plant development. Antioxid Redox Signal 21:1305–1326

Dalio RJD, Pinheiro HP, Sodek L, Haddad CRB (2013) 24-epibrassinolide restores nitrogen metabolism of pigeon pea under saline stress. Bot Stud 54:9

De Grauwe L, Dugardeyn J, Van Der Straeten D (2008) Novel mechanisms of ethylene-gibberellin crosstalk revealed by the gai eto2-1 double mutant. Plant Signal Behav 3:1113–1115

De Michele R, Vurro E, Rigo C, Costa A, Elviri L, di Valentin M, Careri M, Zottini M, di Toppi LS, Schiavo FL (2009) Nitric oxide is involved in cadmium-induced programmed cell death in *Arabidopsis* suspension cultures. Plant Physiol 150:217–228

Demiral T, Türkan I (2004) Does exogenous glycinebetaine affect antioxidative system of rice seedlings under NaCl treatment? J Plant Physiol 161:1089–1100

Du S, Zhang Y, Lin XY, Wang Y, Tang CX (2008) Regulation of nitrate reductase by nitric oxide in Chinese cabbage pakchoi (*Brassica chinensis* L.). Plant Cell Environ 31:195–204

Dugardeyn J, Vandenbussche F, Van Der Straeten D (2008) To grow or not to grow: what can we learn on ethylene-gibberellin cross-talk by in silico gene expression analysis? J Exp Bot 59:1–16

El-Enany AE, Issa AA (2001) Proline alleviates heavy metal stress in *Scenedesmus armatus*. Foliar Microbiol 46:227–230

El-Hamid A, Nabrawy MF, Amera A, Gaber AA, Amal MK (2003) Control the physiological stress resulted from Cd and Pb foliage application on sugar beet plant by using foliar spray with certain growth regulators. J Agric Sci 28:3571–3602

Fan HF, Du CX, Guo SR (2012) Effect of nitric oxide on proline metabolism in cucumber seedlings under salinity stress. J Am Soc Hort Sci 137:127–133

Fan SK, Fang XZ, Guan MY, Ye YQ, Lin XY, Du ST, Jin CW (2014) Exogenous abscisic acid application decreases cadmium accumulation in *Arabidopsis* plants, which is associated with the inhibition of IRT1-mediated cadmium uptake. Front Plant Sci 5:Article 721 | 1

Fediuc E, Lips SH, Erdei L (2005) O-Acetylserine (thiol) lyase activity in *Phragmites* and *Typha* plants under cadmium and NaCl stress conditions and the involvement of ABA in the stress response. J Plant Physiol 162:865–872

Filippou P, Antoniou C, Fotopoulos V (2013) The nitric oxide donor sodium nitroprusside regulates polyamine and proline metabolism in leaves of *Medicago truncatula* plants. Free Rad Biol Med 56:172–183

Finkemeier I, Kluge C, Metwally A, Georgi M, Grotjohann N, Dietz KJ (2003) Alterations in Cd-induced gene expression under nitrogen deficiency in *Hordeum vulgare*. Plant Cell Environ 26:821–833

Fismes J, Vong PC, Guckert A, Frossard E (2000) Influence of sulfur on apparent N-use efficiency, yield and quality of oilseed rape (*Brassica napus* L.) grown on a calcareous soil. Eur J Agron 12:127–141

Foy CD (1998) Plant adaptation to acid, aluminum-toxic soils. Comm Soil Sci Plant Anal 19:959–987

Fuchs Y, Lieberman M (1968) Effects of kinetin, IAA and gibberellin on ethylene production and their interactions in growth of seedlings. Plant Physiol 43:2029–2036

Garcia-Mata C, Gay R, Sokolovski S, Hills A, Lamattina L, Blatt MR (2003) Nitric oxide regulated K^+ and Cl^- channels in guard cells through a subset of abscisic acid-evoked signaling pathways. Proc Natl Acad Sci U S A 100:11116–11121

Garg SK (2013) Role and hormonal regulation of nitrate reductase activity in higher plants: a review. Plant Sci Feed 3:13–20

Gavelienė V, Pakalniškytė L, Novickienė L (2014) Regulation of proline and ethylene levels in rape seedlings for freezing tolerance. Cent Eur J Biol 9:1099–1107

Ghorbanli M, Hadade-Kaveh SH, Sepehr MF (1999) Effects of cadmium and gibberellin on growth and photosynthesis of Glycin max L. Photosynthetica 37:627–631

Gilroy S, Jones RL (1992) Gibberellic acid and abscisic acid coordinately regulated cytoplasmic calcium and secretory activity in barley aleurone protoplasts. Proc Natl Acad Sci 89:3591–3595

Gojon A, Gaymard F (2010) Keeping nitrate in the roots: an unexpected requirement for cadmium tolerance in plants. J Mol Cell Biol 2:299–301

Gómez-Jiménez MC, Matilla AJ, Garrido D (1998) Isolation and characterization of a cDNA encoding an ACC oxidase from *Cicer arietinum* and its expression during embryogenesis and seed germination. Aust J Plant Physiol 25:765–773

Grabov A, Blatt MR (1998) Membrane voltage initiates Ca^{2+} waves and potentiates Ca^{2+} increases with abscisic acid in stomatal guard cells. Proc Natl Acad Sci U S A 95:4778–4783

Gratão PL, Monteiro CC, Carvalho RF, Tezotto T, Piotto FA, Peres LEP, Azevedo RA (2012) Biochemical dissection of diageotropica and never ripe tomato mutants to Cd-stressful conditions. Plant Physiol Biochem 56:79–96

Groppa D, Rosales EP, Iannone MF, Benavides MP (2008) Nitric oxide, polyamines and Cd-induced phytotoxicity in wheat roots. Phytochemistry 69:2609–2615

Guo X, Ma YY, Liu ZH, Liu BH (2008) Effects of exterior abscisic acid on calcium distribution of mesophyll cells and calcium concentration of guard cells in maize seedlings. Agric Sci China 7:438–446

Guo B, Liang Y, Zhu Y (2009) Does salicylic acid regulate antioxidant defense system cell death, cadmium uptake and partitioning to acquire cadmium tolerance in rice? J Plant Physiol 166:20–31

Guo H, Zhu N, Deyholos MK, Liu J, Zhang X, Dong J (2015) Calcium mobilization in salicylic acid-induced *Salvia miltiorrhiza* cell cultures and its effect on the accumulation of rosmarinic acid. Appl Biochem Biotechnol 175:2689–2702

Hare PD, Cress WA, Van Staden J (1999) Proline synthesis and degradation: a model system for elucidating stress-related signal transduction. J Exp Bot 50:413–434

Hasan SA, Fariduddin Q, Ali B, Hayat S, Ahmad A (2009) Cadmium: toxicity and tolerance in plants. J Environ Biol 30:165–174

Hassan MJ, Wang F, Ali S, Zhang GP (2005) Toxic effect of cadmium on rice as affected by nitrogen fertilizer form. Plant Soil 277:359–365

Hayat S, Hasan SA, Hayat Q, Ahmad A (2010) Brassinosteroids protect *Lycopersicon esculentum* from cadmium toxicity applied as shotgun approach. Protoplasma 239:3–14

Hayat Q, Hayat S, Alyemenia MN, Ahmad A (2012) Salicylic acid mediated changes in growth, photosynthesis, nitrogen metabolism and antioxidant defense system in *Cicer arietinum* L. Plant Soil Environ 58:417–423

Hayat S, Hayat Q, Alyemenia MN, Ahmad A (2013) Proline enhances antioxidative enzyme activity, photosynthesis and yield of *Cicer arietinum* L. exposed to cadmium stress. Acta Bot Croat 72:323–335

Hayat S, Hayat Q, Alyemenia MN, Ahmad A (2014) Salicylic acid enhances the efficiency of nitrogen fixation and assimilation in *Cicer arietinum* plants grown under cadmium stress. J Plant Interac 9:35–42

He J, Ren Y, Chen X, Chen H (2014) Protective roles of nitric oxide on seed germination and seedling growth of rice (*Oryza sativa* L.) under cadmium stress. Ecotoxicol Environ Saf 108:114–119

Hollenbach B, Schreiber L, Hartung W, Dietz KJ (1997) Cadmium tends to stimulate expression of lipid transfer protein (ltp) in barley: implications for the involvement of LTP in wax assembly. Planta 203:9–19

Hoque MA, Banu MN, Okuma E, Amako K, Nakamura Y, Shimoishi Y, Murata Y (2007) Exogenous proline and glycinebetaine increase NaCl-induced ascorbate glutathione cycle enzyme activities, and proline improves salt tolerance more than glycinebetaine in tobacco Bright Yellow-2 suspension-cultured cells. J Plant Physiol 164:1457–1468

Hsu YT, Kao CH (2003) Accumulation of ammonium ion in cadmium tolerant and sensitive cultivars of *Oryza sativa*. Plant Growth Regul 39:271–276

Hsu YT, Kao CH (2004) Cadmium toxicity is reduced by nitric oxide in rice leaves. Plant Growth Regul 42:227–238

Iakimova ET, Woltering EJ, Kapchina-Toteva VM, Frans JM, Harren C, Simona MC (2008) Cadmium toxicity in cultured tomato cells – role of ethylene, proteases and oxidative stress in cell death signaling. Cell Biol Int 32:1521–1529

Iqbal N, Nazar R, Syeed S, Masood A, Khan NA (2011) Exogenously-sourced ethylene increases stomatal conductance, photosynthesis, and growth under optimal and deficient nitrogen fertilization in mustard. J Exp Bot 62:4955–4963

Iqbal N, Khan NA, Nazar R, Silva JA (2012) Ethylene-stimulated photosynthesis results from increased nitrogen and sulfur assimilation in mustard types that differ in photosynthetic capacity. Environ Exp Bot 78:84–90

Iqbal N, Trivellini A, Masood A, Ferrante A, Khan NA (2013) Current understanding on ethylene signaling in plants: the influence of nutrient availability. Plant Physiol Biochem 73:128–138

Iqbal N, Umar S, Khan NA, Khan MIR (2014) A new perspective of phytohormones in salinity tolerance: regulation of proline metabolism. Environ Exp Bot 100:34–42

Iqbal N, Umar S, Khan NA (2015) Nitrogen availability regulates proline and ethylene production and alleviates salinity stress in mustard (*Brassica juncea*). J Plant Physiol 178:84–91

Islam MM, Hoque MA, Okuma E, Banu MNA, Shimoishi Y, Nakamura Y, Murata Y (2009) Exogenous proline and glycinebetaine increase antioxidant activities and confer tolerance to cadmium stress in cultured tobacco cells. J Plant Physiol 166:1587–1596

Ismail MA (2008) Involvement of Ca^{2+} in alleviation of Cd^{2+} Toxicity in Common Bean (*Phaseolas vulgaris* L.) plants. Asian J Biol Sci 1:26–32

Jager H, Grill D (1975) Einfluss von SO_2 und HF auf Aminosauren der Fichte (*Picea abies* [L.j Karsten). Eur J For Pathol 5:279–286

Jager H, Pahlich E (1972) Einfluss von SO_2 auf den Aminosaur estoffwechsel von Erbsenkeimlingen. Oecologia (Berl) 9:135–140

Janeczko A, Kocielniak J, Pilipowicz M (2005) Protection of winter rape photosystem 2 by 24-epibrassinolide under cadmium stress. Photosynthetica 43:293–298

Jang SW, Hamayun M, Sohn EY, Shin DH, Kim KU, Lee BH, Lee IJ (2008) Effect of elevated nitrogen levels on endogenous gibberellin and jasmonic acid contents of three rice (*Oryza sativa* L.) cultivars. J Soil Sci Plant Nutr 171:181–186

Jáuregui-Zùñiga D, Ferrer MA, Calderón AA, Muñoz R, Moreno A (2005) Heavy metal stress reduces the deposition of calcium oxalate crystals in leaves of *Phaseolus vulgaris*. J Plant Physiol 162:1183–1187

Jiang Y, Huang B (2001) Effect of calcium on antioxidant activities and water relations associated with heat tolerance in two cool season grasses. J Exp Bot 355:341–349

Jing-hong H, Yang Y, Qing-mao S, Chun-juan D, Zhi-gang Z (2012) Effect of exogenous salicylic acid on nitrogen assimilation of cucumber seedling under drought stress. Acta Hort Sinice 39:81–90

Kalinich JF, Mandava NB, Todhunter JA (1985) Relationship of nucleic acid metabolism to brassinolide induced responses in beans. J Plant Physiol 120:207–214

Kapoor D, Bhardwaj R (2014) Physiological mechanisms of *Brassica Juncea* L. plants exposed to cadmium metal stress. Indian J Appl Res 4:8

Kapoor D, Rattan A, Gautam V, Kapoor N, Bharadwaj R (2014) 24-Epibrassinolide mediated photosynthetic pigments and antioxidative defense systems of radish seedling under cadmium and mercury stress. J Stress Physiol Biochem 10:110–121

Khan NA, Mobin M, Samiullah (2005) The influence of gibberellic acid and sulfur fertilization rate on growth and S-use efficiency of mustard (*Brassica juncea*). Plant Soil 270:269–274

Khan MN, Siddiqui MH, Mohammad F, Naeem M, Khan MMA (2010) Calcium chloride and gibberellic acid protect linseed (*Linum usitatissimum* L.) from NaCl stress by inducing antioxidative defence system and osmoprotectant accumulation. Acta Physiol Plant 32:121–132

Khan MIR, Iqbal N, Masood A, Per TS, Khan NA (2013) Salicylic acid alleviates adverse effects of heat stress on photosynthesis through changes in proline production and ethylene formation. Plant Signal Behav 8, e26374

Khan MIR, Nazir F, Asgher M, Per TS, Khan NA (2014) Selenium and sulfur influence ethylene formation and alleviate cadmium-induced oxidative stress by improving proline and glutathione production in wheat. J Plant Physiol 173:9–18

Khatai L, Goessler W, Lorencova H, Zangger K (2004) Modulation of nitric oxide-mediated metal release from metallothionein by the redox state of glutathione in vitro. Eur J Biochem 271:2408–2416

Kiba T, Kudo T, Kojima M, Sakakibara H (2011) Hormonal control of nitrogen acquisition: roles of auxin, abscisic acid, and cytokinin. J Exp Bot 62:1399–1409

Kim H, Fujiwara T, Hayashi H, Chino M (1997) Effects of exogenous ABA application on sulfate and OAS concentrations, and on composition of seed storage proteins in in vitro cultured soybean immature cotyledons (conference paper). Soil Sci Plant Nutr 43:1119–1123

Kim Y, Yang YY, Lee Y (2002) Pb and Cd uptake in rice toots. Physiol Plant 116:368–372

Koç E, Üstün AS, Çelik N (2013) Effect of exogenously applied salicylic acid on cadmium chloride-induced oxidative stress and nitrogen metabolism in tomato (*Lycopersicon esculentum* L.). Turk J Biol 37:361–369

Koprivova A, North KA, Kopriva S (2008) Complex signaling network in regulation of adenosine 5′- phosphosulfate reductase by salt stress in *Arabidopsis* roots. Plant Physiol 146:1408–1420

Krantev A, Yordanova R, Janda T, Szalai G, Popova L (2008) Treatment with salicylic acid decreases the effect of cadmium on photosynthesis in maize plants. J Plant Physiol 165:920–931

Kumari A, Sheokand S, Swaraj K (2010) Nitric oxide induced alleviation of toxic effects of short term and long term Cd stress on growth, oxidative metabolism and Cd accumulation in chickpea. Braz J Plant Physiol 22:271–284

Kurtyka R, Małkowski E, Kita A, Karcz W (2008) Effect of calcium and cadmium on growth and accumulation of cadmium, calcium, potassium and sodium in maize seedlings. Polish J Environ Stud 17:51–56

Kuthanova A, Gemperlova´ L, Zelenkova S, Eder J, Machácková I, Opatrny Z, Cvikrova M (2004) Cytological changes and alterations in polyamine contents induced by cadmium in tobacco BY-2 cells. Plant Physiol Biochem 42:149–156

Kwak SH, Lee SH (1997) The requirements for Ca^{2+}, protein phosphorylation, and dephosphorylation for ethylene signal transduction in *Pisum sativum* L. Plant Cell Physiol 38:1142–1149

Lamotte O, Gould K, Lecourieux D, Sequeira-Legrand A, Lebrun-Garcia A, Durner J, Pugin A, Wendehenne D (2004) Analysis of nitric oxide signaling functions in tobacco cells challenged by the elicitor cryptogein. Plant Physiol 135:516–529

Laspina NV, Groppa MD, Tomaro ML, Benavides MP (2005) Nitric oxide protects sunflower leaves against Cd-induced oxidative stress. Plant Sci 169:323–330

Lee BR, Jin YL, Avice JC, Cliquet JB, Ourry A, Kim TH (2009) Increased proline loading to phloem and its effects on nitrogen uptake and assimilation in water stressed white clover (*Trifolium repens*). New Phytol 182:654–663

Lin T, Zhu X, Zhang F, Wan X (2011) The detoxification effect of cadmium stress in *Populus yunnanensis*. Res J Bot 4:13–19

Lopez-Carrion AI, Castellano R, Rosales MA, Ruiz JM, Romero L (2008) Role of nitric oxide under saline stress: implications on proline metabolism. Plant Biol 52:587–591

Lu J, Ertl JR, Chen C (1992) Transcriptional regulation of nitrate reductase mRNA levels by cytokinin abscisic acid interactions in etiolated barley leaves. Plant Physiol 98:1255–1260

Lux A, Martinka M, Vaculik M, White PJ (2011) Root responses to cadmium in the rhizosphere: a review. J Exp Bot 62:21–37

Ma W, Xu W, Xu H, Chen Y, He Z, Ma M (2010) Nitric oxide modulates cadmium influx during cadmium-induced programmed cell death in tobacco BY-2 cells. Planta 232:325–335

MacRobbie EAC (1992) Calcium and ABA-induced stomatal closure. Philos Trans R Soc B 338:5–18

Maggio A, Barbieri G, Raimondi G, Pascale SD (2010) Contrasting effects of GA_3 treatments on tomato plants exposed to increasing salinity. J Plant Growth Regul 29:63–72

Mahmood T, Kapuganti J, Gupta WMK (2009) Cadmium stress stimulates nitric oxide production by wheat roots. Pak J Bot 41:1285–1290

Mai Y, Lin S, Zeng X, Ran R (1989) Effect of brassinolide on nitrate reductase activity in rice seedlings. Plant Physiol Commun 2:50–52

Makela P, Munns R, Colmer TD, Peltonen-Sainio P (2003) Growth of tomato and an ABA-deficient mutant (sitiens) under saline conditions. Physiol Plant 117:58–63

Manaa A, Gharbi E, Mimouni H, Wasti S, Aschi-Smiti S, Lutts S, Ben Ahmed H (2014) Simultaneous application of salicylic acid and calcium improves salt tolerance in two contrasting tomato (*Solanum lycopersicum*) cultivars. South Afr J Bot 95:32–39

Marefat E, Shabani L, Amooaghaie R (2012) Interaction effect of cadmium and salicylic acid on proline and antioxidant enzyme activity in soybean. Malaysian Appl Biol 41:25–30

Marschner H (1995) Mineral nutrition of higher plants, 2nd edn. Academic, San Diego

Masood A, Khan NA (2013) Ethylene and gibberellic acid interplay in regulation of photosynthetic capacity inhibition by cadmium. J Plant Biochem Physiol 1:3

Masood A, Iqbal N, Khan NA (2012) Role of ethylene in alleviation of cadmium-induced photosynthetic capacity inhibition by sulphur in mustard. Plant Cell Environ 35:524–533

McAinsh MR, Brownlee C, Hetherington AM (1997) Calcium ions as second messengers in guard cell signal transduction. Physiol Plant 100:16–29

Meher HC, Gajbhiye VT, Singh G (2011) Salicylic acid-induced glutathione status in tomato crop and resistance to root-knot nematode, *Meloidogyne incognita* (Kofoid & White). Chitwood J Xenobiotics 1, e5

Mehta SK, Gaur JP (1999) Heavy-metal-induced proline accumulation and its role in ameliorating metal toxicity in *Chlorella vulgaris*. New Phytol 143:253–259

Mendoza-Cózatl DG, Jobe TO, Hauser F, Schroeder JI (2011) Long-distance transport, vacuolar sequestration, tolerance, and transcriptional responses induced by cadmium and arsenic. Curr Opin Plant Biol 14:554–562

Meng H, Hua S, Shamsi IH, Jilani G, Li Y et al (2009) Cadmium-induced stress on the seed germination and seedling growth of *Brassica napus* L., and its alleviation through exogenous plant growth regulators. Plant Growth Regul 58:47–59

Metwally A, Safronova VI, Belimov AA, Dietz KJ (2005) Genotypic variation of the response to cadmium toxicity in *Pisum sativum* L. J Exp Bot 56:167–178

Misra N, Gupta AK (2006) Interactive effects of sodium and calcium on proline metabolism in salt tolerant green gram cultivar. J Plant Physiol 1:1–12

Misra N, Saxena P (2009) Effect of salicylic acid on proline metabolism in lentil grown under salinity stress. Plant Sci 177:181–189

Moat AG, Foster JW, Spector MP (2003) Biosynthesis and metabolism of amino acids. In: Moat AG, Foster JW, Spector MP (eds) Microbial physiology. Wiley, New York, pp 503–544

Molinari HBC, Marur CJ, Daros E, de Campos MKF, de Carvalho JFRP, Filho JCB (2007) Evaluation of the stress-inducible production of praline in transgenic sugarcane (*Saccharum* spp.): osmotic adjustment, chlorophyll fluorescence and oxidative stress. Physiol Plant 130:218–229

Moradkhani S, Nejad RAK, Dilmaghani K, Chaparzadeh N (2013) Salicylic acid decreases Cd toxicity in sunflower plants. Ann Biol Res 4:135–141

Mori IC, Schroeder JI (2004) Reactive oxygen species activation of plant Ca^{2+} channels. A signaling mechanism in polar growth, hormone transduction, stress signaling and hypothetically mechanotransduction. Plant Physiol 135:702–708

Moussa HR, El-Gamal SM (2010) Role of salicylic acid in regulation of cadmium toxicity in wheat (*Triticum aestivum* L.). J Plant Nutr 33(10):1460–1471

Murata Y, Pei ZM, Mori IC, Schroeder J (2001) Abscisic acid activation of plasma membrane Ca^{2+} channels in guard cells requires cytosolic NAD(P)H and is differentially disrupted upstream and downstream of reactive oxygen species production in abi1-1 and abi2-1 protein phosphatase 2C mutants. Plant Cell 13:2513–2523

Nazar R, Iqbal N, Syeed S, Khan NA (2011) Salicylic acid alleviates decreases in photo-synthesis under salt stress by enhancing nitrogen and sulfur assimilation and antioxidant metabolism differentially in two mungbean cultivars. J Plant Physiol 168:807–815

Nazar R, Iqbal N, Masood A, Khan MIR, Khan NA (2012) Cadmium toxicity in plants and role of mineral nutrients in its alleviation. J Plant Sci 3:1476–1489

Nazar R, Khan MIR, Iqbal N, Masood A, Khan NA (2014) Involvement of ethylene in reversal of salt-inhibited photosynthesis by sulphur in mustard. Physiol Plant 152:331–344

Nazar R, Umar S, Khan NA, Sareer O (2015) Salicylic acid supplementation improves photosynthesis and growth in mustard through changes in proline accumulation and ethylene formation under drought stress. S Afr J Bot 98:84–94

Nedjimi B, Daoud Y (2009) Cadmium accumulation in *Atriplex halimus* subsp. *schweinfurthii* and its influence on growth, proline, root hydraulic conductivity and nutrient uptake. Flora-Mor Dist Func Ecol Plant 204:316–324

Neill SJ, Desikan R, Hancock JT (2003) Nitric oxide signaling in plants. New Phytol 159:11–35

Nelson MT (1986) Interactions of divalent cations with single calcium channels from rat brain synaptosomes. J Gen Physiol 87:201–222

Nies DH (1999) Microbial heavy-metal resistance. Appl Microbiol Biotechnol 51:730–750

Nikiforova VJ, Daub CO, Hesse H, Willmitzer L, Hoefgen R (2005) Integrative gene-metabolite network with implemented causality deciphers informational fluxes of sulphur stress response. J Exp Bot 56:1887–1896

Oh M-H, Kim HS, Wu X, Clouse SD, Zielinski RE, Huber SC (2012) Calcium/calmodulin inhibition of the *Arabidopsis* BRASSINOSTEROID-INSENSITIVE 1 receptor kinase provides a possible link between calcium and brassinosteroid signaling. Biochem J 443:515–523

Okuma E, Murakami Y, Shimoishi Y, Tada M, Murata Y (2004) Effects of exogenous application of proline and betaine on the growth of tobacco cultured cells under saline conditions. Soil Sci Plant Nutr 50:1301–1305

Ouariti O, Boussama N, Zarrouk M, Cherif A, Ghorbal MH (1997) Cadmium- and copper-induced changes in tomato membrane lipids. Phytochemistry 45:1343–1350

Oulhajd A, Kuschk P, Humback K (2006) Heavy metal stress and leaf senescence induce the barley gene HvC2d1 encoding a calcium-dependent novel C2 domain-like protein. New Phytol 170:261–273

Pal UR, Gossett DR, Sims JL, Legett JE (1976) Molybdenum and sulfur nutrition effects on nitrate reductase in barley tobacco. Can J Bot 54:2014–2022

Paleg LG, Doughlas TJ, Vandaal A, Keech DB (1981) Proline and betaine protect enzymes against heat inactivation. Aust J Plant Physiol 8:107–114

Panda P, Nath S, Chanu TT, Sharma GD, Panda SK (2011) Cadmium stress-induced oxidative stress and role of nitric oxide in rice (Oryza sativa L.). Acta Physiol Plant 33:1737–1747

Pandey S (2001) Response of nitrate assimilation to putrescine and glutamine supply in subabul (*Leucaena leucocephala*) cotyledons. Indian J Agric Sci 71:341–343

Pankovic D, Plesnicar M, Arsenijeevic-Maksimovic I, Petrovic N, Sakac Z, Kastori R (2000) Effects of nitrogen nutrition on photosynthesis in Cd-treated sunflower plants. Ann Bot 86:841–847

Parmar P, Kumari N, Sharma V (2013) Structural and functional alterations in photosynthetic apparatus of plants under cadmium stress. Bot Stud 54:45

Perfus-Barbeoch L, Leonhardt N, Vavasseur A, Forestier C (2002) Heavy metal toxicity: cadmium permeates through calcium channels and disturbs the plant water status. Plant J 32:539–548

Picazo I, Ros R, Moya JL (2007) Heavy metal hormone interactions in rice plants: effect on growth, net photosynthesis and carbohydrate distribution. J Plant Growth Regul 14:61–67

Popova LP, Maslenkova LT, Yordanova RY, Ivanova AP, Krantev AP, Szalai G, Janda T (2009) Exogenous treatment with salicylic acid attenuates cadmium toxicity in pea seedlings. Plant Physiol Biochem 47:224–231

Potters G, Pasternak TP, Guisez Y, Palme KJ, Jansen MA (2007) Stress-induced morphogenic responses: growing out of trouble? Trends Plant Sci 12:98–105

Pustovoitova TN, Bavrina TV, Zhdanova NE (2000) Drought tolerance of transgenic tobacco plants carrying the iaaM and iaaH genes of auxin biosynthesis. Russ J Plant Physiol 47:380e385

Qiao X, Wang P, Shi G, Yang H (2015) Zinc conferred cadmium tolerance in *Lemna minor* L. via modulating polyamines and proline metabolism. Plant Growth Regul. doi:10.1007/s10725-015-0027-0

Radic S, Babic M, Skobic D, Roje V, Pevalek-Kozlina B (2010) Ecotoxicological effects of aluminum and zinc on growth and antioxidants in *Lemna minor* L. Ecotoxicol Environ Saf 73:336–342

Rady MM (2011) Effect of 24-epibrassinolide on growth, yield, antioxidant system and cadmium content of bean (*Phaseolus vulgaris* L.) plants under salinity and cadmium stress. Sci Hortic 129:232–237

Rais L, Masood A, Inam A, Khan NA (2013) Sulfur and nitrogen co-ordinately improve photosynthetic efficiency, growth and proline accumulation in two cultivars of mustard under salt stress. J Plant Biochem Physiol 1:1

Raz V, Fluhr R (1992) Calcium requirement for ethylene-dependent responses. Plant Cell 4:1123–1130

Rivetta A, Negrini N, Cocucci M (1997) Involvement of Ca^{2+} calmodulin in Cd^{2+} toxicity during the early phases of radish (*Raphanus sativus* L.) seed germination. Plant Cell Environ 20:600–608

Rodríguez-Serrano M, Romero-Puertas MC, Pazmiño DM, Testillano PS, Risueño MC, del Río LA, Sandalio LM (2009) Cellular response of pea plants to cadmium toxicity: cross talk between reactive oxygen species, nitric oxide, and calcium. Plant Physiol 150:229–243

Rubio M, Escrig I, Martinez-Cortina C, Lopez-Benet F, Sanz A (1994) Cadmium and nickel accumulation in rice plants. Effects on mineral nutrition and possible interactions of abscisic and gibberellic acids. Plant Growth Regul 14:151–157

Sakakibara H (2003) Nitrate-specific and cytokinin-mediated nitrogen signaling pathways in plants. J Plant Res 116:253–257

Sánchez E, Ávila-Quezada G, Gardea AA, Ruiz JM, Romero L (2007) Biosynthesis of proline in fruits of green bean plants: deficiency versus toxicity of nitrogen. Int J Exp Bot 76:143–152

Sandalio LM, Dalurzo HC, Gómez M, Romero-Puertas MC, del Río LA (2001) Cadmium-induced changes in the growth and oxidative metabolism of pea plants. J Exp Bot 52:2115–2126

Sankla M, Huber W (1975) Effect of salt and abscisic acid on in vivo activity of nitrate reductase in seedlings of *Phaseolus aconitifolius*. Z Pflanzenphysiol 76:467–470

Schwartz A (1985) Role of calcium and EGTA on stomatal movements in Commelina communis. Plant Physiol 79:1003–1005

Seki M, Umezawa T, Urano K (2007) Regulatory metabolic networks in drought stress responses. Curr Opin Plant Biol 10:296–302

Schroeder JI, Hagiwara S (1990) Repetitive increases in cytosolic Ca^{2+} of guard cells by abscisic acid activation of nonselective Ca^{2+} permeable channels. Proc Natl Acad Sci U S A 87:9305–9309

Shah K, Dubey RS (1998) Effect of cadmium on proline accumulation and ribonuclease activity in rice seedlings: role of proline as a possible enzyme protectant. Biol Plant 40:121–130

Shah SH, Wainwright SJ, Merrett MJ (1990) The interaction of sodium and calcium chlorides and light on growth, potassium nutrition and proline accumulation in callus culture of *Medicago sativa* L. New Phytol 116:37–45

Shah SH, Tobita S, Shono M (2001) Supplemental calcium regulates proline accumulation in NaCl-stressed suspension culture of *Oryza sativa* L. at the level of mRNA translation. Pak J Biol Sci 4:707–710

Sharma S, Dietz K (2006) The significance of amino acids and amino acid-derived molecules in plant responses and adaptation to heavy metal stress. J Exp Bot 57:711–726

Sharma SS, Schat H, Vooijs R (1998) In vitro alleviation of heavy metal induced enzyme inhibition by proline. Phytochemistry 46:1531–1535

Shi GR, Cai QS, Liu QQ, Wu L (2009) Salicylic acid-mediated alleviation of cadmium toxicity in hemp plants in relation to cadmium uptake, photosynthesis and antioxidant enzymes. Acta Physiol Plant 31:969–977

Siddiqui MH, Khan MN, Mohammad F, Khan MMA (2008) Role of nitrogen and gibberellins (GA_3) in the regulation of enzyme activities and in osmoprotectant accumulation in *Brassica juncea* L. under salt stress. J Agron Crop Sci 194:214–224

Siddiqui MH, Al-Whaibi MH, Basalah MO (2011) Interactive effect of calcium and gibberellin on nickel tolerance in relation to antioxidant systems in *Triticum aestivum* L. Protoplasma 248:503–511

Signora L, De Smet I, Foyer CH, Zhang H (2001) ABA plays a central role in mediating the regulatory effects of nitrate on root branching in *Arabidopsis*. Plant J 28:655–662

Simaei M, Khavarinejad RA, Saadatmand S, Bernard F, Fahimi H (2011) Interactive effects of salicylic acid and nitric oxide on soybean plants under NaCl salinity. Russ J Plant Physiol 58:783–790

Singh PK, Chaturvedi VK (2012) Effects of salicylic acid on seedling growth and nitrogen use efficiency in cucumber (*Cucumis sativus* L.). Plant Biosyst 146:302–308

Singh I, Shah K (2015) Evidences for suppression of cadmium induced oxidative stress in presence of sulphosalicylic acid in rice seedlings. Plant Growth Regul 76:12p

Singla B, Chugh A, Khurana JP, Khurana P (2006) An early auxin-responsive Aux/IAA gene from wheat (*Triticum aestivum*) is induced by epibrassinolide and differentially regulated by light and calcium. J Exp Bot 57:4059–4070

Siripornadulsil S, Traina S, Verma DP, Sayre RT (2002) Molecular mechanisms of proline-mediated tolerance to toxic heavy metals in transgenic microalgae. Plant Cell 14:2837–2847

Stroiński A, giżewska K, zielezińska M (2013) Abscisic acid is required in transduction of cadmium signal to potato roots. Biol Plant 57:121–127

Sun JY, Shen ZG (2007) Effects of Cd stress on photosynthetic characteristics and nutrient uptake of cabbages with different Cd-tolerance. Chin J Appl Ecol 18:2605–2610

Sun Q, Yec ZH, Wang XR, Wong MH (2007) Cadmium hyperaccumulation leads to an increase of glutathione rather than phytochelatins in the cadmium hyperaccumulator *Sedum alfredii*. J Plant Physiol 164:1489–1498

Suzuki N (2005) Alleviation by calcium of cadmium-induced root growth inhibition in *Arabidopsis* seedlings. Plant Biotechnol 22:19–25

Talanova VV, Titov AF, Boeva NP (2000) Effect of increasing concentrations of lead and cadmium on cucumber seedlings. Biol Plant 43:441–444

Tarighaleslami M, Zarghami R, Boojar MMA, Oveysi M (2012) Effects of drought stress and different nitrogen levels on morphological traits of proline in leaf and protein of corn seed (*Zea mays* L.). Am-Eurasian J Agric Environ Sci 12:49–56

Terrile MC, Tonon CV, Iglesias MJ, Lamattina L, Casalongue CA (2010) Extracellular ATP and nitric oxide signaling pathways regulate redox-dependent responses associated to root hair growth in etiolated *Arabidopsis* seedlings. Plant Signal Behav 5:698–701

Upreti KK, Murti GSR (2004) Effects of brassinosteroids on growth, nodulation, phytohormone content and nitrogenase activity in French bean under water stress. Biol Plant 48:407–411

Verma K, Mehta SK, Shekhawat GS (2013) Nitric oxide (NO) counteracts cadmium induced cytotoxic processes mediated by reactive oxygen species (ROS) in *Brassica juncea*: cross-talk between ROS, NO and antioxidant responses. BioMetals 26:255–269

Villiers F, Jourdain A, Bastien O, Leonhardt N et al (2012) Evidence for functional interaction between brassinosteroids and cadmium response in *Arabidopsis thaliana*. J Exp Bot 63:1185–1200

Wang QC, Song H (2009) Calcium protects *Trifolium repens* L. seedlings against cadmium stress. Plant Cell Rep 28:1341–1349

Wang LJ, Chen SJ, Kong WF, Li SH, Archbold DD (2006) Salicylic acid pretreatment alleviates chilling injury and affects the antioxidant system and heat shock proteins of peaches during cold storage. Postharvest Biol Technol 41:244–225

Wang Q, Liang X, Dong Y, Xu L, Zhang X, Kong J, Liu S (2013) Effects of exogenous salicylic acid and nitric oxide on physiological characteristics of perennial ryegrass under cadmium stress. J Plant Growth Regul 32:721–731

Wen X-P, Ban Y, Inoue H, Matsuda N, Kita M et al (2011) Antisense inhibition of a spermidine synthase gene highlights the role of polyamines for stress alleviation in pear shoots subjected to salinity and cadmium. Environ Exp Bot 72:157–166

Wu JT, Chang SJ, Chou TL (1995) Intracellular proline accumulation in some algae exposed to copper and cadmium. Bot Bull Acad Sinica 36:89–93

Xie HL, Jiang RF, Zhang FS, McGrath SP, Zhao FJ (2009) Effect of nitrogen form on the rhizosphere dynamics and uptake of cadmium and zinc by the hyperaccumulator Thlaspi caerulescens. Plant and Soil 318:205–215

Xiong J, Fu G, Tao L, Zhu C (2010) Roles of nitric oxide in alleviating heavy metal toxicity in plants. Arch Biochem Biophys 497:13–20

Xu J, Yin HX, Li X (2009) Protective effects of proline against cadmium toxicity in micropropagated hyperaccumulator, *Solanum nigrum* L. Plant Cell Rep 28:325–333

Xu J, Wang W, Yin H, Liu X, Sun H, Mi Q (2010) Exogenous nitric oxide improves antioxidative capacity and reduces auxin degradation in roots of *Medicago truncatula* seedlings under cadmium stress. Plant Soil 326:321–330

Xu X, Shi G, Ding C, Xu Y, Zhao J, Yang H, Pan Q (2011) Regulation of exogenous spermidine on the reactive oxygen species level and polyamine metabolism in *Alternanthera philoxeroides* (Mart.) Griseb under copper stress. Plant Growth Regul 63:251–258

Xu C, Li X, Zhang L (2013a) The effect of calcium chloride on growth, photosynthesis, and antioxidant responses of *Zoysia japonica* under drought conditions. PLoS One 8, e68214

Xu L, Dong Y, Fan Z, Kong J, Liu S, Bai S (2013b) Effects of the application of exogenous NO at different growth stage on the physiological characteristics of peanut grown in Cd-contaminated soil. J Plant Interact 9:285–296

Yadav S, Hayat S, Wani AS, Irfan M, Ahmad A (2012) Homobrassinolide and 24-epibrassinolide on nitrate reductase activity, proline content, and antioxidative enzymes of tomato. Intl J Veg Sci 18:161–170

Yamagami K, Nishimura S, Sorimachi M (1998) Cd^{2+} and Co^{2+} at micromolar concentrations mobilize Ca^{2+} via the generation of 1,4,5-triphosphate in bovine chromaffin cells. Brain Res 798:316–319

Yang C-S, Tzou B-C, Liu YP, Tsai M-J, Shye SQ, Tzeng S-F (2008) Inhibition of cadmium induced oxidative injury in rat primary astrocytes by the addition of antioxidants and the reduction of intracellular calcium. J Cell Biochem 103:825–834

Ye J-L, Mao W-P, Wu A-L, Zhang N-N, Zhang C, Yu Y-J, Zhou L, Wei C-L (2007) Cadmium-induced apoptosis in human normal liver L-02 cells by acting on mitochondria and regulating Ca^{2+} signals. Environ Toxicol Phar 24:45–54

Yendrek CR, Lee YC, Morris V, Liang Y, Pislariu CI, Burkart G, Meckfessel MH, Salehin M, Kessler H, Wessler H, Lloyd M, Lutton H, Teillet A, Sherrier DJ, Journet EP, Harris JM, Dickstein R (2010) A putative transporter is essential for integrating nutrient and hormone signaling with lateral root growth and nodule development in *Medicago truncatula*. Plant J 62:100–112

Yilmaz R, Sakcali S, Yarci C, Aksoy A, Ozturk M (2006) Use of *Aesculus hippocastanum* L. as a biomonitor of heavy metal pollution. Pak J Bot 38:1519–1527

Yoo JH, Park CY, Kim JC, Heo WD, Cheong MS, Park HC, Kim MC, Moon BC, Choi MS, Kang YH, Lee JH, Kim HS, Lee SM, Yoon HW, Lim CO, Yun DJ, Lee

SY, Chung WS, Cho MJ (2005) Direct interaction of a divergent CaM isoform and the transcription factor, MYB2, enhances salt tolerance in *Arabidopsis*. J Biol Chem 280:3697–3706

Youm JW, Jeon JH, Choi D, Yi SY, Joung H, Kim HS (2008) Ectopic expression of pepper CaPF1 in potato enhances multiple stresses tolerance and delays initiation of in vitro tuberization. Planta 228:701–708

Yuan L, Yuan Y, Du J, Sun J, Guo S (2012) Effects of 24-epibrassinolide on nitrogen metabolism in cucumber seedlings under $Ca(NO_3)^2$ stress. Plant Physiol Biochem 61:29–35

Yusuf M, Fariduddin Q, Ahmad I, Ahmad A (2014) Brassinosteroid-mediated evaluation of antioxidant system and nitrogen metabolism in two contrasting cultivars of *Vigna radiata* under different levels of nickel. Physiol Mol Biol Plant 20:449–460

Zafar S, Nasri M, Moghadam HRT, Zahedi H (2014) Effect of zinc and sulfur foliar applications on physiological characteristics of sunflower (*Helianthus annuus* L.) under water deficit stress. Int J Biosci 5:87–96

Zakery-Asl MA, Bolandnazara S, Oustanb S (2014) Effect of salinity and nitrogen on growth, sodium, potassium accumulation, and osmotic adjustment of halophyte *Suaedaa egyptiaca* (Hasselq.). Zoh Arch Agron Soil Sci 60:785–792

Zawoznik MS, Groppa MD, Tomaro ML, Benavides MP (2007) Endogenous salicylic acid potentiates cadmium-induced oxidative stress in *Arabidopsis thaliana*. Plant Sci 173:190–197

Zengin FK, Munzuroglu O (2005) Effects of some heavy metals on content of chlorophyll, proline and some antioxidant chemicals in bean (*Phaseolus vulgaris* L.) seedlings. Acta Biol Cracov Bot 47:157–164

Zhang WN, Chen WL (2011) Role of salicylic acid in alleviating photochemical damage and autophagic cell death induction of cadmium stress in *Arabidopsis thaliana*. Photochem Photobiol Sci 10:947–955

Zhang LP, Mehta SK, Liu ZP, Yang ZM (2008) Copper induced proline synthesis is associated with nitric oxide generation in *Chlamydomonas reinhardtii*. Plant Cell Physiol 49:411–419

Zhang Z, Li F, Li D, Zhang H, Huang R (2010) Expression of ethylene response factor JERF1 in rice improves tolerance to drought. Planta 232:765e774

Zhang Y, Xu S, Yang S, Chen Y (2015) Salicylic acid alleviates cadmium-induced inhibition of growth and photosynthesis through upregulating antioxidant defense system in two melon cultivars (*Cucumis melo* L.). Protoplasma 252:911–924

Zhao Y, Qi Z, Berkowitz GA (2013) Teaching an old hormone new tricks: cytosolic Ca^{2+} elevation involvement in plant brassinosteroid signal transduction cascades. Plant Physiol 163:555–565

Zhenyan HE, Jiangchuan LI, Zhang H, Ma MI (2005) different effects of calcium and lanthanum on the expression of phytochelatin synthase gene and cadmium absorption in *Lactuca sativa*. Plant Sci 168:309–318

Zhou ZS, Guo K, Elbaz AA, Yang ZM (2009) Salicylic acid alleviates mercury toxicity by preventing oxidative stress in roots of *Medicago sativa*. Environ Exp Bot 65:27–34

Zhu E, Liu D, Li JG, Li TQ, Yang XE, He ZL, Stoffella PJ (2011) Effect of nitrogen fertilizer on growth and cadmium accumulation in *Sedum alfredii* Hance. J P Nutr 34:115–126

Zhu XF, Jiang T, Wang ZW, Lei GJ, Shi YZ, Li GX, Zheng SJ (2012) Gibberellic acid alleviates cadmium toxicity by reducing nitric oxide accumulation and expression of IRT1 in *Arabidopsis thaliana*. J Hazard Mat 239–240:302–307

Proline Accumulation in Plants: Roles in Stress Tolerance and Plant Development

9

Mudasir Irfan Dar, Mohd Irfan Naikoo, Farha Rehman, Fauzia Naushin, and Fareed Ahmad Khan

Abstract

Proline accumulation occurs in a wide range of plant species in response to various kinds of environmental stresses. A large body of evidence suggests that a positive correlation occurs between proline accumulation and plant stress tolerance. In this chapter, we will discuss the metabolism of proline accumulation and its role in stress tolerance in plants. Existing literature indicates that despite acting as an osmolyte, proline also plays important roles during stress as a metal chelator and an antioxidative defence molecule. Moreover, when applied exogenously at low concentrations, proline enhanced stress tolerance in plants. However, some reports point out adverse effects of proline when applied at higher doses. Role of proline in seed germination, flowering and other developmental programmes is also presented in this chapter.

9.1 Introduction

Plants are subjected to various kinds of abiotic and biotic stresses throughout their life cycles which include salinity, drought, temperature extremes, heavy metal, infection by pathogens, nutrient deficiency and UV radiation (Hare and Cress 1997; Saradhi et al. 1995; Siripornadulsil et al. 2002). A general response of plants to various kinds of stresses is the accumulation of compatible osmolytes such as proline, glycine betaine, proline betaine, glycerol, mannitol and sorbitol, etc. which protect cells against damage caused by stress (Hare and Cress 1997). These small uncharged molecules at physiological pH are highly soluble in water, which allows them to accumulate at

M.I. Dar (✉) • M.I. Naikoo • F.A. Khan
Environmental Botany Section, Department of Botany, Aligarh Muslim University, Aligarh, UP 202002, India
e-mail: irfanmudasir@gmail.com

F. Rehman
Department of Botany, Faculty of Science, Mohammad Ali Jauhar University, Rampur, UP 24490, India

F. Naushin
Department of Botany, Women's College, Aligarh Muslim University, Aligarh, UP 202002, India

N. Iqbal et al. (eds.), *Osmolytes and Plants Acclimation to Changing Environment: Emerging Omics Technologies*, DOI 10.1007/978-81-322-2616-1_9

elevated concentration in the cytosol of plant cells without harming cellular structures (Low 1985; Trovato et al. 2008). These osmolytes help plants to cope up with stress conditions by adjusting cellular osmotic pressure, contributing in ROS detoxification, protection of membrane integrity and stabilization of enzymes/proteins (Ashraf and Foolad 2007; Bohnert and Jensen 1996; Hayat et al. 2012). Among them, proline plays a pivotal role (Hare and Cress 1997) and accumulates in a large number of species under salinity (Yoshiba et al. 1995), drought (Choudhary et al. 2005), cold, nutrient deficiency (Hare and Cress 1997), heavy metals (Schat et al. 1997), pathogen attack (Fabro et al. 2004; Haudecoeur et al. 2009; Sreedevi et al. 2013; Rehman et al. 2014) and high acidity (Hare and Cress 1997). Proline is a proteinogenic amino acid with the α-amino group present as a secondary amine and is essential for primary metabolism (Szabados and Savoure 2009; Verslues and Sharma 2010). Different kinds of organisms including protozoa, eubacteria, marine invertebrates and also various plant species under stress conditions have been found to accumulate the amino acid proline (Kaneshiro et al. 1969; Poulin et al. 1987; Csonka 1989; Burton 1991; Brown and Hellebust 1978; Verbruggen et al. 1993; Yoshiba et al. 1995; Peng et al. 1996; Nakashima et al. 1998; Mattioli et al. 2008).

9.2 Proline Metabolism

Intracellular proline levels in plants are governed by biosynthesis, catabolism and transport between cells and different compartments of cell (Szabados and Savoure 2009). In higher plants, proline biosynthesis occurs either via the glutamate or the ornithine pathway. The former involves the synthesis of proline from glutamic acid via intermediate pyrroline-5-carboxylate (P5C) in the cytoplasm or chloroplast and is the key biochemical pathway under physiological conditions and under nitrogen insufficiency or osmotic stress (Delanauney and Verma 1993; Trovato et al. 2008; Verslues and Sharma 2010). The biosynthesis of proline from this pathway was first found in bacteria (Leisinger 1987) and subsequently in numerous prokaryotic and eukaryotic organisms (Krishna and Leisinger 1979; Adams and Frank 1980; Csonka 1989). The core enzymes in this reaction are pyrroline-5-carboxylate synthetase (P5CS) and pyrroline-5-carboxylate reductase (P5CR) (Sekhar et al. 2007; Fig. 9.1). P5CS is encoded by two genes, whilst P5CR is encoded by only one in the majority of plant species (Strizhov et al. 1997; Armengaud et al. 2004; Verbruggen et al. 1993). In another pathway, proline synthesis occurs via deamination of ornithine which is transaminated to P5C by ornithine-delta-aminotransferase (OAT) (Verbruggen and Hermans 2008; Fig. 9.1). It has been reported that the ornithine pathway occurs during supra-optimal nitrogen conditions and seedling development (Roosens et al. 1998; Armengaud et al. 2004).

Plant cells have a potential to accumulate proline speedily and degrade it quickly when needed (Trovato et al. 2008). During proline biosynthesis, conversion product P5C may cause the production of reactive oxygen species (ROS) and induction of apoptosis and consequently a programmed cell death when undue amounts are accumulated (Hellmann et al. 2000; Székely et al. 2008). Therefore, plants need to degrade it quickly as soon as the stress is relieved. Proline catabolism occurs in mitochondria via the incessant action of proline dehydrogenase (PDH) or proline oxidase (POX) producing P5C from proline and P5C dehydrogenase (P5CDH), which converts P5C to glutamate (Szabados and Savoure 2009; Fig. 9.1). PDH is encoded by two genes, whereas only one P5CDH gene is known in *Arabidopsis* and tobacco (*Nicotiana tabacum*) (Kiyosue et al. 1996; Verbruggen et al. 1996; Ribarits et al. 2007). The interconversion of proline and glutamate is sometimes known as the "proline cycle" (Verslues and Sharma 2010). Unlike proline biosynthesis, the catabolic pathway involving PDH and P5CDH seems to be the only route for plants to degrade excess proline. *Arabidopsis pdh* mutants (lacking a functional PDH) are not capable of eliminating excess proline when supplied exogenously (Nanjo et al. 2003). Similarly, *p5cdh* mutants (lacking a functional P5CDH) are hypersensitive to exoge-

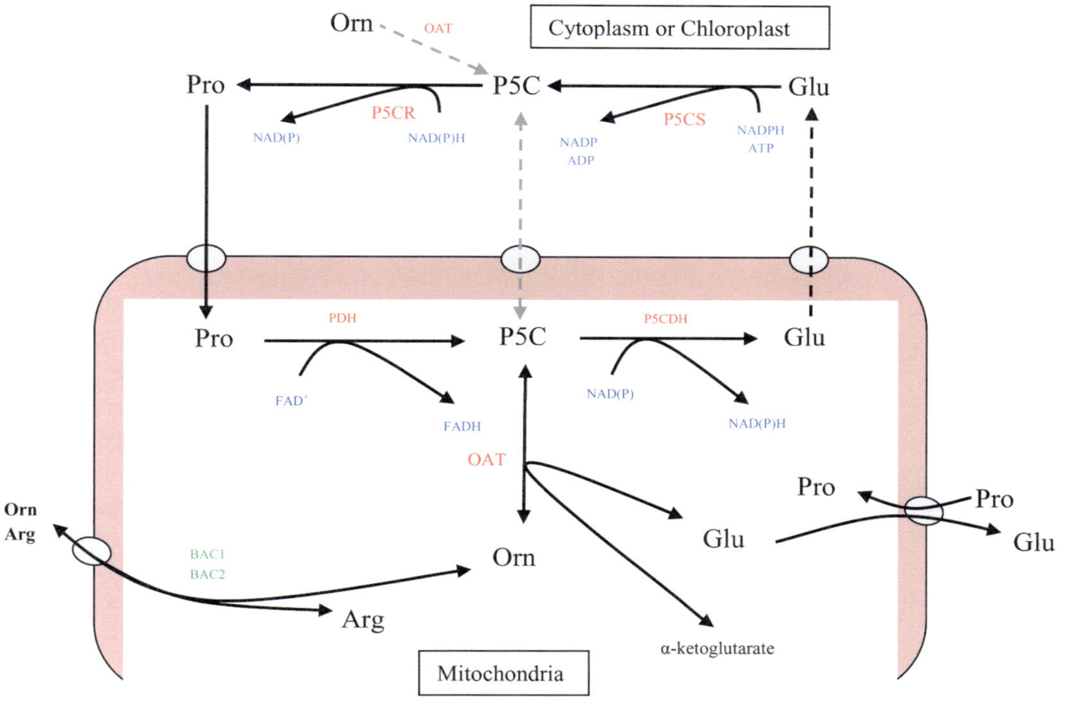

Fig. 9.1 The core pathways of proline metabolism in higher plants Pro biosynthesis from Glu occurs in the cytoplasm and/or chloroplast via two enzymatic steps. Pro catabolism to Glu occurs in the mitochondria also by two enzymatic steps. Both Pro biosynthesis and catabolism use common intermediate P5C (formed by spontaneous cyclization of glutamic-5-semialdehyde produced by P5CS or PDH). *Solid lines* represent known metabolic or transport steps, whilst *dashed lines* represent proposed but not demonstrated steps. Carriers BAC1 and BAC2 may also influence proline metabolism by movement of Arg or Orn. OAT had been proposed to be a cytoplasmic enzyme that functioned as an alternative route to proline, but recent evidence has placed it in the mitochondria. Arg, arginine; BAC, basic amino acid carrier; Glu, glutamate; Orn, ornithine; Pro, proline (Adapted from Verslues and Sharma (2010)

nously supplied proline, whilst plants overexpressing *P5CDH* are less sensitive to proline (Deuschle et al. 2001) as overproduction of P5CDH boosts the rate of degradation of P5C thus thwarting cell death (Hellmann et al. 2000; Deuschle et al. 2001).

Depending upon environmental conditions, proline can be synthesized in different subcellular compartments (Szabados and Savoure 2009). Housekeeping biosynthesis of proline takes place in the cytosol, and in *Arabidopsis* it is controlled by the *P5CS2* gene (Székely et al. 2008; Szabados and Savoure 2009), which is active in dividing meristematic tissues, such as shoot and root tips, and inflorescences (Madan et al. 1995; Deuschle et al. 2001; Tripathi and Gaur 2004). Both P5CS genes are active in floral shoot apical meristems and contribute proline for flower development (Csonka and Hanson 1991). Proline synthesis in chloroplasts is controlled by the stress-induced *P5CS1* gene in *Arabidopsis* (Székely et al. 2008; Savoure et al. 1995; Strizhov et al. 1997; Szabados and Savoure 2009) (Fig. 9.1).

9.3 Role in Stress Tolerance

Although plants exhibit species-specific differences in proline accumulation, there have been mixed results regarding the relationship between proline accumulation and stress tolerance in plants. In some cases, proline accumulation was found higher in stress-tolerant than in stress-sensitive plants. For example, in the roots of salt-tolerant alfalfa plants, proline accumulation increased dramatically under saline stress, whilst salt-sensitive plants responded slowly (Fougere et al. 1991; Petrusa and Winicov 1997). Similarly, salt-tolerant ecotypes of *Agrostis stolonifera* concentrated more proline under salt-stressed conditions than salt-sensitive ecotypes (Ahmad et al. 1981). Significant enhancement in salt tolerance was also found at the cellular level in some plant species. For example, in vitro studies with *Brassica juncea* revealed that elevated levels of free proline were found in salt-tolerant calli under salt stress compared with non-stressed calli (Madan et al. 1995; Gangopadhyay et al. 1997). Similarly, salt-adapted calli of *Citrus limon* had considerably greater proline concentration compared with salt-sensitive calli (Piqueras et al. 1996).

Proline accumulation also occurs in plants in response to drought stress. For example, water deficit rice plants accumulated high amounts of proline in leaves (Hsu et al. 2003) which was attributed to improved contents of the precursors for proline biosynthesis, including glutamic acid, ornithine and arginine. In the case of wheat, rate of proline accumulation and utilization was also considerably higher in the drought-tolerant cultivar than drought-sensitive cultivar under the conditions of drought stress (Nayyar and Walia 2003). Moreover, activities of proline biosynthetic enzymes P5CR and OAT increased primarily in tolerant lines in *B. juncea* plants grown under stress environment; however, the activity of enzyme which degrades proline "proline oxidase" decreased in all lines (Madan et al. 1995).

Drought, cold and salt stress causes excessive accumulation of salts in the soil which directly (because of salinization) or indirectly (because of water loss) results in osmotic stress (Trovato et al. 2008). As a result, the soil water potential gradually decreases, which hinders and finally stops flow of water gradient from roots to apical shoot. The resulting osmotic stress causes closure of stomata, reduction in photosynthesis rate and growth inhibition (Trovato et al. 2008). Within the cell, osmotic stress results in the production of ROS and the accumulation of toxic ions such as Na^+ or Cl^-, which can severely damage the membrane structures and other cell components (Apel and Hirt 2004). To maintain low levels of Na^+ ions in the cytoplasm, plants possess specific transporters such as the plasma membrane SOS1 Na^+/H^+ antiporter which pump out Na^+ ions into the vacuole or outside the cell (Shi et al. 2000). The mRNA of this antiporter has been shown to be specifically stabilized during osmotic stress (Chung et al. 2008) and is escorted by the accumulation of compatible solutes in the cytosol, to balance the osmotic potential of cytosol and vacuole. This was further supported by the cytosolic localization of P5CS1 in *Arabidopsis* mesophyll leaves under non-stress conditions, whilst under salt stress, P5CS1 is translocated to and mainly spotted in chloroplasts (Székely et al. 2008).

Moreover, the rate of the Calvin cycle is reduced during stress conditions, preventing oxidation of NADPH and restoration of $NADP^+$. The flow of electrons in the electron transport chain gets suppressed because of inadequate electron acceptor $NADP^+$, which leads to produce singlet oxygen in the PSI reaction centre and results in the accumulation of ROS (Chaves et al. 2009). Proline biosynthesis is a reductive pathway and uses NADPH to reduce glutamate to P5C and P5C to proline and generates $NADP^+$ that can be further used as electron acceptor in the oxidative pentose phosphate pathway (Trovato et al. 2008). Furthermore, during the phosphorylation of glutamate, ADP is produced from ATP. An increasing rate of proline biosynthesis in chloroplasts during stress conditions causes to maintain the low $NADPH/NADP^+$ ratio, which results to continue the flow of electrons between photosynthetic excitation centres, stabilizes the redox balance, and protects an important photosynthetic apparatus from photoinhibition (Hare and Cress 1997). During the time of stress recovery, proline oxidation leads to oxidative respiration which provides energy to the cell: the oxidation of one molecule of proline capitulates 30 ATP equivalents (Atkinson 1977) and is therefore well suitable to keep up high-energy-requiring processes (Trovato et al. 2008). The important role of proline biosynthesis for maintaining $NADP^+$ pools during stress conditions was further confirmed when in transgenic soybean plants, inhibition of proline biosynthesis and NADPH-NADP$^+$ conversion by antisense P5CR led to drought hypersensitivity, whilst overexpression of P5CR resulted in drought tolerance (De Ronde et al. 2004).

In mitochondria, proline performs distinctive protective functions. After relief from stress, proline pools contribute a reducing potential for mitochondria through the oxidation of proline by PDH and P5CDH, providing electrons for the respiratory chain and consequently supplying energy to resume growth (Hare and Cress 1997; Kishor et al. 2005). Proline stabilizes mitochondrial respiration by protecting Complex II of the mitochondrial electron transport chain during salt stress (Hamilton and Heckathorn 2001). The P5C-proline cycle discovered in recent times can send electrons to the electron transport chain of mitochondria without producing glutamate and, under some conditions, can cause generation of more ROS in the mitochondria (Miller et al. 2009). Thus, proline catabolism plays a key role in regulating cellular ROS balance and can also control various other regulatory pathways (Hayat et al. 2012).

Proline accumulation has also been suggested to activate alternative detoxification pathway by protecting and stabilizing ROS scavenging enzymes. In salt-stressed tobacco cells, proline has been shown to increase the activities of methylglyoxal detoxification enzymes; enhance peroxidase, glutathione-S-transferase, superoxide dismutase and catalase activities; and increase the glutathione redox state (Hoque et al. 2008; Islam et al. 2010). Accumulation of P5CS1 and P5CR in chloroplasts during the conditions of salt stress indicates that under such adverse conditions, glutamate-derived proline biosynthesis is elevated in plastids, where photosynthesis takes place (Székely et al. 2008).

However, the correlation between proline accumulation and abiotic stress tolerance in plants is not always clear. Accumulation of proline in the leaves was believed to be a symptom of salt injury rather than a sign of salt tolerance in rice plants grown under salinity (Lutts et al. 1999). Similarly, assessment of proline accumulation and distribution in two sorghum genotypes differing in salt tolerance indicated that proline accumulation occurred due to salt stress and not because of salt tolerance (de-Lacerda et al. 2003).

9.4 Response of Plants to Exogenously Applied Proline

A large body of evidence suggested that exogenous application of proline to plants can also increase their stress tolerance. For example, exogenously applied proline provided osmoprotection and promoted growth in various plant species grown under salt-stressed conditions (Csonka and Hanson 1991; Yancey 1994). Exogenous application of proline (30 mM) on salt-stressed rice seedlings neutralized the adverse effects of salinity on the growth of premature plants (Roy et al. 1993). Ali et al. (2007) reported that the exogenous proline sprayed at seedling and/or at vegetative stage of *Zea mays* enhanced its growth under the conditions of water deficiency conditions.

Application of exogenous proline protected cell membranes from salt-induced oxidative stress by increasing activities of different antioxidative enzymes (Yan et al. 2000). Exogenous application of 10 mM proline has been reported to facilitate the growth of salt-stressed tobacco suspension cells (Okuma et al. 2000). Exogenous application of proline enhanced the activities of superoxide dismutase and peroxidase in soybean cell cultures maintained under salinity (Yan et al. 2000; Hua and Guo 2002). The increase in the activities of superoxide dismutase and peroxidase is believed to contribute increased salt tolerance in plants (Yan et al. 2000; Hua and Guo 2002). Kaul et al. (2008) using in vitro studies showed that exogenously applied l-proline proved to be an effective free radical (particularly ROS) scavenger. Hong et al. (2000) suggested that the role of proline as a free radical scavenger in relieving stress is more appreciable than its role as an osmolyte.

Under the conditions of trace metal stress, proline accumulation induces the formation of phytochelatins which chelate with metals like cadmium (Cd), thus alleviating metal toxicity (De Knecht et al. 1994). Xu et al. (2009) found that proline treatment caused reduction in ROS level and protected the membrane integrity of callus cells of *Solanum nigrum* when exposed to high concentrations of Cd, eventually regenerating its shoots. Exogenously applied proline proved to protect the activity of glucose-6-phosphate dehydrogenase and nitrate reductase against inhibition by Cd and Zn (Sharma et al. 1998). This protection was due to the formation of a proline-metal complex (Sharma et al. 1998). Similar complex-forming characters of proline were reported by Farago and Mullen (1979) where complex formation occurred between proline and Cu in metal-tolerant *Armeria maritima*. Proline pretreatment also minimized the toxicity of Hg^{2+} in rice (*Oryza sativa*) by scavenging ROS, such as H_2O_2 (Doke 1997).

Despite the fact that exogenous application of proline to plants subjected to abiotic stresses usually prevents or recuperates the deleterious effects caused by stress, on the other hand, application of high concentrations of proline to plants may inhibit their growth or may have detrimental effects on cellular metabolisms of plants (Ehsanpour and Fatahian 2003; Nanjo et al. 2003). For example, exogenous application of proline caused damage to ultrastructures of chloroplast and mitochondria in *Arabidopsis* plants (Hare et al. 2002), which may subsequently cause a significant increase in ROS in these organelles resulting a significant decrease in photosynthetic electron acceptor pools (Hare et al. 2002). Thus, the broadly accepted hypothesis that proline as an inert compatible solute is accumulated to high levels with least effects on cellular metabolism was questioned (Hare et al. 2002). Application of 20–33 Mm proline to cell cultures of *Vigna radiata* under salt (NaCl) stress alleviated the adverse effects, but concentrations of 50 mM or above proved inhibitory for the growth of both salt-stressed and non-stressed cell cultures (Kumar and Sharma 1989). In this study, contents of Na^+ and Cl^- declined in cells when the concentration of proline reached optimal, but they started increasing with further addition in proline concentration. Similarly, Roy et al. in 1993 determined in rice that 30 mM proline was the most helpful concentration in improving germination and seedling growth under saline conditions, whilst higher concentrations (40 or 50 mM) caused reduced seedling growth and lowered K^+/Na^+ ratio in the leaves. Thus, despite of its much accepted positive role, the toxic effects of proline at high concentrations may cause problems (Ashraf and Foolad 2007). From the above examples, it is therefore very important to determine optimal concentrations of exogenous proline when using as a tolerance-inducing agent.

9.5 Role in Flowering and Development

An increasing body of evidence indicates that proline is also involved in flowering and development (Mauro et al. 1996; Nanjo et al. 1999; Samach et al. 2000; Trovato et al. 2001; Székely et al. 2008; Mattioli et al. 2008, 2009; Lehmann et al. 2010), and it is suggested that the developmental role of proline is not coupled with its role in stress response (Mattioli et al. 2008; Trovato et al. 2008). Existing literature suggests that proline may play a role in flowering and development both as a metabolite and as a signal molecule (Mattioli et al. 2009).

Accumulation of proline in undeveloped seeds of *Vicia faba* indicates that proline might be playing an important role in the development of generative organs (Venekamp and Koot 1984). Numerous studies reported high-proline contents in *Arabidopsis* seeds (Chiang and Dandekar 1995; Schmidt et al. 2007), although data from other plant species are meagre. However, the genes which encode enzymes of proline metabolism are expressed during seed development of different species of plants, including *Arabidopsis*, *Solanum lycopersicum*, *Medicago truncatula* and *Oryza sativa* (Verbruggen et al. 1993; Fujita et al. 1998; Armengaud et al. 2004; Hur et al. 2004). The antisense repression of *AtP5CS1* postponed the emergence of radicle during the seed germination of *Arabidopsis*, indicating that proline synthesis restores the $NADP^+$ pool and consequently activates the oxidative pentose phosphate pathway (Hare et al. 2003).

Various scientists also reported considerable levels of proline in the reproductive organs of different plant species (Chiang and Dandekar 1995; Schwacke et al. 1999; Mattioli et al. 2012), which creates the possibility that this amino acid could be also accumulated under non-stressed conditions for developmental functions. For example, proline accounts about 26 % of the total amino acid pool in the reproductive tissues such as florets, pollen, siliques and seeds of *Arabidopsis*, whilst in the case of vegetative tissues, it only represents 1–3 % (Chiang and Dandekar 1995). Schwacke et al. (1999) analysed various organs of tomato for proline accumulation and found that the amount of this amino acid in tomato flowers was 60 times higher than other organs. Similar physiological accumulations of free proline have been reported at different concentrations, in reproductive organs of other plant species (Chiang and Dandekar 1995), and in most cases the amount of this amino acid is higher than required demand of protein synthesis. Mattioli et al. (2012) found that the development of the male gametophyte in *P5CS1* and *P5CS2* mutants of *Arabidopsis thaliana* was severely compromised, thereby providing genetic evidence that proline is required for pollen development and fertility.

The findings of Mattioli et al. (2008, 2009) raised the possibility that low proline levels contained in apical meristems may signal the most favourable conditions for the plant to flower, whilst higher levels of proline might be read as a stress signal by the plant and induce adaptive responses. In a similar way, low concentration of glucose prompted growth stimulation in yeast whilst at high concentration suppressed the growth (Cho et al. 2006).

The possible role of proline in flower and reproductive development is supported by upregulation of genes involved in proline biosynthesis (*P5CS*, *P5CR*) and transport (*PT*) in reproductive tissues of various plant species (Verbruggen et al. 1993; Savoure et al. 1995; Schwacke et al. 1999). It has been reported that constitutive overexpression of *P5CS1* caused early flowering in transgenic plants under normal (Mattioli et al. 2008) or salt stress conditions (Kishor et al. 1995), whilst Nanjo et al. (1999) found that decreased expression of *P5CS1* in antisense *Arabidopsis* plants inhibits bolting and delayed flowering. Besides bolting, there are other developmental processes in which proline is suggested to play a role, including elongation of pollen tube. In petunia and tomato, the amino acid proline has been found in abundance and suggested as the key nitrogen and carbon source for carrying pollen tube elongation (Zhang et al. 1982; Schwacke et al. 1999), the elongation of the hairy roots in dicotyledonous plants infected by *Agrobacterium rhizogenes* (Trovato et al. 2001) and the elongation of the maize primary root in which proline was found to accumulate particularly in the elongation zone of the maize root at low water potential (Voetberg and Sharp 1991; Spollen et al. 2008), indicating that proline might be generally used by the plant cell in developmental processes involving rapid cell growth as an energy source (Mattioli et al. 2009).

A positive correlation between proline and cell elongation can also be described in terms of protein synthesis, as hydroxyproline-rich glycoproteins (HRGPs, extensins and arabinogalactan proteins) are important structural components of the plant cell wall which play an important role in the regulation of cell division, cell wall self-assembly and cell extension (Snowalter 1993; Majewska-Sawka and Nothnagel 2000). Nanjo et al. (1999) found declined proline and hydroxyproline levels in the cell wall protein fraction of antisense *P5CS1*transgenic *Arabidopsis* impaired in bolting, further supporting the hypothesis that proline content and cell elongation in plants are positively related with each other.

Although the accumulation of proline in reproductive organs has been frequently reported in various plant species, its exact role is still not clear. It is proposed that proline protects developing cells from osmotic damages, especially during pollen development and embryogenesis where tissues experience spontaneous water loss (Trovato et al. 2008; Mattioli et al. 2009). Similarly, the desiccation process that occurs in reproductive tissues due to osmotic stress caused by environmental factors may critically damage the plant cell, and it is likely to be thwarted by proline accumulation (Trovato et al. 2008). Accordingly, higher levels of proline have been found (Chiang and Dandekar 1995) in tissues with low water content as compared to tissues with high water content. The correlation between proline accumulation and water content is not always tight. Florets, for example, have been depicted by Chiang and Dandekar (1995) as the organs with the highest proline content, in spite of their quite high water content.

The oxidation of a single molecule of proline provides 30 ATP equivalents (Atkinson 1977), making this amino acid well fit for sustaining high-energy-requiring processes. Therefore, it is believed that proline has a major role in sustaining the plant reproduction, a highly metabolically requiring process, by providing energy (Micheu et al. 2000).

9.6 Conclusion

Proline accumulation in response to various kinds of environmental stresses and in some developmental processes is a well-established fact, and its role in such processes has been confirmed by various transgenic approaches. Various kinds of functions for proline as an osmolyte and/or as an energy source and/or as an ROS scavenger and/or stress reliever have been proposed. However, the very (adaptative) function of proline accumulation has been questioned. The balance between proline biosynthesis and its degradation is suggested to be very important in determining its osmoprotective and developmental role. When proline was applied exogenously to plants exposed to stress environment, it resulted in improved growth and other physiological attributes of plants. Exogenously applied proline scavenges the ROS generated in plants subjected to various environmental stresses. Low doses of exogenous proline proved to protect plants from salinity, drought, heavy metal and temperature stress. However, higher concentrations may cause toxicity.

An involvement of proline in developmental processes, viz. seed germination, flower transition, flower development and other developmental processes, has been recognized. Proline may act as a signal molecule during flower transition, whilst during flowering and other developmental processes it provides energetic needs for rapidly dividing or elongating cells.

References

Adams E, Frank L (1980) Metabolism of proline and the hydroxyprolines. Annu Rev Biochem 49:1005–1061

Ahmad I, Wainwright SJ, Stewart GR (1981) The solute and water relations of *Agrostis stolonifera* ecotypes differing in their salt tolerance. New Phytol 87:615–629

Ali Q, Ashraf M, Athar HUR (2007) Exogenously applied proline at different growth stages enhances growth of two maize cultivars grown under water deficit conditions. Pak J Bot 39:1133–1144

Apel K, Hirt H (2004) Reactive oxygen species: metabolism, oxidative stress, and signal transduction. Ann Rev Plant Biol 55:373–399

Armengaud P, Thiery L, Buhot N, Grenier-De March G, Savouré A (2004) Transcriptional regulation of proline biosynthesis in *Medicago truncatula* reveals developmental and environmental specific features. Physiol Plant 120:442–450

Ashraf M, Foolad MR (2007) Roles of glycine betaine and proline in improving plant abiotic stress resistance. Environ Exp Bot 59:206–216

Atkinson DE (1977) Cellular energy metabolism and its regulation. Academic, New York

Bohnert HJ, Jensen RG (1996) Strategies for engineering water-stress tolerance in plants. Trends Biotechnol 14:89–97

Brown LM, Hellebust JA (1978) Sorbitol and proline as intracellular osmotic solutes in the green alga *Stichococcus bacillaris*. Can J Bot 56:676–679

Burton RS (1991) Regulation of proline synthesis in osmotic response: effects of protein synthesis inhibitors. J Exp Zool 259:272–277

Chaves MM et al (2009) Photosynthesis under drought and salt stress: regulation mechanisms from whole plant to cell. Ann Bot 103:551–560

Chiang H-H, Dandekar AM (1995) Regulation of proline accumulation in *Arabidopsis thaliana* (L) Heynh. during development and in response to desiccation. Plant Cell Environ 18:1280–1290

Cho YH, Yoo SD, Sheen J (2006) Regulatory functions of nuclear hexokinase1 complex in glucose signaling. Cell 127:579–589

Choudhary NL, Sairam RK, Tyagi A (2005) Expression of Δ^1-pyrroline-5- carboxylate synthetase gene during drought in rice (*Oryza sativa* L.). Indian J Biochem Biophys 42:366–370

Chung JS, Zhu JK, Bressan RA, Hasegawa PM, Shi H (2008) Reactive oxygen species mediate Na^{+}-induced SOS1 mRNA stability in *Arabidopsis*. Plant J 53:554–565

Csonka LN (1989) Physiological and genetic responses of bacteria to osmotic stress. Microbiol Rev 53:121–147

Csonka LN, Hanson AD (1991) Prokaryotic osmoregulation: genetics and physiology. Annu Rev Microbiol 45:569–606

De Knecht JA, Van Dillen M, Koevoets PLM, Schat H, Verkleij JAC, Ernst WHO (1994) Phytochelatins in cadmium-sensitive and cadmium-tolerant *Silene vulgaris*. Plant Physiol 104:255–261

De Ronde JAD, Cress WA, Kruger GHJ, Strasser RJ, Staden JV (2004) Photosynthetic response of transgenic soybean plants, containing an *Arabidopsis P5CR* gene, during heat and drought stress. J Plant Physiol 161:1211–1224

de-Lacerda CF, Cambraia J, Oliva MA, Ruiz HA, Prisco JT (2003) Solute accumulation and distribution during shoot and leaf development in two sorghum genotypes under salt stress. Environ Exp Bot 49:107–120

Delanauney AJ, Verma DPS (1993) Proline biosynthesis and osmo-regulation in plants. Plant J 4:215–223

Deuschle K, Funck D, Hellmann H, Däschner K, Binder S, Frommer WB (2001) A nuclear gene encoding mitochondrial delta 1-pyrroline-5-carboxylate dehydrogenase and its potential role in protection from proline toxicity. Plant J 27:345–356

Doke N (1997) The oxidative burst: role in signal transduction and plant stress. In: Scandalios JG (ed) Oxidative stress and the molecular biology of antioxidant defenses. Cold Spring Harbor Press, New York, pp 785–813

Ehsanpour AA, Fatahian N (2003) Effects of salt and proline on *Medicago sativa* callus. Plant Cell Tiss Org Cult 73:53–56

Fabro G, Kovacs I, Pavet V, Szabados L, Alvarez ME (2004) Proline accumulation and *AtP5CS2* gene activation are induced plant pathogen incompatible interactions in *Arabidopsis*. Mol Plant Microbe Interact 17:343–350

Farago ME, Mullen WA (1979) Plants which accumulate metals. Part IV. A possible copper-proline complex from the roots of *Armeria maritima*. Inorg Chim Acta 32:L93–L94

Fougere F, Le-Rudulier D, Streeter JG (1991) Effects of salt stress on amino acid, organic acid, and carbohydrate composition of roots, bacteroids, and cytosol of alfalfa (*Medicago sativa* L.). Plant Physiol 96:1228–1236

Fujita T, Maggio A, García-Ríos M, Bressan RA, Csonka LN (1998) Comparative analysis of the regulation of expression and structures of two evolutionarily divergent genes for D1-pyrroline- 5-carboxylate synthetase from tomato. Plant Physiol 118:661–674

Gangopadhyay G, Basu S, Gupta S (1997) In vitro selection and physiological characterization of NaCl and mannitol-adapted callus lines in *Brassica juncea*. Plant Cell Tiss Org Cult 50:161–169

Hamilton EW 3rd, Heckathorn SA (2001) Mitochondrial adaptations to NaCl complex I is protected by antioxidants and small heat shock proteins, whereas complex II is protected by proline and betaine. Plant Physiol 126:1266–1274

Hare PD, Cress WA (1997) Metabolic implications of stress-induced proline accumulation in plants. Plant Growth Regul 21:79–102

Hare PD, Cress WA, Staden JV (2002) Disruptive effects of exogenous proline on chloroplast and mitochondrial ultrastructure in *Arabidopsis* leaves. S Afr J Bot 68:393–396

Hare PD, Cress WA, van Staden J (2003) A regulatory role for proline metabolism in stimulating Arabidopsis thaliana seed germination. Plant Growth Regul 39:41–50

Haudecoeur E, Planamente S, Cirou A, Tannieres M, Shelp BJ, Morera S, Faure D (2009) Proline antagonizes GABA-induced quenching of quorum-sensing in *Agrobacterium tumefaciens*. Proc Natl Acad Sci U S A 106:14587–14592

Hayat S, Hayat Q, Alyemeni MN, Wani AS, Pichtel J, Ahmad A (2012) Role of proline under changing environments: a review. Plant Signal Behav 7:1456–1466

Hellmann H, Funk D, Rentsch D, Frommer WB (2000) Hypersensitivity of an *Arabidopsis* sugar signaling mutant towards exogenous proline application. Plant Physiol 123:779–790

Hong Z, Lakkineni K, Zhang Z, Verma DP (2000) Removal of feedback inhibition of delta(1)-pyrroline-5-carboxylate synthetase results in increased proline accumulation and protection of plants from osmotic stress. Plant Physiol 122:1129–1136

Hoque MA, Banu MNA, Nakamura Y, Shimoishi Y, Murata Y (2008) Proline and glycinebetaine enhance antioxidant defense and methylglyoxal detoxification systems and reduce NaCl-induced damage in cultured tobacco cells. J Plant Physiol 165:813–824

Hsu SY, Hsu YT, Kao CH (2003) The effect of polyethylene glycol on proline accumulation in rice leaves. Biol Plant 46:73–78

Hua B, Guo WY (2002) Effect of exogenous proline on SOD and POD activity of soybean callus under salt stress. Acta Agric Boreali Sinica 17:37–40

Hur J, Jung KH, Lee CH, An G (2004) Stress-inducible *OsP5CS2* gene is essential for salt and cold tolerance in rice. Plant Sci 167:417–426

Islam MM, Hoque MA, Okuma E, Banu MNA, Shimoishi Y, Nakamura Y, Murata Y (2010) Exogenous proline and glycinebetaine increase antioxidant enzyme activities and confer tolerance to cadmium stress in cultured tobacco cells. J Plant Physiol 166:1587–1597

Kaneshiro ES, Holz GG Jr, Dunham PB (1969) Osmoregulation in a marine ciliate, *Miamiensis avi-*

dus. II. Regulation of intracellular free amino acids. Biol Bull 137:161–169

Kaul S, Sharma SS, Mehta IK (2008) Free radical scavenging potential of L-proline: evidence from in vitro assays. Amino Acids 34:315–320

Kishor PBK, Hong Z, Miao GH, Hu CAA, Verma DPS (1995) Overexpression of Δ1-pyrroline-5-carboxylate synthetase increases proline production and confers osmotolerance in transgenic plants. Plant Physiol 108:1387–1394

Kishor PBK, Sangam S, Amrutha RN, Laxmi PS, Naidu KR, Rao KRSS, Rao S, Reddy KJ, Theriappan P, Sreenivasulu N (2005) Regulation of proline biosynthesis, degradation, uptake and transport in higher plants: its implications in plant growth and abiotic stress tolerance. Curr Sci 88:424–438

Kiyosue T, Yoshiba Y, Yamaguchi-Shinozaki K, Shinozaki K (1996) A nuclear gene encoding mitochondrial proline dehydrogenase, an enzyme involved in proline metabolism, is upregulated by proline but downregulated by dehydration in *Arabidopsis*. Plant Cell 8:1323–1335

Krishna RV, Leisinger T (1979) Biosynthesis of proline in Pseudomonas aeruginosa. Partial purification and characterization of gamma-glutamyl kinase. Biochem J 181:215–222

Kumar V, Sharma DR (1989) Effect of exogenous proline on growth and ion content in NaCl stressed and non-stressed cells of mung bean, *Vigna radiata* var. radiata. Indian J Exp Biol 27:813–815

Lehmann S, Funck D, Szabados L, Rentsch D (2010) Proline metabolism and transport in plant development. Amino Acids 39:949–962

Leisinger T (1987) Biosynthesis of proline in *Escherichia coli* and *Salmonella typhimurium*. In: Neidhart FC, Ingraham JL, Low KB, Magasanik B, Schaechter M, Umbarger HE (eds) Cellular and molecular biology. American Society for Microbiology, Washington, DC, pp 345–351

Low PS (1985) Molecular basis of the biological compatibility of nature's osmolytes. In: Gilles R, Gilles-Baillien M (eds) Transport processes, iono- and osmoregulation. Springer, Berlin, pp 469–477

Lutts S, Majerus V, Kinet J-M (1999) NaCl effects on proline metabolism in rice (*Oryza sativa*) seedlings. Physiol Plant 105:450–458

Madan S, Nainawatee HS, Jain RK, Chowdhury JB (1995) Proline and proline metabolising enzymes in in-vitro selected NaCl-tolerant *Brassica juncea* L. under salt stress. Ann Bot (Lond) 76:51–57

Majewska-Sawka A, Nothnagel EA (2000) The multiple roles of arabinogalactan proteins in plant development. Plant Physiol 122:3–9

Mattioli R, Marchese D, D'Angeli S, Altamura MM, Costantino P, Trovato M (2008) Modulation of intracellular proline levels affects flowering time and inflorescence architecture in *Arabidopsis*. Plant Mol Biol 66:277–288

Mattioli R, Falasca G, Sabatini S, Costantino P, Altamura MM, Trovato M (2009) The proline biosynthetic genes *P5CS1* and *P5CS2* play overlapping roles in *Arabidopsis* flower transition but not in embryo development. Physiol Plant 137:72–85

Mattioli R, Biancucci M, Lonoce C, Costantino P, Trovato M (2012) Proline is required for male gametophyte development in *Arabidopsis*. BMC Plant Biol 12:236

Mauro ML, Trovato M, De Paolis A, Gallelli A, Costantino P, Altamura MM (1996) The plant oncogene *rolD* stimulates flowering in transgenic tobacco plants. Dev Biol 180:693–700

Micheu S, Crailsheim K, Leonhard B (2000) Importance of proline and other amino acids during honeybee flight (*Apis mellifera carnica* POLLMANN). Amino Acids 18:157–175

Miller G, Honig A, Stein H, Suzuki N, Mittler R, Zilberstein A (2009) Unraveling delta1-pyrroline-5-carboxylateproline cycle in plants by uncoupled expression of proline oxidation enzymes. J Biol Chem 284:26482–26492

Nakashima K, Satoh R, Kiyosue T, Yamaguchi-Shinozaki K, Shinozaki K (1998) A gene encoding proline dehydrogenase is not only induced by proline and hypo osmolarity, but is also developmentally regulated in the reproductive organs of Arabidopsis. Plant Physiol 118:1233–1241

Nanjo T, Kobayashi M, Yoshiba Y, Sanada Y, Wada K, Tukaya H et al (1999) Biological functions of proline in morphogenesis and osmotolerance revealed in antisense transgenic *Arabidopsis*. Plant J 18:185–193

Nanjo T, Fujita M, Seki M, Kato T, Tabata S, Shinozaki K (2003) Toxicity of free proline revealed in an *Arabidopsis* T-DNA-tagged mutant deficient in proline dehydrogenase. Plant Cell Physiol 44:541–548

Nayyar H, Walia DP (2003) Water stress induced proline accumulation in contrasting wheat genotypes as affected by calcium and abscisic acid. Biol Plant 46:275–279

Okuma E, Soeda K, Tada M, Murata Y (2000) Exogenous proline mitigates the inhibition of growth of *Nicotiana tabacum* cultured cells under saline conditions. Soil Sci Plant Nutr 46:257–263

Peng Z, Lu Q, Verma DP (1996) Reciprocal regulation of delta-1- pyrroline-5-carboxylate synthetase and proline dehydrogenase genes controls proline levels during and after osmotic stress in plants. Mol Gen Genet 253:334–341

Petrusa LM, Winicov I (1997) Proline status in salt tolerant and salt sensitive alfalfa cell lines and plants in response to NaCl. Plant Physiol Biochem 35:303–310

Piqueras A, Hernandez JM, Olmos E, Hellin E, Sevilla F (1996) Changes in antioxidant enzymes and organic solutes associated with adaptation of citrus cells to salt stress. Plant Cell Tiss Org Cult 45:53–60

Poulin R, Larochelle J, Hellebust JA (1987) The regulation of amino acid metabolism during hyperosmotic stress in *Acanthamoeba castellanii*. J Exp Zool 243:365–378

Rehman F, Khan FA, Anis SB (2014) Assessment of aphid infestation levels in some cultivars of mustard with varying defensive traits. Arch Phytopathol Plant Protect 47:1866–1874

Ribarits A, Abdullaev A, Tashpulatov A, Richter A, Heberle-Bors E, Touraev A (2007) Two tobacco proline dehydrogenases are differentially regulated and play a role in early plant development. Planta 225:1313–1324

Roosens NH, Thu TT, Iskandar HM, Jacobs M (1998) Isolation of the ornithine-delta-amino transferase cDNA and effect of salt stress on its expression in *Arabidopsis thaliana*. Plant Physiol 117:263–271

Roy D, Basu N, Bhunia A, Banerjee SK (1993) Counteraction of exogenous l-proline with NaCl in salt-sensitive cultivar of rice. Biol Plant 35:69–72

Samach A, Onouchi H, Gold SE, Ditta GS, Schwarz-Sommer S, Yanofsky MF, Coupland G (2000) Distinct roles of *CONSTANS* target genes in reproductive development of *Arabidopsis*. Science 288:1613–1616

Saradhi P, Alia P, Arora S, Prasad KV (1995) Proline accumulates in plants exposed to UV radiation and protects them against UV induced peroxidation. Biochem Biophys Res Commun 209:1–5

Savoure A, Jaoua S, Hua X-E, Ardiles W, Van Montagu M, Verbruggen N (1995) Isolation, characterization, and chromosomal location of a gene encoding the delta 1-pyrroline-5-carboxylate synthetase in *Arabidopsis thaliana*. FEBS Lett 372:13–19

Schat H, Sharma SS, Vooijs R (1997) Heavy metal induced accumulation of free proline in a metal tolerant and a non tolerant ecotype of *Silene vulgaris*. Physiol Plant 101:477–482

Schmidt R, Stransky H, Koch W (2007) The amino acid permease AAP8 is important for early seed development in *Arabidopsis thaliana*. Planta 226:805–813

Schwacke R, Grallath S, Breitkreuz KE, Stransky H, Frommer WB, Rentsch D (1999) LeProT1, a transporter for proline, glycine betaine, and γ -amino butyric acid in tomato pollen. Plant Cell 11:377–391

Sekhar PN, Amrutha RN, Sangam S, Verma DP, Kishor PB (2007) Biochemical characterization, homology modelling and docking studies of ornithine delta aminotransferase- an important enzyme in proline biosynthesis of plants. J Mol Graph Model 26:709–719

Sharma SS, Schat H, Vooijs R (1998) In vitro alleviation of heavy metal-induced enzyme inhibition by proline. Phytochemistry 1998(49):1531–1535

Shi H, Ishitani M, Kim C, Zhu J-K (2000) The *Arabidopsis thaliana* salt tolerance gene *SOS1* encodes a putative Na^+/H^+ antiporter. Proc Natl Acad Sci Lett U S A 97:6896–6901

Siripornadulsil S, Train S, Verma DPS, Sayre RT (2002) Molecular mechanisms of proline-mediated tolerance to toxic heavy metals in transgenic microalgae. Plant Cell 14:2837–2847

Snowalter AM (1993) Structure and function of plant cell wall proteins. Plant Cell 5:9–23

Spollen WG, Tao W, Valliyodan B, Chen K, Hejlek LG, Kim JJ, Noble ME, Zhu J, Bohnert HJ, Henderson D, Schachtman DP, Davis GE et al (2008) Spatial distribution of transcript changes in the maize primary root elongation zone at low water potential. BMC Plant Biol 8:1–32

Sreedevi S, Remani KN, Benjamin S (2013) Biotic stress induced biochemical and isozyme variations in ginger and tomato by *Ralstonia solanacearum*. Am J Plant Sci 4:1601–1610

Strizhov N, Abraham E, Okresz L, Blickling S, Zilberstein A, Schell J, Koncz C, Szabados L (1997) Differential expression of two *P5CS* genes controlling proline accumulation during salt-stress requires ABA and is regulated by *ABA1*, *ABI1* and *AXR2* in Arabidopsis. Plant J 12:557–569

Szabados L, Savoure A (2009) Proline: a multifunctional amino acid. Trends Plant Sci 15:89–97

Székely G, Ábrahám E, Cséplő A, Rigó G, Zsigmond L, Csiszár J, Ayaydin F et al (2008) Duplicated *P5CS* genes of *Arabidopsis* play distinct roles in stress regulation and developmental control of proline biosynthesis. Plant J 53:11–28

Tripathi BN, Gaur JP (2004) Relationship between copper and zinc-induced oxidative stress and proline accumulation in *Scenedesmus* sp. Planta 219:397–404

Trovato M, Maras B, Linhares F, Costantino P (2001) The plant oncogene *rolD* encodes a functional ornithine cyclodeaminase. Proc Natl Acad Sci U S A 98:13449–13453

Trovato M, Mattioli R, Costantino P (2008) Multiple roles of proline in plant stress tolerance and development. Rendiconti Lincei 19:325–346

Venekamp JH, Koot JTM (1984) The distribution of free amino acids, especially of proline, in the organs of field bean plants. *Vicia faba* L., during development in the field. J Plant Physiol 116:343–349

Verbruggen N, Hermans C (2008) Proline accumulation in plants: a review. Amino Acids 35:753–759

Verbruggen N, Villarroel R, Van Montagu M (1993) Osmoregulation of a pyrroline-5-carboxylate reductase gene in *Arabidopsis thaliana*. Plant Physiol 103:771–781

Verbruggen N, Hua XJ, May M, Van Montagu M (1996) Environmental and developmental signals modulate proline homeostasis: evidence for a negative transcriptional regulator. Proc Natl Acad Sci U S A 93:8787–8791

Verslues PE, Sharma S (2010) Proline metabolism and its implications for plant-environment interaction. Arabidopsis Book 8, e0140

Voetberg GS, Sharp RE (1991) Growth of the maize primary root tip at low water potentials. III. Role of increased proline deposition in osmotic adjustment. Plant Physiol 96:1125–1130

Xu J, Yin H, Li X (2009) Protective effects of proline against cadmium toxicity in micropropagated hyperaccumulator, *Solanum nigrum* L. Plant Cell Rep 28:325–333

Yan H, Gang LZ, Zhao CY, Guo WY (2000) Effects of exogenous proline on the physiology of soybean plantlets regenerated from embryos in vitro and on the ultrastructure of their mitochondria under NaCl stress. Soybean Sci 19:314–319

Yancey PH (1994) Compatible and counteracting solutes. In: Strange K (ed) Cellular and molecular physiology of cell volume regulation. CRC Press, Boca Raton, pp 81–109

Yoshiba Y et al (1995) Correlation between the induction of a gene for D1- pyrroline-5-carboxylate synthetase and the accumulation of proline in *Arabidopsis thaliana* under osmotic stress. Plant J 7:751–776

Zhang HQ, Croes A, Linskens H (1982) Protein synthesis in germinating pollen of *Petunia*: role of proline. Planta 154:199–203

Index

N. Iqbal et al. (eds.), *Osmolytes and Plants Acclimation to Changing Environment: Emerging Omics Technologies*, DOI 10.1007/978-81-322-2616-1

MIX
Papier aus verantwortungsvollen Quellen
Paper from responsible sources
FSC® C105338

If you have any concerns about our products,
you can contact us on
ProductSafety@springernature.com

In case Publisher is established outside the EU,
the EU authorized representative is:
Springer Nature Customer Service Center GmbH
Europaplatz 3, 69115 Heidelberg, Germany

Printed by Libri Plureos GmbH
in Hamburg, Germany